NUTRITION OF GRAZING RUMINANTS IN WARM CLIMATES

ANIMAL FEEDING AND NUTRITION

A Series of Monographs and Treatises

Tony J. Cunha, Editor

Distinguished Service Professor Emeritus
University of Florida
Gainesville, Florida

and

Dean Emeritus, School of Agriculture
California State Polytechnic University
Pomona, California

Tony J. Cunha, SWINE FEEDING AND NUTRITION, 1977

W. J. Miller, DAIRY CATTLE FEEDING AND NUTRITION, 1979

Tilden Wayne Perry, BEEF CATTLE FEEDING AND NUTRITION, 1980

Tony J. Cunha, HORSE FEEDING AND NUTRITION, 1980

Charles T. Robbins, WILDLIFE FEEDING AND NUTRITION, 1983

T. W. Perry, ANIMAL LIFE-CYCLE FEEDING AND NUTRITION, 1984

Lee Russell McDowell, NUTRITION OF GRAZING RUMINANTS
IN WARM CLIMATES, 1985

Raymond L. Shirley, NITROGEN AND ENERGY NUTRITION
OF RUMINANTS (in press)

NUTRITION OF GRAZING RUMINANTS IN WARM CLIMATES

Edited by

Lee Russell McDowell
Department of Animal Science
University of Florida
Gainesville, Florida

1985

ACADEMIC PRESS, INC.
Harcourt Brace Jovanovich, Publishers

Orlando   San Diego   New York
Austin   London   Montreal   Sydney
Tokyo   Toronto

COPYRIGHT © 1985 BY ACADEMIC PRESS, INC.
ALL RIGHTS RESERVED.
NO PART OF THIS PUBLICATION MAY BE REPRODUCED OR
TRANSMITTED IN ANY FORM OR BY ANY MEANS, ELECTRONIC
OR MECHANICAL, INCLUDING PHOTOCOPY, RECORDING, OR
ANY INFORMATION STORAGE AND RETRIEVAL SYSTEM, WITHOUT
PERMISSION IN WRITING FROM THE PUBLISHER.

ACADEMIC PRESS, INC.
Orlando, Florida 32887

United Kingdom Edition published by
ACADEMIC PRESS INC. (LONDON) LTD.
24–28 Oval Road, London NW1 7DX

Library of Congress Cataloging in Publication Data
Main entry under title.

Nutrition of grazing ruminants in warm climates.

   (Animal feeding and nutrition)
   Includes index.
   1. Animal nutrition–Topics.   2. Ruminants–Feeding
and feeds–Tropics.   I. McDowell, L. R., Date.
II. Series.
SF95.N88   1985   636.2   85-6120
ISBN 0–12–483370–5 (alk. paper)
ISBN 0–12–483371–3 (paperback)

PRINTED IN THE UNITED STATES OF AMERICA

85 86 87 88   9 8 7 6 5 4 3 2 1

In dedication to
animal agriculturalists and researchers
involved with grazing livestock in tropical regions

and to

Dr. Raymond B. Becker
the pioneer of mineral research for
grazing livestock in Florida

Contents

Contributors	xiii
Foreword	xv
Preface	xvii

1 The Role of Ruminants in Warm Climates
J. K. LOOSLI AND L. R. McDOWELL

I.	Introduction	1
II.	What Are the Ruminants?	3
III.	Contributions of Domestic Ruminants to Human Welfare	12
IV.	Improving the Efficiency of Tropical Ruminants	17
	References	18

2 Nutrient Requirements of Ruminants
L. R. McDOWELL

I.	General Requirements	21
II.	Energy–Protein Requirements	22
III.	Factors Influencing Mineral Requirements	28
IV.	Nutritional Relationships to Diseases and Parasites	32
V.	Environment and Stress	33
VI.	Nonnutrient Factors Affecting Requirements	34
	References	34

3 Water Requirements for Grazing Ruminants and Water as a Source of Minerals
RAY L. SHIRLEY

I.	Introduction	37
II.	Water Turnover in Ruminants	39
III.	Voluntary Intake of Water	40
IV.	Effect of Dry Matter Intake on Water Consumption	40

V.	Effect of Water Deprivation on Feed Intake and Utilization	42
VI.	Effect of Temperature on Water Intake and Digestibility of Nutrients	44
VII.	Effect of Salinity on Water Consumption	45
VIII.	Water and Electrolyte Metabolism	48
IX.	Nutrient Elements in Water	49
X.	Toxic Elements and Substances in Drinking Water	50
XI.	Summary	54
	References	55

4 Thermal Stress as a Factor Associated with Nutrient Requirements and Interrelationships
R. J. COLLIER AND D. K. BEEDE

I.	Introduction	59
II.	Physiological Responses of Ruminants to Thermal Stress	60
III.	Metabolic and Hormonal Responses to Thermal Stress	61
IV.	Effects of Thermal Stress on Nutrient Requirements	62
	References	70

5 Forages for Grazing Systems in Warm Climates
PAUL MISLEVY

I.	Introduction	73
II.	Selected Warm-Climate Perennial Grass Types	75
III.	Selected Warm-Climate Perennial Legumes	91
	References	101

6 Pasture Management for Optimum Ruminant Production
L. V. CROWDER

I.	Pasture Maintenance and Renovation	104
II.	Pasture Type and Grazing Management	109
III.	Grass–Legume Mixtures	112
IV.	Grazing Management Systems	115
V.	Herbage Quality	119
VI.	Pasture Use for Animal Production	122
	References	125

7 Providing Energy–Protein Supplementation during the Dry Season
E. J. GOLDING

I.	Development of the Production System	130
II.	Improving Ruminant Production Potential of Dry-Season Forage	135
III.	Sources of Supplemental Crude Protein or Energy	139
IV.	Management for Forage Conservation and Efficient Dry-Season Feeding	154
	References	158

8 Contribution of Tropical Forages and Soil toward Meeting Mineral Requirements of Grazing Ruminants
L. R. McDOWELL

I.	Introduction	165
II.	Tropical Forages as Sources of Minerals	166
III.	Soils as Sources of Minerals	176
	References	185

9 Calcium, Phosphorous, and Fluorine
L. R. McDOWELL

I.	General	189
II.	Calcium and Phosphorus	190
III.	Fluorine	204
	References	210

10 Common Salt (Sodium and Chlorine), Magnesium, and Potassium
L. R. McDOWELL

I.	General	214
II.	Common Salt (Sodium and Chlorine)	214
III.	Magnesium	220
IV.	Potassium	228
	References	233

11 Copper, Molybdenum, and Sulfur
L. R. McDOWELL

I.	Introduction	237
II.	Metabolism of Copper and Molybdenum	238
III.	Copper and Molybdenum Requirements	239
IV.	Copper Deficiency	240
V.	Prevention and Control of Copper Deficiency	248
VI.	Toxicity of Copper and Molybdenum	249
VII.	Sulfur	251
	References	255

12 Cobalt, Iodine, and Selenium
L. R. McDOWELL

I.	General	259
II.	Cobalt	260
III.	Iodine	268
IV.	Selenium	275
	References	287

13 Iron, Manganese, and Zinc
L. R. McDOWELL

I.	General	291
II.	Iron	292
III.	Manganese	297
IV.	Zinc	304
	References	311

14 Newly Discovered and Toxic Elements
SCOT N. WILLIAMS AND L. R. McDOWELL

I.	Introduction	317
II.	Toxic Elements	318
III.	Newly Discovered Trace Elements	327
IV.	Significance of Newly Discovered and Toxic Elements for Grazing Livestock	334
	References	335

15 Detection of Mineral Status of Grazing Ruminants
L. R. McDOWELL

I.	Introduction	339
II.	Clinical and Pathological Evaluation	340
III.	Analysis of Water, Soil, and Forage	341
IV.	Examination of Tissues and Fluids	343
V.	Response to Supplementation	346
VI.	Analyses Most Indicative of Mineral Status	347
VII.	A Mapping Technique for Determining Mineral Deficiencies and Toxicities	354
	References	355

16 Incidence of Nutrient Deficiencies and Excesses in Tropical Regions and Beneficial Results of Mineral Supplementation
L. R. McDOWELL

I.	Introduction	359
II.	Geographical Distribution of Nutritional Deficiencies and Toxicities	360
III.	Energy–Protein Deficiencies in Ruminants	361
IV.	Incidence of Mineral Deficiencies and Toxicities	364
V.	Mineral Supplementation Results	369
VI.	Disease Conditions Related to Minerals	372
VII.	Seasonal Needs for Supplemental Minerals	376
VIII.	Economic Benefits from Mineral Supplementation	377
	References	378

17 Free-Choice Mineral Supplementation and Methods of Mineral Evaluation
L. R. McDOWELL

I.	Introduction	383
II.	Methods of Providing Minerals to Grazing Livestock	384
III.	Free-Choice Mineral Supplementation	385
IV.	Factors Affecting Mineral Consumption	386
V.	Selecting a Free-Choice Mineral Supplement	395
VI.	Information Required for Mineral Supplement Formulation	396
VII.	Calculations Required for Mineral Supplement Formulation	400
VIII.	Mineral Supplement Evaluation	402
	References	406

18 Vitamin Nutrition for Ruminants
L. R. McDOWELL

I.	Introduction	409
II.	Vitamin A	410
III.	Vitamin D	419
IV.	Vitamin E	421
V.	Vitamin K	422
VI.	B-Complex Vitamins	423
VII.	Providing Vitamin Supplements	427
	References	428

Appendix 431

Index 435

Contributors

Numbers in parentheses indicate the pages on which the authors' contributions begin.

D. K. BEEDE (59), Dairy Science Department, University of Florida, Gainesville, Florida 32611

R. J. COLLIER (59), Dairy Science Department, University of Florida, Gainesville, Florida 32611

L. V. CROWDER (103), The Rockefeller Foundation, and Department of Agronomy, University of Florida, Gainesville, Florida 32611

E. J. GOLDING (129), Department of Animal Science, University of Florida, North Florida Research and Education Center, Quincy, Florida 32351

J. K. LOOSLI (1), Department of Animal Science, University of Florida, Gainesville, Florida 32611

L. R. MCDOWELL (1, 21, 165, 189, 213, 237, 259, 291, 317, 339, 359, 383, 409), Department of Animal Science, University of Florida, Gainesville, Florida 32611

PAUL MISLEVY (73), Department of Agronomy, University of Florida, Agricultural Research Center, Ona, Florida 33865

RAY L. SHIRLEY (37), Department of Animal Science, University of Florida, Gainesville, Florida 32611

SCOT N. WILLIAMS (317), Department of Animal Science, University of Florida, Gainesville, Florida 32611

Foreword

This is the seventh in a series of books on animal feeding and nutrition. These seven books, and others to follow in this series, are designed to keep the reader abreast of the rapid developments in animal feeding and nutrition that have occurred in recent years. This new technology has resulted in many changes in domestic animal diets and in more supplementation with minerals, vitamins, amino acids, nonprotein nitrogen compounds, and proper feed additives. Moreover, new developments improve feed processing and preservation methods, and provide for better use of by-product feeds and improved stability and availability of nutrients.

Farmers are also developing animals that produce and reproduce at an increasingly higher rate, which increases nutrient needs. Moreover, increasing attention is being given to consumer needs and desires, which means animal production programs are changing to produce carcasses that are leaner and have less fat and other animal products with better eating and keeping quality. Since feed constitutes the major share of animal production costs, proper feeding and nutrition become increasingly important as farmers strive to become more efficient. One objective of the books in this series is, therefore, to collate and interpret the vast and complex literature for specific animal production programs.

The author of this book, Dr. Lee R. McDowell, has been working in Latin America, Africa, and Asia with hundreds of other collaborating animal scientists for the last 20 years. His interests have included developing feed analysis techniques for tropical and subtropical feeds, feed composition tables, and mineral and nutrient supplementation programs and locating mineral and other nutrient-deficient areas and mapping them. This book thus will be especially important to researchers and practitioners in Latin America, Africa, Asia, Australia, and other warm-climate areas of the world. But its contents, well illustrated with photographs and tables, will have application throughout all ruminant grazing areas of the world.

Many scientists and others think there is a need to double the world's animal protein food production in the next 15–20 years as a means of improving the protein quality of the world's human diet. This is a critical need, since well over one billion people suffer from chronic malnutrition, with over one-half of them being children under 5 years of age. In addition, every $2\frac{1}{2}$–3 years, the world's population increases by 200–250 million people, which is the equivalent of another United States to feed. The developing countries of the world already have two-thirds of the world's animals but produce only one-fifth of the world's meat, milk, and eggs. This indicates the need to improve their animal production efficiency, which this book is designed to help accomplish. It will be very useful to researchers, students, teachers, farmers, extension specialists, feed manufacturers, and all others in the United States and abroad who are concerned with the feeding, nutrition, and production of ruminant animals under grazing conditions and who are directly or indirectly involved with the world's food problems.

Tony J. Cunha

Preface

The objective of this book is to review the basic knowledge and methodology of feeding grazing ruminants in tropical and semitropical countries. It is hoped this information will be of use to farmers, research specialists, teachers, students, extension specialists, feed manufacturers, and others throughout the world concerned with the nutrition of grazing ruminants. A unique feature is the identification of nutritional limitations of grazing ruminants in the tropics, which will be beneficial for increasing animal production efficiency through the application of improved nutrition. A large number of photographs illustrate nutritional deficiencies and conditions in tropical countries.

This book contains 18 chapters concerned with the nutrition of grazing ruminants. The first chapter deals with the contributions, locations, and various types of ruminants and their importance to human welfare in the tropics and subtropics. Chapters 2–4 progress through nutrient requirements of grazing ruminants in warm climates, the effects of tropical heat on these requirements, and water requirements for ruminant species. Chapters 5–7 discuss grass and legume forage species suitable for tropical regions, pasture management procedures, and energy–protein supplementation programs needed during the extensive dry periods. The importance of tropical forages and soils toward meeting mineral requirements is discussed in Chapter 8. Chapters 9–14 contain concise, up-to-date summaries of minerals emphasizing mineral deficiencies and imbalances for grazing livestock. Evaluation of mineral status, incidence of mineral deficiencies and excesses in tropical regions, and benefits and methods of mineral supplementation for grazing ruminants are discussed in Chapters 15–17. Chapter 18 reviews vitamin nutrition considerations for ruminants consuming tropical forages.

In addition to the outstanding and authoritative contributions of fellow contributors for 7 of the 18 chapters, in preparing this book I have also obtained numerous helpful suggestions from eminent scientists and livestock producers in both the United States and the tropical countries of Latin America, Africa, Australia, and Southeast Asia. I wish to express my sincere appreciation to them and to those who supplied photographs and other material used. I am especially grateful to the following: C. B. Ammerman, R. B. Becker, D. K. Beede, B. Bock, J. H. Conrad, G. L. Ellis, C. E. Fenton, J. K. Loosli, P. G. Mallonée, M. McGlothen, J. E. Moore, G. O. Mott, F. M. Pate, Hugh Popenoe, R. L. Shirley, and H. D. Wallace (Florida); O. A. Beath (Wyoming); J. A. Boling (Kentucky); Bernardo J. Carrillo (Argentina); D. C. Church (Oregon); Jürgen Döbereiner (Brazil); I. A. Dyer (Washington); Mariano Echevarría (Peru); Nelson dos Santos Fernández (Brazil); Carlos García B. (Mexico); U. S. Garrigus (Illinois); S. L. Hansard (Tennessee); O. M. Mahmoud (Sudan); W. J. Miller (Georgia); Francisco Megale (Brazil); David Morillo (Venezuela); O. H. Muth (Oregon); M. W. Neathery (Georgia); G. Patterson (Pfizer Co.); R. L. Preston (Missouri); J. L. Shupe (Utah); S. E. Smith (New York); Carlos H. Tokarnia (Brazil); B. D. H. Van Niekerk (South Africa); and Eliecer Alberto Velasco (Venezuela).

I am particularly grateful to Lorraine M. McDowell and G. L. Ellis for their useful suggestions and assistance in the editing of the entire book. Likewise, I wish to acknowledge with thanks and appreciation of the skill and care of Patricia Joyce for overseeing the typing and proofing of chapters and also that of Sarah McKee, Patricia French, and Vera Hartsuch for valuable assistance. Also, I am indebted to the Animal Science Department and the Center for Tropical Agriculture of the University of Florida for providing the opportunity for this project. I also want to thank US/AID for their support in the mineral research phase of this book, which was supported in 25 tropical countries of Latin America, Africa, and Asia. Their support has made it possible to help these three continents in livestock and food production. Finally, I thank Tony J. Cunha for encouraging me to undertake the responsibility of serving as editor of this book.

Lee Russell McDowell

NUTRITION OF GRAZING RUMINANTS IN WARM CLIMATES

1

The Role of Ruminants in Warm Climates

J. K. LOOSLI AND L. R. McDOWELL
Department of Animal Science
University of Florida
Gainesville, Florida

I.	Introduction.	1
II.	What Are the Ruminants?.	3
	A. Cattle.	5
	B. Buffaloes.	5
	C. Sheep.	6
	D. Goats.	7
	E. Camelids.	8
	F. Wild Ruminants.	10
III.	Contributions of Domestic Ruminants to Human Welfare.	12
	A. Meat and Milk.	12
	B. Work.	14
	C. Other Products and Services.	16
IV.	Improving the Efficiency of Tropical Ruminants.	17
	References.	18

I. INTRODUCTION

The objective of this chapter is to define ruminant animals, both domestic and wild, to review their locations and numbers, and to note the important contributions they make to human welfare in the tropics and subtropics in comparison to temperate zones. Other chapters outline the requirements of the different ruminant species for specific nutrients and the effects of deficiencies of these nutrients and of heat on the requirements of grazing ruminants. Detailed data on water requirements of various species are presented as well as water representing a source of minerals.

Grass and legume forage species as they are best adapted to diverse conditions of warm tropical regions are illustrated. Systems of pasture improvement and management procedures are outlined. Utilization of forages, of crop residues, and of by-products and systems for conserving and storing feed supplies for dry seasons are presented. Research results are summarized on methods of detecting mineral deficiencies and on excesses of ruminants and effective supplementation to ensure optimum utilization of natural resources for the benefit of mankind.

Most tropical and subtropical countries that lie within the 30-degree north–south latitudes have warm temperatures throughout the year, except in higher elevations (above 2000 m). There are wide variations in rainfall within this zone from the humid lowland tropics to the deserts, not only in yearly rain but also between the rainy and the dry seasons. These climatic differences have resulted in vast variations in plant vegetation and in animal life. Animal species, and even ruminants, differ markedly from the humid rain forests to the dry deserts and from the lowland plains to the high altitudes of the mountains. Nonruminant herbivores, which vary in size from larger to smaller than any ruminant, are distributed about as widely as ruminants, and they use the same food resources. Some omnivores, such as the pig and birds, also contribute to mankind's food supplies and welfare.

Most of the less developed countries lie within 30 degrees of the equator. Some of the developed or industrialized countries also are within this zone. For example, the northern two-thirds of Australia, Florida, the gulf coast, and southern Texas in the United States, most of Egypt, parts of the Middle East, and southern China are within the warm-climate zone. Data on human population and animal numbers are tabulated by countries and not by climatic zones, and the same is true for production of animal products. It is perfectly clear, however, that most famines occur in tropical countries and most of the underfed people live there. Loosli *et al.* (1974) pointed out that the tropics could be the most food-productive areas on earth, far outdoing the temperate zones, but they are the part of the world with the least to eat. Improving productivity of ruminants could help relieve the problem.

For a long time some European and other countries have tried to help tropical countries improve food and animal production. In spite of these and more recent assistance programs, food production does not meet the requirements to sustain the human population. In the last decade, international symposia have reviewed the problems causing low production of domestic animals in the tropics and proposed solutions (Gilchrist and

Mackie, 1984; Simpson and Evangelou, 1984; Smith, 1976; Yousef, 1982). International Research Institutes has made important progress in advancing production of rice, wheat, corn, root crops, and grain legumes. Only recently has International Research Institutes started to study animal production and diseases in tropical Africa.

II. WHAT ARE THE RUMINANTS?

The larger ruminants have a special role because of their wide distribution throughout the world. They have greater digestive capacity to convert cellulose and other fibrous materials into useful products than do nonruminant herbivores.

The ruminant species are shown in Table 1.1. Of these families, the Antilocapridae, of which the pronghorn antelope is an example, occur only in the United States and Canada, north of the warm-climate zone. Most of the Camelidae are found in warm-climate zones, but the alpaca and llama are in the high altitudes of the Andes of South America, where there is no heat stress. Most of the camels are in the dry areas of Africa, the Middle East, and Asia, including the deserts, with extremely high daytime temperatures during the warmer season. Species of Bovidae are found in all temperature zones of the earth, except the arctic and antarctic. The yak, bison, and some cattle are well adapted to the cold climates, but antelope, buffaloes, tropical cattle, gazelle, eland, and other wild ruminants live mostly in the tropical zone. Cattle, goats, and sheep are found in both the temperate and tropical zones. The Giraffidae and Tragulidae are found only in the tropics. Of the 111 species of Bovidae (Table

TABLE 1.1

Families of Suborder Ruminantia with Numbers of Genera and Species[a]

Family	Number of:		Examples
	Genera	Species	
Antilocapridae	1	1	Pronghorn antelope
Bovidae	44	111	Antelope, bison, buffalo, cattle, eland, gazelle, goat, sheep, yak
Camelidae	2	4	Alpaca, llama, camel
Cervidae	16	37	Caribou, deer, elk, moose
Giraffidae	2	2	Giraffe, okapia
Tragulidae	2	4	Chevrotain, mousedeer

[a] Modified from McDowell (1977).

TABLE 1.2

Distribution of Domestic Ruminants (1000 head)[a]

Area	Cattle	Buffaloes	Sheep	Goats
World	1,213,731	131,146	1,116,449	454,667
Asia	369,342	127,636	325,085	365,171
China	64,130	30,080	120,880	80,448
India	182,408	61,296	4,108	71,644
Indonesia	6,482	2,321	3,611	8,051
Japan	4,136	—	11	80
Philippines	1,983	3,037	31	1,450
Thailand	4,963	5,583	62	31
Africa	169,527	2,380	182,091	146,125
Australia	26,321	1	135,706	—
Europe	134,411	457	134,309	11,599
North and Central America	176,306	8	21,786	11,207
South America	231,751	315	197,863	18,582
USA	110,961	—	12,513	1,400
USSR	115,900	350	143,599	5,824
Developed	424,582	807	524,490	24,410
Developing	789,149	130,340	591,959	430,257
Developing, %	65	99	53	95

[a] Modified from FAO (1981).

1.1), only the buffalo, camel, cattle, goat, sheep, and yak have been domesticated. The reindeer is the only member of the Cervidae family that is domestic. It is not clear why more of the ruminant species and other herbivores have not been domesticated.

Of the domesticated ruminants, the buffalo, camel, cattle, goat, and sheep are the only ones that are found in the warm tropics. The reindeer and caribou occur only in the cold, northern areas of the world. The alpaca and llama are distributed in the higher regions of the Andes of South America. The yak and its crosses exist in the high mountains of Asia. A number of nonruminant herbivores compete with ruminants for feed supplies and serve as human food to some extent. These include the elephant, hippopotamus, rhinoceros, horse, donkey, and zebra along with many smaller herbivorous animals.

The numbers and world distribution of domestic grazing cattle, buffaloes, sheep, and goats are shown in Table 1.2, based on the Food and Agriculture Organization of the United Nations (FAO, 1981) estimates. Cattle and sheep are the most numerous ruminants, followed by goats and buffaloes. Some 99% of the buffaloes and 95% of the goats are in the developing countries, largely in the warm climate zones. In contrast, developing countries have only 65% of the cattle and 53% of the sheep.

A. Cattle

There are several hundred breeds and types of cattle in the world. These have developed as a result of natural selection to the environment or by willful selection by cattle breeders. The most important breeds and types have been described by Payne (1970). Cattle are divided into three main species: *Bos indicus,* developed in the Indian subcontinent; *Bos taurus,* developed in Europe; and *Bos javanicus,* developed in Bali, Indonesia. The European cattle are not very tolerant to high temperatures but adapt to the cold European and American winters. There are some strains that were moved to tropical Africa, South America, and the Philippines centuries ago that are as adapted to high temperatures as Indian or Zebu breeds of cattle. These are the Criollo cattle of Latin America and the West African Shorthorn and Ndama breeds (Starkey, 1984). The latter are also resistant to trypanosomiasis, carried by the tsetse fly in wide areas of Africa. The other two species of cattle are tolerant to the tropical climate but suffer in cold weather.

The domestic Mithan (*Bos frontalis*) is thought to be a type of gaur. It crossbreeds readily with cattle. The yak (*Bos grunniens*), used for packing goods, is a small animal found in the high mountains of Asia. It mates readily with cattle but is not found in the warm climates.

B. Buffaloes

There are between 130 and 150 million buffaloes in the world (Table 1.2). Some 99% of these are found in developing countries. Only small numbers exist outside of the warm-climate countries. At least 95% are found on small farms. Cockrill's book (1974) gives information about the different breeds and types of buffaloes in various countries. There are three main divisions within buffaloes. The domestic ones are the river type found in India, the Middle East, Egypt, Italy, Brazil, and Trinidad; and the Swamp, or Carabao, found in the Philippines and Southeast Asia (Fig. 1.1). The third type is the wild Cape Buffalo of Central Africa (*Syncerus caffer*). There are also two species of wild buffalo-like animals: the Tamaraw (*Bubalus mindorensis*) in the Philippines and the much smaller Anoas (*Bub. depressecornis*) of Indonesia, which is about the size of a goat. These two are in danger of becoming extinct [National Research Council (NRC), 1983].

Fig. 1.1. Swamp type of buffalo in Malaysia.

C. Sheep

There are about 1116.4 million sheep in the world (Table 1.2). Some 450 different breeds and types have been identified. Most sheep in the tropics are found grazing with the nomadic cattle herds or as small flocks or single animals eating the grass, weeds, leaves, and waste residues on farms, on roadways or streets of towns and cities, and in unfarmed areas. They are useful as scavengers, cleaning up all edible plant materials. The book by Devendra and McLeroy (1982) describes the most important tropical breeds of sheep and goats.

Most of the world's sheep produce wool, but some tropical breeds grow hair instead of wool. Sheep usually thrive best outside of the humid areas; however, the West African Dwarf sheep are adapted to the higher-rainfall regions of West Africa. The small sheep recognized as the Florida Native do well in the high humidity of Florida and do not suffer unduly from intestinal parasites without treatment. Some of the more common breeds of sheep are listed in Table 1.3.

Some of the temperate breeds of sheep have been crossbred with tropical sheep to increase meat or wool production. The Finnish Landrace sheep have been used to improve fertility of both tropical and temperate sheep.

TABLE 1.3
Common Breeds and Types of Sheep

Hair sheep	Wool	Meat and wool
Awasi	Merino	Suffolk
Masai	Corriedale	Columbia
Ouda	Romney	Mutton Merino
Fattail	Targhee	Columbia

D. Goats

The FAO (1981) data indicate there are about 455 million goats in the world (Fig. 1.2). Some 95% of them are found in developing countries. Most of the goats are in the drier regions of the warm-climate countries and relatively few in the high-rainfall, humid regions. In Africa, the nomadic cattle owners usually have some goats and sheep grazing with their cattle. These provide a source of meat since cattle are sold and not used as meat for the family. Goats are also found on small farms and in the

Fig. 1.2. Goats grazing signalgrass (*Brachiaria spp.*) at Ebini Experiment Station, Guyana. (Courtesy of G. O. Mott, University of Florida, Gainesville, Florida.)

TABLE 1.4

Important Breeds and Types of Goats

Breed	Locality	Breed	Locality
Dairy type		Milk and meat type	
Alpine	Europe	Barbari	India, Pakistan
Anglo-nubian	England, Sudan	Damascus	Syria
La Mancha	USA	Jamnapari	India, Pakistan
Toggenburg	Europe	Nubian	Sudan
Saanen	Europe		
		Fiber	
Meat type		Angora	Turkey, South Africa, Texas
Boer	South Africa		
Ma Tou	China	Cashmere	Iran, China
Kambing Kajang	Malaysia		
Sapel	Northern Africa	Skins	
West African dwarf	West Africa	Mubende	Uganda
		Red Sokoto	West Africa

villages where they serve as scavengers, consuming the citrus and banana peels, corn husks, and other plant materials scattered along the roadways and paths as well as the weeds, bushes, tree leaves and fallen leaves, and surplus grass. In the tropics, goats serve mostly as a source of meat and fiber (Devendra and Owen, 1983), while in the temperate zone they give milk and meat. Everywhere goat skins are prized as a source of fine leather. There are many different breeds and types of goats, some of which are listed in Table 1.4.

E. Camelids

There are six different species of camels (McDowell, 1977). The larger camels found in Africa, the Middle East, and southern Asia are either the one-humped *Camelus dromedarius* or the two-humped *C. bactrianus*. The smaller Camelidae of Latin America consist of four species, as shown in Table 1.5. Most of them are domestic, but a few wild animals still exist. The camels are the most adaptable of all ruminants. The large camels are able to exist in dry desert areas where cattle cannot thrive. They are unusual in their ability to conserve water loss from the body (Schmidt-Nielson, 1969). They have been known to travel in the summer heat of the Sahara for a week or more without water and then be able to replace water losses amounting to 20–25% of total body weight within 30 min without ill effects. The large camels, in addition to their service as pack animals and for farm traction, provide useful amounts of milk and the

TABLE 1.5

Distribution of Latin American Camelidae Species (Thousands)[a,b]

Species	Argentina	Bolivia	Chile	Peru
Alpaca	Some	300	20	289
Llama	500	2500	70	915
Vicuña	—	2	—	30
Guanaco	100	Some	Some	Some

[a] Modified from McDowell (1977).
[b] In 1978, FAO estimated there were 15 million camels in the world, 12 million in the dry parts of Africa.

main meat source in some places. Leather from camel skins and fine-quality brushes from camel hair also deserve mention.

The smaller Camelidae are found in the high mountains of South America (Fig. 1.3). Like the large camels, they are noted for their ability to serve as pack animals. They produce a fine-quality wool which is used to make warm clothing. They also provide important amounts of meat in the

Fig. 1.3. Alpaca and llama in the highlands (altiplano) of Bolivia.

Andes. Llamas and alpacas, given their proven productive superiority over European livestock, should play an increasingly important role in the economy of Andean nations. Conception and reproductive success rates are low for both sheep and cattle, and altitude-related diseases can still produce significant mortality in sheep and cattle, even after almost five centuries.

F. Wild Ruminants

Spinage (1962) has described the most common larger wild ruminants found in Africa as listed in Table 1.6. The number of these is not known in most countries, but outside of zoos they are mostly in tropical Africa. Wild ruminants are important to African countries as a source of income from tourists and as sources of meat, skins, and trophies.

Yousef (1982) reviewed briefly the information on ranching of wild ungulates in Africa. These wild animals store less body fat than domestic cattle and represent better sources of protein. Some species are better adapted than cattle to utilizing the drier, less productive land areas and, therefore, these animals should be exploited. The problem, however, of

TABLE 1.6

The Larger Wild Ruminants of Africa and Asia[a]

Common name	Scientific name	Common name	Scientific name
Cape buffalo	*Syncerus caffer*	Oribis	
Dikdik		Oryx	*Oryx beisa*
Eland	*Taurotragus oryx*	Reedbuck	
Gerenuk	*Litocranius walleri*	Steinboks	
Giraffe	*Giraffa Camelopardalis*	Sunis	
Hartebeest or Kongoni	*Alcelaphus buselaphus*	Topi	*Damaliscus korrigum*
		Waterbuck	*Kobus ellipsiprymnus*
Impala	*Aepyceros melampus*	Wildebeest or gnu	*Gorgon taurinus*
Klipspringer			
Kob	*Rhynchotragus kirkii*	Zebra	*Equus grevyi, burchelli*
Okapi			
Little-known Asian ruminants			
Banteng or Bali	*Bos javanicus*	Tamaraw	*Bubalus mendorensis*
		Anoas	*Bubalus depressicornis*
Gaur	*Bos gaurus*		
Kouprey	*Bos sauveli*		

[a] Modified from Spinage (1962) and NRC (1983).

harvesting and marketing remains a challenge. Utilizing wildlife for meat production not only makes it possible to use marginal land efficiently, but also helps to preserve the wildlife. Harvesting some of the animals will help to keep the population healthier and to maintain a higher level of reproduction. A recent study in Kenya (McDowell et al., 1983) suggests that a combination of cattle and wild ruminants may be the most economical way to increase the production of animal protein in tropical Africa.

Omololu (1974) stressed the deplorable state of nutrition in Nigeria and the great need to increase energy and protein supplies. About half of the meat supplies come from wild animals. Nigeria was relatively well off compared with some Sahelial countries during the 1970s drought and with Ethiopia in the more recent famine. The critical need to improve the utilization of the drier areas, which have potential only as grazing lands, is evident. The great desire for meat has resulted in the destruction of almost all wild ruminants in West Africa. Their reintroduction could add to meat supplies in some areas.

Some wild ruminants are being produced as sources of meat. The Red Deer is farmed in Great Britain and New Zealand. The eland is grown on farms in the USSR. The American bison is reared on farms in the United States and Canada. Direct harvest of the native animals may be more economic. In the Transvaal, South Africa, over 3000 ranches market more than 3,000,000 kg of fresh meat annually from wild ruminants. There are some 27 million wild ruminants in the United States, Canada, Europe, and the USSR. The surplus animals are hunted yearly for meat, skins, and trophies. Estimated game meat production in selected African countries is presented in Table 1.7. This valuable source of protein with adequate management, could be greatly increased in many countries. In Texas,

TABLE 1.7

Estimated Game Meat Production per Year in Selected African Countries[a]

Country	Per capita production (kg)	Country	Per capita production (kg)
Botswana	7.5	Kenya	0.5
Chad	0.7	Liberia	3.0
Congo	4.0	Nigeria	1.2
Ethiopia	0.2	Sudan	0.3
Ghana	2.7	Zaire	2.6

[a] Modified from Simpson and Evangelow (1984).

290,000 deer harvested yearly yield over 12,000 t of meat. Over 63,000 deer produce over 4000 t of carcass meat annually in Utah.

III. CONTRIBUTIONS OF DOMESTIC RUMINANTS TO HUMAN WELFARE

A. Meat and Milk

Number one on the list of human needs is a dependable food supply. Domestic ruminants are the primary suppliers of meat and milk. Table 1.8 shows the world sources of meat. Developing countries have 65% of the cattle and 99% of the buffaloes (Table 1.2), but they produce only 33% of the total meat produced from cattle and buffaloes. The meat from these animals is recorded together because it cannot be easily distinguished.

TABLE 1.8

Meat Production (1000 t)[a]

Area	Beef and buffalo	Mutton and goat	Pig meat	Poultry	Hen eggs
World	46,518	7,679	54,967	30,109	27,248
Asia	5,158	2,769	20,149	6,814	8,523
China	2,331	747	16,580	3,781	4,465
India	194	399	69	108	87
Indonesia	168	60	86	105	90
Japan	398	—	1,510	1,020	1,990
Philippines	122	—	399	187	188
Thailand	215	—	240	97	172
Africa	3,047	1,240	358	1,188	908
Australia	1,557	539	217	344	190
Europe	10,542	1,206	18,337	7,160	7,002
North and Central America	12,149	194	9,065	10,106	5,383
South America	6,878	319	1,770	2,312	1,285
USA	9,951	137	7,521	8,713	4,094
USSR	6,700	853	5,000	2,158	3,790
Developed	31,203	3,466	33,613	20,508	17,712
Developing	15,195	4,213	21,354	9,601	9,535
Developing, %	33	58	39	32	35

[a] Modified from FAO (1981).

TABLE 1.9
World Milk Production[a]

		Cow's milk		Other milk		
Area	Milk cows	Per animal (kr/yr)	Total (t)	Buffalo (t)	Sheep (t)	Goat (t)
World	186,785	1,969	426,706	24,779	7,576	7,042
Asia	42,228	705	33,919	23,426	3,335	3,247
China	7,811	601	5,395	1,390	489	110
India	20,000	504	10,073	14,576	—	743
Israel	103	6,733	694	—	19	24
Japan	1,218	5,336	6,500	—	—	—
Philippines	—	—	—	18	—	—
USA	10,801	5,358	57,923	—	—	—
Africa	19,050	483	10,365	1,255	671	1,405
North and Central America	24,112	3,022	76,269	—	—	309
South America	17,919	1,012	23,254	—	34	135
Europe	48,576	3,467	180,272	98	8,436	1,546
Oceania	1,546	3,196	12,427	—	—	—
USSR	39,754	2,083	19,200	—	100	400
Developed	110,393	3,128	358,379	98	3,555	1,970
Developing	86,393	667	68,327	24,618	4,021	5,072
Developing, %	44	21	16	99.6	53	72

[a] Modified from FAO (1981).

Developing countries have 53% of the sheep and 95% of the goats, yet they produce only 58% of the total mutton and goat meat. The many reasons why meat production is less efficient in tropical than in the temperate zones are discussed later. Domestic ruminants contribute 39% of the world's meat, pigs about an equal amount, and poultry about 22% (Table 1.8).

Milk production by cattle in warm climates is even less efficient than meat production. Developing countries have 44% of the milk cows but produce only 16% of cow's milk (Table 1.9). Milk produced per cow is only 20% as much in developing countries as in the rest of the world. Most of the buffalo milk is produced in Asia, especially in India. Buffaloes produce more total milk in India and Egypt than cattle produce. A majority of the world's buffaloes are used for work and meat rather than primarily for milk. The total milk produced by sheep and goats is small, yet it is important in many places. Of the total milk supply, 91.5% comes from

TABLE 1.10

The Importance of Goats as a Source of Meat and Milk in Selected Countries (Percentage from All Ruminants)[a]

Country	Meat (%)	Milk (%)	Country	Meat (%)	Milk (%)
India	47.1	2.9	Greece	20.2	25.0
Cyprus	26.7	39.4	Bangladesh	19.2	33.1
Nepal	26.2	4.3	Turkey	17.0	12.1
Pakistan	26.0	3.7	Haiti	16.7	37.1
Nigeria	25.4	?	Ethiopia	15.8	12.1

[a] Modified from Devendra (Yousef, 1982).

cattle, 5.3% from buffaloes, and the remaining 3.2% about equally from sheep and goats. Camels, yaks, and reindeer produce important amounts of milk in their limited areas, but the amounts are not recorded. The importance of goats in supplying milk and meat in selected countries is illustrated in Table 1.10.

B. Work

There are only three practical power sources for field operations: human, animal, or engine. Cattle, buffaloes, and other large animals are used for various kinds of work in many countries (Figs. 1.4 and 1.5). The use of animals for draft in the cultivation of the soil and in hauling farm produce continues to be a dominant factor in Southeast Asia, the Near East, and some Mediterranean countries. In many areas of Africa, crop production is done only with hand tools. An adult can work 0.5 ha of land and a family is limited to not more than 2 ha. The family's entire effort is usually needed to produce food for their own use, with only small extras for sale in the best years. About one-quarter of the cultivated area in developing countries falls into this type of farming.

With a good pair of work oxen, a man can cultivate 3–4 ha of land. With tractor power much more land could be cultivated, but the cost and skill involved in importing, maintaining, and operating this type of machinery makes this impractical for most small farmers. The effective use of animal power appears to be the best way to increase farm crop output per capita. Table 1.11 shows the comparative draft potential of some animals found in most tropical countries. Larger animals could produce more draft power than shown. Cattle and buffaloes are more widely used than horses or mules, because they also supply acceptable milk and meat. Animals are also used to thresh grains by treading and to provide power for pumping

1. The Role of Ruminants in Warm Climates

Fig. 1.4. Draft animals in Bolivia.

Fig. 1.5. Cattle plowing near Addis Ababa, Ethiopia.

TABLE 1.11

Draft Power of Animals[a]

Animal	Body weight (kg)	Approx. draught (kg)	Speed (m/sec)	Power developed (kg/sec)	(Hp)
Buffalo	400	50	0.8	55	0.75
Cattle	500	60	0.6	56	0.75
Horse	400	60	1.0	75	1.00
Mule	350	50	0.9	52	0.70
Donkey	200	30	0.7	25	0.35

[a] Modified from Cochrill (1974).

water and performing other tasks. Use of animals for work of whatever kind increases their requirements for energy. Adequate feeding is necessary for sustained work. Goe (1983) has summarized the requirements of animals used for traction. Inns (1980) reviewed the research in order to improve the type of harness or yoke for cattle and buffaloes. Use of a harness with a collar or breast strap increased output by at least 25%. Improved equipment for farm cultivation and hauling has been designed. Production of the necessary equipment needs to be expanded in various countries to effectively enhance food production.

The buffalo has greater capacity for work in hot, marshy, flooded fields than any other farm animal. The camel is still common in the Sahara for riding and packing goods where other animals could not survive. The llama and alpaca do much of the hauling and packing of goods and produce in the higher Andes, where horses and cattle have not succeeded.

C. Other Products and Services

The use and importance of wool from sheep and hair from goats and Camelids is widely recognized. Animals are an important way to accumulate and store capital. They also contribute to recreation and entertainment in various ways.

By-products from the slaughter of an animal for meat include the hide, pelt, horns, hooves, and bones. These are used to produce various goods. Inedible organs, bones, and excess fatty tissue are recycled into food supplements. Among other by-products from animal wastes are glue, gelatin, oils, and some pharmaceuticals, as reviewed by McDowell (1977).

Animal manure is widely used as a fertilizer to help maintain crop yields. In West Africa, farmers pay nomadic cattle herders to bring cattle onto their land to eat the crop residues and leave the droppings on the land. Manure is hauled on pack animals or small carts from cattle holding areas to be spread on nearby crop lands. In India cattle manure, sometimes mixed with straw, is dried in cakes and stored to be used as fuel to cook the food for the family. A fresh manure–straw mixture is used as plaster to seal the cracks in adobe walls of buildings. Manure is recycled into animal feeds. Alone or with crop residues it is fermented to produce methane gas for cooking and heating.

IV. IMPROVING THE EFFICIENCY OF TROPICAL RUMINANTS

The productivity of animals in the tropics is considerably lower than that of similar breeds in temperate climates. Some reasons for this are heat stress, lower quality of feeds, deficiencies of specific nutrients, poor management, more losses from diseases and parasites, and lower genetic merit for rapid growth and milk yields.

Yousef (1982) pointed out that 50% of the world's livestock are in the tropics, yet they contribute less than 25% of the animal protein for humans. Meat production by tropical cattle is very low, from one-third to one-twentieth of that in the United States (Smith, 1976). Birth weight of calves is 15–20% lower in the tropics than for the same breeds in temperate climates. Growth rates are slower and the age of females at first calving is 4–6 years in the tropics compared to 2–3 years in temperate climates. Calving percentages are 40–50% versus about 85%. Age of steers at slaughter is 4–6 years compared to 1.5–2 years, and the carcasses are only 50–70% as heavy, and the slaughter rate is about 12% compared to 35% in the United States.

Young cattle gain weight during the rainy season and then lose at least half or more of the gain during the next dry season. Studies have shown that the weight losses can be prevented by feeding supplements, but in many cases the feeds are not available, or they are too expensive. There are two ways to starve, one is a lack of feed and the other is a deficiency of a specific nutrient. Both animals and people suffer in both of these ways.

A lack of feed is usually the result of too little rainfall. Heady (1984) pointed out that about 24% of the world's land area in the rain forests has from 0–2.4 months when lack of moisture limits plant growth. The 49% of the land in savanna grasslands and deciduous forests has dry periods of 2.5–7 months, and 27% of the land in thorn tree, short grass, and desert shrub areas has 7.5–12 months without rain. The variability of the yearly rainfall is even more of a problem than the low rainfall. The way to overcome a lack of moisture for forage production is to irrigate or to provide management to utilize all of the feed produced in the rainy season and to adjust animal numbers to fit the feed supply. Another method is to develop forages that require less moisture. As an example, grain sorghums require about one-half the moisture level of corn. In certain areas of Africa, grain sorghums are grown and used for bread making, etc., since the moisture level is too low to grow other grains. Better management and disease control can reduce death losses of animals. Selection of animals that are adapted to the tropical environment and have a greater resistance to the diseases and parasites can effect improvement, but selection is a slow process.

REFERENCES

Cockrill, W. R. (1974). "The Husbandry and Health of the Domestic Buffalo." FAO, Rome.
Devendra, C., and McLeroy, G. B. (1982). "Goat and Sheep Production in the Tropics." Longman, London.
Devendra, C., and Owen, J. E. (1983). *World Anim. Rev.* **47,** 19–29.
Food and Agriculture Organization of the United Nations (FAO) (1981). "Food and Agriculture Organization Yearbook." Rome.
Gilchrist, F. M. C. and Mackie, R. I. (eds.) (1984). "Symposium on Herbivore Nutrition in the Sub-tropics and Tropics—Problems and Prospects." The Science Press, Craighall, South Africa.
Goe, M. R. (1983). *World Anim. Rev.* **45,** 2–17.
Heady, H. F., (1984). *In* "Herbivore Nutrition in the Subtropics and Tropics" (F. M. C. Gilchrist and R. I. Mackie, eds.), pp. 29–47. The Science Press, Craighall, South Africa.
Inns, F. M. (1980). *World Anim. Rev.* **34,** 2–10.
Loosli, J. K., Oyenuga, V. A., and Babatunde, G. M. (eds.) (1974). "Animal Production in the Tropics," p. 1–12. Heinemann Educational Books, Ibadan, Nigeria.
McDowell, R. E. (1977). "Ruminant Products: More Than Meat." Winrock International Livestock Research and Training Center, Petit Jean Mountain, Morrilton, Arkansas.
McDowell, R. E., Sisler, D. G., Schermerhorn, E. C., Reed, J. D., and Bauer, R. P. (1983). "Game or Cattle for Meat Production on Kenya Rangelands." Cornell International Agriculture Mimeo 101. Cornell Univ., Ithaca, New York.
National Research Council (NRC) (1983). "Little Known Asian Animals." National Academy Press, Washington, D.C.
Omololu, A. (1974). *In* "Animal Production in the Tropics" (J. K. Loosli, V. A. Oyenuga, and G. M. Babatunde, eds.), pp. 13–17. Heineman Educational Books, Ibadan, Nigeria.
Payne, W. J. A. (1970). "Cattle Production in the Tropics," Vol. 1. Longman, London.
Schmidt-Nielson, K. (1969). *Sci. Am.* **201,** 140–151.

Simpson, J. R., and Evangelou, P. (eds.) (1984). "Livestock Development in Subsaharan Africa." Westview Press, Boulder, Colorado.
Smith, A. J. (ed.) (1976). "Beef Cattle Production in Developing Countries." Univ. of Edinburgh, Edinburgh.
Spinage, C. A. (1962). "Animals of East Africa." Collins, St. James's Place, London.
Starkey, P. H. (1984). *World Anim. Rev.* **50,** 2–15.
Yousef, M. K. (ed.) (1982). "Animal Production in the Tropics." Praeger, New York.

2

Nutrient Requirements of Ruminants

L. R. McDOWELL

Department of Animal Science
University of Florida
Gainesville, Florida

I.	General Requirements.	21
II.	Energy–Protein Requirements.	22
	A. Introduction.	22
	B. Forage Intake.	22
	C. Maintenance Requirements.	26
	D. Production Requirements.	26
	E. Nonprotein Nitrogen.	27
	F. Animal Traction Requirements.	27
III.	Factors Influencing Mineral Requirements.	28
	A. Introduction.	28
	B. Physiological State and Level of Production.	28
	C. Breed and Adaptation.	30
	D. Age Variation.	31
	E. Biological Availability and Interrelationships.	31
	F. Intake and Seasonal Effects.	32
IV.	Nutritional Relationships to Diseases and Parasites.	32
V.	Environment and Stress.	33
VI.	Nonnutrient Factors Affecting Requirements.	34
	References.	34

I. GENERAL REQUIREMENTS

Nutritional disorders including deficiencies, toxicities, and imbalances are severely inhibiting the grazing ruminant livestock industries of many countries. The nutrients required for ruminants can generally be grouped into the following categories: water, energy, protein, vitamins, and minerals. Ruminants are known to definitely require at least 15 minerals, with 7 in greater quantities referred to as macrominerals: calcium (Ca), chlorine

(Cl), magnesium (Mg), phosphorus (P), potassium (K), sodium (Na), and sulfur (S). Minerals required in smaller quantities are the trace elements: cobalt (Co), copper (Cu), Iron (Fe), iodine (I), manganese (Mn), molybdenum (Mo), selenium (Se), and zinc (Zn).

Quantitative nutrient requirements are given in publications by the U.S. National Research Council for beef cattle (NRC, 1984), dairy cattle (NRC, 1978), sheep (NRC, 1975) and goats (NRC, 1981). The other leading source of nutrient requirements for ruminants is the British Agricultural Research Council (ARC, 1980). A recent publication dealing with the nutrient requirements of ruminants in developing countries has been published (Kearl, 1982) which includes recommendations not only for cattle, sheep, and goats but also for domestic water buffalo. National Research Council requirements of energy, protein, Ca, P, Vitamin A, and Vitamin D for various classes of ruminants are presented in Table 2.1 (beef cattle), Table 2.2 (dairy cattle), and Table 2.3 (sheep). Scientific literature on goat nutrient requirements (NRC, 1981) is not nearly as extensive and comprehensive as it is for cattle and sheep, and thus recommendations have relied heavily on extrapolation of values derived from these species.

The present chapter will illustrate nutrient requirements for specific classes of beef cattle, dairy cattle, and sheep. The discussion, where appropriate, will emphasize requirements for grazing ruminants. Water requirements are discussed in Chapter 3, with the majority of references to vitamins presented in Chapter 18. The effects of thermal stress on the nutrient requirements of grazing livestock is illustrated in Chapter 4.

II. ENERGY–PROTEIN REQUIREMENTS

A. Introduction

Ruminant livestock production in tropical regions is largely dependent on unimproved grasslands, which often supply inadequate energy and protein. Energy–protein shortages are often found in the semiarid grazing areas and in grazing areas characterized by dry and wet seasons, owing to inadequate feed reserves for the dry season. Both the quantity and the quality of available herbage closely follows the rainfall pattern, which is strictly seasonal.

B. Forage Intake

To determine whether nutrient requirements of grazing livestock are being provided, the quantity and quality of the diet that is actually con-

TABLE 2.1

Nutrient Requirements for Selected Classes of Beef Cattle (Nutrient Concentration in Diet Dry Matter)[a,b]

Weight (kg)	Daily gain (kg)	Dry matter intake (kg)	Protein (%)	NE_m[c] (Mcal/kg)	NE_g[c] (Mcal/kg)	TDN[c] (%)	Calcium (%)	Phosphorus (%)
Medium-frame steer calves								
136	0.23	3.5	9.6	1.10	0.55	54.0	0.31	0.20
	1.36	3.6	19.9	2.09	1.41	85.0	1.13	0.47
364	0.23	7.5	7.7	1.10	0.55	54.0	0.22	0.17
	1.36	7.6	10.8	2.09	1.41	85.0	0.42	0.25
Large-frame heifer calves and compensating medium-frame yearling heifers								
136	0.23	3.5	9.5	1.10	0.55	54.0	0.31	0.20
	0.91	4.0	14.6	1.63	1.01	69.5	0.69	0.30
364	0.23	7.5	7.7	1.10	0.55	54.0	0.20	0.17
	0.91	8.5	9.0	1.63	1.01	69.5	0.28	0.19
Pregnant yearling heifers—last third of pregnancy								
318	0.64	7.2	9.0	1.32	0.75	60.3	0.33	0.21
409	0.64	8.6	8.5	1.28	0.70	59.1	0.30	0.21
Dry pregnant mature cows—last third of pregnancy								
364	0.41	7.6	8.2	1.12	NA	54.5	0.26	0.20
546	0.41	10.1	7.8	1.08	NA	52.9	0.26	0.21
Cows nursing calves—average milking ability first 3–4 months postpartum 10 kg milk/day								
455	0.0	9.2	9.6	1.21	NA	56.6	0.28	0.22
591	0.0	11.0	9.1	1.14	NA	55.1	0.27	0.22

[a] Modified from NRC (1984).
[b] Vitamin A requirements for growing steers and heifers are 2200 IU per kilogram of diet. Vitamin A requirements per kilogram of diet are 2800 IU for pregnant heifers and cows and 3900 IU for lactating cows. The Vitamin D requirement of beef cattle is 275 IU per kilogram of diet.
[c] Abbreviations: (NE_m), net energy for maintenance; (NE_g), net energy for gain; (TDN), total digestible nutrients.

TABLE 2.2
Nutrient Requirements for Selected Classes of Dairy Cattle (Nutrient Concentration in Diet Dry Matter)[a,b]

Nutrients (Concentration in the feed dry matter)	Cow weight (kg)	Lactating cow rations - Daily milk yields (kg)				Nonlactating cattle rations		
						Dry pregnant cows	Mature bulls	Growing heifers and bulls
	≤400	<8	8–13	13–18	>18			
	500	<11	11–17	17–23	>23			
	600	<14	14–21	21–29	>29			
	≥700	<18	18–26	26–35	>35			
Ration number		I	II	III	IV	V	VI	VII
Crude protein (%)		13.0	14.0	15.0	16.0	11.0	8.5	12.0
Energy[c]								
NE_l (Mcal/kg)[c]		1.42	1.52	1.62	1.72	1.35	—	—
NE_m (Mcal/kg)		—	—	—	—	—	1.20	1.26
NE_g (Mcal/kg)		—	—	—	—	—	—	0.60
ME (Mcal/kg)		2.36	2.53	2.71	2.89	2.23	2.04	2.23
DE (Mcal/kg)		2.78	2.95	3.13	3.31	2.65	2.47	2.65
TDN (%)		63	67	71	75	60	56	60
Crude fiber (%)		17	17	17	17[a]	17	15	15
Acid detergent fiber (%)		21	21	21	21	21	19	19
Ether extract (%)		2	2	2	2	2	2	2
Minerals								
Calcium (%)		0.43	0.48	0.54	0.60	0.37	0.24	0.40
Phosphorus (%)		0.31	0.34	0.38	0.40	0.26	0.18	0.26
Vitamins								
Vitamin A (IU/kg)		3200	3200	3200	3200	3200	3200	2200
Vitamin D (IU/kg)		300	300	300	300	300	300	300

[a] Modified from NRC (1978).

[b] It is difficult to formulate high-energy rations with a minimum of 17% crude fiber. However, fat percentage depression may occur when rations with less than 17% crude fiber or 21% acid detergent fiber are fed to lactating cows.

[c] Abbreviations: (NE_l), net energy for lactation; (NE_m), net energy for maintenance; (NE_g), net energy for gain; (ME), metabolizable energy; (DE), digestible energy; (TDN), total digestible nutrients.

TABLE 2.3

Nutrient Requirements for Selected Classes of Sheep (Nutrient Concentration in Diet Dry Matter)[a]

Weight (kg)	Daily gain (g)	Daily dry matter (kg)	TDN[b] (%)	DE[b] (Mcal/kg)	ME[b] (Mcal/kg)	Crude protein (%)	Calcium (%)	Phosphorus (%)	Vitamin A (IU/kg)	Vitamin D (IU/kg)
Ewes maintenance[c]										
50	10	1.0	55	2.4	2.0	8.9	0.30	0.28	1275	278
80	10	1.3	55	2.4	2.0	8.9	0.25	0.24	1569	342
Nonlactating and first 15 weeks of gestation										
50	30	1.1	55	2.4	2.0	9.0	0.27	0.25	1159	253
80	30	1.5	55	2.4	2.0	9.0	0.22	0.21	1360	296
First 8 weeks of lactation with sucking singles or last 8 weeks of lactation with sucking twins[d]										
50	−25(+80)	2.1	65	2.9	2.4	10.4	0.52	0.37	2024	132
80	−25(+80)	2.6	65	2.9	2.4	10.4	0.48	0.34	2615	171
Replacement lambs and yearlings[e]										
30	180	1.3	62	2.7	2.2	10.0	0.45	0.25	981	128
60	40	1.5	55	2.4	2.0	8.9	0.43	0.24	1700	222
Lambs finishing[f]										
30	200	1.3	64	2.8	2.3	11.0	0.37	0.23	588	128
50	220	1.8	70	3.1	2.5	11.0	0.28	0.17	708	154

[a] Modified from NRC (1975).
[b] Abbreviations: (TDN), total digestible nutrients; (DE), digestible energy; (ME), metabolizable energy.
[c] Values are for ewes in modern condition, not excessively fat or thin.
[d] Values in parentheses are for ewes with sucking twins last 8 weeks of lactation.
[e] Requirements for replacement lambs (ewe and ram) starting when lambs are weaned.
[f] Maximum gains expected.

sumed must be known. When animals are maintained under normal conditions, dry matter intake (DMI) is influenced primarily by body size (Tables 2.1, 2.2, and 2.3), energy density of the diet, and rate of digestion or fermentation. The DMI will drop significantly below expected levels when animals are subjected to grazing poor-quality forage. Low amounts of digestible energy (DE) and protein content of the diet imposes a severe physical restriction on the amount of feed an animal can consume. Data utilizing native *Hyparrhenia rufa* pasture showed that *Bos indicus* steers consumed dry matter equivalent to 1.2% of their body weight when the herbage contained 50% digestible organic matter (DOM), but as the dry season progressed, intake of forage fell to 0.8% of body weight when DOM dropped to 38% (Smith, 1962). Intake of forage is sharply reduced when the total protein content is less than 6–7% (Milford and Haydock, 1965).

C. Maintenance Requirements

Maintenance requirements can be defined as the amount of feed energy and protein that will result in no loss or gain in body energy and protein. There are variations in maintenance requirements based on sex, breed, and physiological age. In general, breeds and individuals maturing at heavier weights may have a higher maintenance requirement, and *Bos indicus* breeds, which are more adapted to warm climates, may require less. With grazing livestock, the "activity increment" of movement may raise energy maintenance needs by 10–20%, or more, above those of the animal at rest. Kearl (1982) suggests adding 25% above basic maintenance requirements for animals grazing pastures where limited physical exercise is required, 50% for open ranges with long distances to available water, and 75% when extremely rugged mountain or harsh desert environments prevail.

D. Production Requirements

For production functions such as growth and milk secretion, a substantial part of the animal's total feed is used for maintenance, the proportion so used decreasing with rising rate of production. Comparing the protein and energy requirements of a medium-frame steer gaining either 0.23 or 1.36 kg daily would be 9.6 and 19.9% protein and 54.0 and 85.0% total digestible nutrients (TDN), respectively (Table 2.1). The lactating ability of the ruminant dictates the increased need for dietary energy and protein. Table 2.2 illustrates the greatly increased energy and protein requirements of dairy cattle as milk production increases.

Dietary needs of nonpregnant ruminants increase after pregnancy, especially during the later phase of gestation. Kearl (1982) averaged six reports on energy needs of sheep during the last 6 weeks of pregnancy and found requirements to be 171% of maintenance requirement. The NRC (1975) suggests multiplying by a factor of 1.5 times the maintenance requirement for single lambs and 2 times for twin lambs to derive the metabolizable energy (ME) requirement during the last 6 weeks of pregnancy.

E. Nonprotein Nitrogen

Ruminants have the ability to use both protein and nonprotein sources of nitrogen. The most common source of nonprotein nitrogen (NPN) fed to ruminants is urea. Urea is rapidly hydrolyzed to ammonia by microbial urease in the rumen. The extent of synthesis of ammonia into microbial protein is largely dependent on the energy concentration of the diet.

Precautions are necessary when NPN is fed both to prevent ammonia toxicity and to avoid reduction in feed intake. Single doses of urea at 0.3–0.8 g per kilogram of body weight have toxic effects (NRC, 1984). Toxicity can be prevented by thoroughly mixing urea with the diet and setting the maximum concentration of 1% of the diet dry matter or one-third of the total dietary protein. Special care in mineral supplementation must be exercised, since most sources of protein provide substantial amounts of S, K, and P, which are absent in NPN sources.

F. Animal Traction Requirements

Nitis (1980) reported that 45% of the large ruminant population in Indonesia and 85% of that in Thailand are raised primarily for land preparation and transport purposes. Bullocks are the main source of draft power in India, while swamp buffaloes have become the "beast of burden" throughout the rice production areas of the world.

Age, sex, breed, species, tractive effort(s) produced, and duration of work will determine energy needs. Mature oxen or buffaloes have requirements for maintenance and work, while immature animals and pregnant or lactating cows employed for draft have additional needs. In general, energy expenditure of working animals, such as oxen and water buffaloes, is three to eight times that of basal metabolism, but actual relative metabolic rates will be affected by frequency and level of work (Goe, 1983). The general conclusion is that an increase in protein for work above maintenance levels is not needed, provided the maintenance ration is adequately balanced.

Draft up a grade is performed less efficiently in relation to energy requirements than draft on level ground. Estimates of the net energy cost of carrying loads on the horizontal were 3.3. and 4.8 per kilogram carried per meter traveled and for raising loads against gravity (6% slope), 23 and 36 J per kilogram carried per vertical meter for Brahman and crossbreds, respectively (Smith and Pearson, 1984). At 30°C, animals walking on the 6%+ slope showed large increases in rectal temperature and respiration rate within 50 min of starting work, with the crossbred animals having significantly higher respiration rates than the purebred Brahmans.

In warm climate regions of the world, it is often difficult during the greater part of the year for ruminants to obtain needed energy while grazing on natural pasture alone. Therefore, for most of the year, energy supplementation will be needed for peak working performance.

III. FACTORS INFLUENCING MINERAL REQUIREMENTS

A. Introduction

Approximate mineral requirements for various ruminant livestock are presented in Table 2.4. Many factors affect mineral requirements, including nature and level of production, age, level and chemical form of elements, interrelationship with other nutrients, mineral intake, breed, and animal adaptation. The criterion for adequacy is important, as illustrated by the fact that minimum Zn requirements for spermatogenesis and testicular development in male sheep are higher than for growth, and Mn requirements are, similarly, lower for growth than for fertility (Underwood, 1981).

B. Physiological State and Level of Production

Mineral requirements are highly dependent on the level of productivity and physiological state of the animal (NRC, 1984). As an example, a young, pregnant beef cow during her first lactation would have substantially higher mineral requirements than a mature dry cow. High-yielding milking cows obviously require much more dietary Ca and P than low-yielding cows because of the richness of milk in those elements (Table 2.2).

Improved management practices that lead to improved milk production and growth rates for grazing ruminants will necessitate more attention to mineral nutrition. Mineral deficiencies, perhaps only marginal until this time, are likely to become important, and previously unsuspected nutritional deficiency signs may occur as production level increases.

TABLE 2.4
Suggested Mineral Requirements for Ruminants (Dry Basis)

Required elements	Beef cattle[a]		Lactating dairy cows[b]		Sheep[c]		Goats[d,e]	
	Suggested value	Range[f]	Suggested value	Range	Suggested value	Range	Suggested value	Range
Macroelements								
Calcium (%)	(Table 2.1)	—	(Table 2.2)	0.43–0.60	—	0.21–0.52	—	—
Phosphorus (%)	(Table 2.1)	—	(Table 2.2)	0.31–0.40	—	0.16–0.37	—	—
Magnesium (%)	0.10	0.05–0.25	0.20	—	—	0.04–0.08	—	—
Potassium (%)	0.65	0.50–0.70	0.80	0.80–1.20	0.50	—	—	0.50–0.80
Sodium (%)	0.08	0.06–0.10	0.18	—	—	0.04–0.10	—	—
Sulfur (%)	0.10	0.08–0.15	0.2	—	—	0.14–0.26	—	0.16–0.32
Microelements								
Cobalt (ppm)	0.10	0.07–0.11	0.1	—	0.1	—	0.1	—
Copper (ppm)	8.0	4–10	10.0	—	5.0	—	—	—
Iodine (ppm)	0.50	0.20–2.00	0.5	—	—	0.10–0.80	—	—
Iron (ppm)	50.0	50–100	50	—	—	30–50	—	—
Manganese (ppm)	40.0	20–50	40	—	—	20–40	>5.5	—
Molybdenum (ppm)	—	—	—	—	>0.5	—	—	—
Selenium (ppm)	0.20	0.05–0.30	0.1	—	0.1	—	—	—
Zinc (ppm)	30.0	20–40	40	—	—	35–50	>10.0	—
Toxic elements[g]								
Copper (ppm)	115		80		8–25		?	
Fluorine (ppm)	20–100		30		60–200		?	
Molybdenum (ppm)	6		6		5–20		?	
Selenium (ppm)	5[h]		5		>2.0		?	
Zinc (ppm)	500		500		1000		1000	

[a] NRC (1984).
[b] NRC (1978).
[c] NRC (1975).
[d] NRC (1981).
[e] Mineral requirements for goats have not been studied in detail. Lactating dairy goats have requirements similar to lactating dairy cattle. Other goats have mineral requirements similar to sheep (Haenlin, 1980).
[f] The listing of a range recognizes that requirements for most minerals are affected by a variety of dietary and animal factors.
[g] NRC (1980).
[h] McDowell et al. (1984).

C. Breed and Adaptation

Important differences in mineral metabolism for tropical livestock can be attributed to breed and adaptation. A most important aspect of adaptability is concerned with local breeds and types that, through the centuries, have accustomed themselves to a certain dietary and management regime. Any sudden change, such as increased size due to grading up, will result in a higher maintenance and production requirement. This gives rise to a need for more nutrients, which, if not supplied, will cause a general upset in the animal's metabolism as it attempts to meet these demands. In an area around Salinas, Mexico, lambs have died of Co deficiency (Phillips, 1956). This malady only appeared as the local sheep type was being replaced by the more robust Rambouillet, through a grading-up program.

It is not unusual for cattle introduced into an area to show deficiency signs, while the indigenous breeds that are slow-growing and late-maturing do not exhibit the deficiencies to the same degree. In Brasil, *Bos inducus* exhibited clinical Co deficiency when fed forage containing 0.08 ppm Co while the more indigenous breeds of *Bos taurus* were not affected until the Co level dropped to 0.05 ppm or lower (Correa, 1957). Payne (1966) suggests the possibility that unacclimatized cattle that sweat profusely in hot temperatures and lose saliva and mucus from the mouth may lose significant quantities of minerals, particularly in the arid tropics.

The effect of breed differences on mineral requirements has often been observed in ruminants. A far higher instance of milk fever among Jerseys than among Holsteins reflects such an effect (Miller, 1979). Likewise, Blackface sheep require and tolerate substantially more Cu than Welsh sheep (Wiener and Field, 1969).

There is also extensive evidence showing marked variations within breeds in the efficiency of absorption of minerals from the diet: 3–35% for Mg in dairy cows and 40–80% for P and 2–10% for Cu in adult sheep (Field, 1984). When different breeds of sheep grazed certain pastures in Scotland, one breed exhibited signs of Cu poisoning whereas another showed signs of Cu deficiency (Wiener *et al.*, 1977). The most probable cause for this apparent variation in dietary requirements for some microelements between breeds could be genetic differences in the efficiency of absorption of the mineral in the diet (Field, 1984). Wiener *et al.* (1977) found variations in the absorption of dietary Cu between different breeds of growing lambs, and Field (1984) found much greater variation in the absorption of Mg between monozygotic twin cows. It is, therefore, suggested that future recommendations of dietary requirements, particularly minerals, should take into account the breed when determining the needs of animals. (Field, 1984).

D. Age Variation

Animal age is an important consideration for mineral requirements. Dietary requirements may decline with age because the major requirement for growth often remains constant (for a given liveweight gain), while appetite increases in proportion to body size. These changes in the need for the trace elements Fe and Zn are greatest since their requirements for growth are particularly high (Suttle, 1979). Likewise, Se and I deficiencies occur in the offspring of clinically normal appearing mothers. Thus, when other factors (particularly absorption) are constant, the young ruminant is more vulnerable to trace-element deficiencies. The older animal also has the advantage of being more resistant to mineral excesses. In mature cows, homeostatic control mechanisms that regulate Zn content in tissues are much more effective than in calves; therefore, mature cows probably are able to tolerate higher concentrations of dietary Zn (Kincaid et al., 1976).

Young animals have the advantage of being more efficient in metabolizing specific nutrients than mature animals, particularly in regard to absorption efficiency. Jorgensen (1983) reported that above 14 months of age absorption efficiency decreased so that aged dairy cows absorbed P less efficiently than growing heifers. Mineral mobilization is also more efficient in young animals. As indicated by Rook and Storry (1962), about 30% of skeletal Mg in young ruminants can be mobilized under conditions of Mg deprivation while in adult animals, only 2% of bone Mg can be used for physiological needs. This is most apparent in the observation that older cattle are much more susceptible to grass tetany.

E. Biological Availability and Interrelationships

Specific mineral requirements are difficult to pinpoint since exact needs depend on chemical form and numerous mineral interrelationships. The relative biological availability of the desired element in a compound or supplement is one of the major considerations in the selection of a suitable source of the element (Ammerman and Miller, 1972). Numerous dietary factors, including protein source and level, interrelationships among the mineral ions, and certain chelating agents, influence the utilization of mineral ions. With some elements, the chemical form has a major impact on the availability of the element. For instance, Fe is far more available as ferrous sulfate than as ferric oxide (Miller, 1979).

Other constituents of the diet often have a major impact on the amount of minerals needed and tolerated. For instance, Cu requirement and tolerance are very closely related to dietary Mo. As the Mo increases, the need and tolerance for Cu also increase. Even the form of the Mo seems to have an influence, with that in natural forages having more than added inorganic Mo (Miller, 1979).

In many respects, the dietary requirements for minerals are more difficult to accurately define than those for the organic nutrients because many factors determine the utilization of minerals. For example, interrelationships among minerals or relationships between minerals and organic fractions may result in enhanced or decreased mineral utilization. Numerous mineral interrelationships that affect requirements and mineral status of the animal include Ca–P, Ca–Zn, Cu–Mo–S, Cu–Fe, Se–As, Se–S, Fe–P, Na–K, and Mg–K. The organic constituents of the diet can have a major impact on the amounts of different mineral elements needed and tolerated. A good illustration is the relationship between vitamin E and Se and between vitamin B_{12} and Co; also, the effect of vitamin D on Ca and P metabolism is well known. Goitrogenic substances and chelates such as oxalic acid and phytic acid each influence specific mineral requirements (McDowell, 1976).

F. Intake and Seasonal Effects

Adequate intake of forages by grazing ruminants is essential in meeting mineral requirements. Factors that greatly reduce forage intake, such as low protein (< 7.0%) content and increased degree of lignification, likewise reduce the total minerals consumed.

Since tropical forages contain lesser amounts of minerals during the dry season, it is logical to assume that grazing livestock would most likely suffer mineral inadequacies during this time. On the contrary, numerous reports, including those from Kenya, Brazil, and South Africa, have noted that specific mineral deficiencies are more prevalent during the wet season (McDowell *et al.*, 1984). During the wet season, livestock gain weight rapidly since energy and protein supplies are adequate. Associated with the rapid growth during the wet season, mineral requirements are high, while during the dry season, inadequate protein and energy result in weight loss, thereby greatly reducing mineral requirements (see Chapter 16).

IV. NUTRITIONAL RELATIONSHIPS TO DISEASES AND PARASITES

Adequate nutrition is essential for providing a greater resistance to infections and for counteracting parasitic infestations. An animal's ability to produce antibodies depends on adequate intakes of protein and other essential nutrients. If any one of the essential nutrients is lacking, growth stops, reproduction may cease, and eventually the animals will die of complications with diseases from lowered resistance to infections (Loosli, 1974). Adequate intakes of protein and vitamin A are particularly essential to provide resistance to infections and are necessary to recovery from disease and parasites. White Fulani cattle thrive in the presence of mild infestations with trypanosomiasis, if they are fed well (Hill, 1964). Work oxen of Ibadan, Nigeria, if receiving supplemental feed, show no ill effects from trypanosomiasis but will die from the combined stress of work and infection if restricted to poor dry season grazing (Loosli, 1974).

Where nutritional factors are limiting production, any effect of parasites on the true utilization of nutrients will be of considerable importance. Evidence for a reduction in the digestibility of nitrogen due to chronic infection with *Trichostrongylus colubriformis* in sheep has been reported (Barger, 1973). Likewise, in sheep, with the same organism, Sykes and Coop (1976) report a 50% reduction in the efficiency of the utilization of ME for maintenance and growth. Sykes and Coop (1976) administered the intestinal parasite *T. colubriformis* to growing lambs and concluded that an induced Ca and P deficiency resulted from parasitic damage in the small intestine. Paynter et al. (1979) found that fecal egg counts in lambs that had been treated with Se 10 weeks before sampling were significantly lower than those of control lambs, suggesting Se played a role in alleviating "ill-thrift" resulting, in part, from enhanced immune responses to helminth infaunation.

V. ENVIRONMENT AND STRESS

Stated requirements commonly have been established in an environment protected from climatic extremes and are less appropriate when animals are exposed to stressful environments. Of the many stresses affecting rate and efficiency of animal productivity, more is known of the consequences of the thermal environment and associated factors of humidity, radiation, and air movement than of factors such as altitude, sound, animal density, confinement, chemical or biological contamination, and specific nutrient deficiency or toxicity. Studies show environmental stress generally alters animal performance in one or more of five ways (Curtis, 1984): (1) it alters internal functions; (2) it diverts nutrients to use in higher-priority maintenance processes; (3) it reduces productivity directly; (4) it increases variability; or (5) it impairs disease resistance.

Heat, cold, crowding, regrouping, weaning, limit feeding, noise, and movement restraint are eight stressors that can play a central role in altering resistance to infection. The single most important reaction affecting the performance of heat-stressed ruminants is the reduction of voluntary feed intake (see Chapter 4).

VI. Nonnutrient Factors Affecting Requirements

Nonnutrient factors including antibiotics, hormones, and buffers have been used for a number of years and have in many instances provided a favorable return through improved animal performance, improved utilization of feed, and reduced incidence of disease and other disorders. A listing and use of feed additives, drugs, and implants for cattle has been prepared by Perry (1980).

Antibiotics, in addition to reducing the incidence of liver abscesses and shipping fever, have improved daily gain and feed efficiency generally in excess of 2–5%, with greatest response in the early part of the finishing period. Hormone or hormonelike compounds generally improve gains in feedlot cattle 10–15% and increase feed efficiency 7–10%. Changes in feeding and management by substitution of grain for forages has created a need for dietary buffers. Buffers counteract the low pH resulting from grain diets and aid in preventing metabolic disturbances in both high-producing dairy cattle and finishing beef cattle.

A new aspect of energy metabolism was brought about by the introduction of the ionophores, characterized by monensin and lasalocid. Much of the evidence today indicates the use of ionophores improves the efficiency of feedstuffs by 10%. High-roughage diets hold great potential for increased gains and feed efficiency in response to ionophores. The major limitation to use of ionophore compounds for grazing cattle is lack of a suitable method of administration. However, promising results have been reported by adding lasalocid to free-choice mineral mixtures (Pitman and Pate, 1984). Increased gains for the grazing period of 4.79% and 12.3% for the low- and high-lasalocid intakes, respectively, were obtained over the no-lasalocid controls.

REFERENCES

Agricultural Research Council (ARC) (1980). "The Nutrient Requirements of Ruminant Livestock." Agricultural Research Council—Commonwealth Agricultural Bureaux, Slough, England.
Ammerman, C. B., and Miller, S. M. (1972). *J. Anim. Sci.* **35,** 681–694.
Barger, I. A. (1973). *Aust. J. Exp. Agric. Anim. Husb.* **13,** 42–47.
Correa, R. (1957). *Arq. Inst. Biol.* **24**(15), 199–227.

Curtis, S. E. (1984). *Feedstuffs.* **56**(40), 44–45.
Field, A. C. (1984). In "IMC Mineral Conference," pp. 71–93. International Minerals and Chemical Corporation, Mundelein, Illinois.
Goe, M. R. (1983). *World Anim. Rev.* **45,** 2–17.
Haenlein, G. F. W. (1980). *J. Dairy Sci.* **63,** 1729–1748.
Hill, D. H. (1964). *Outlook on Agric.* **4,** 80–85.
Jorgensen, N. A. (1983). *Proc. Minn. Nutr. Conf.* **44,** 15–27.
Kearl, L. C. (1982). "Nutrient Requirements of Ruminants in Developing Countries." Utah State Univ., Logan.
Kincaid, R. L., Miller, W. J., Fowler, P. R., Gentry, R. P., Hampton, D. L., and Neathery, M. W. (1976). *J. Dairy Sci.* **59,** 1580–1584.
Loosli, J. K. (1974). *Proc. Niger. Soc. Anim. Prod.* **1,** 74–82.
McDowell, L. R. (1976). In "Beef Cattle Production in Developing Countries" (T. Smith, ed.), pp. 216–241. Centre for Tropical Veterinary Medicine, Univ. of Edinburgh, Edinburgh.
McDowell, L. R., Conrad, J. H., Ellis, G. L., and Loosli, J. K. (1983). "Minerals for Grazing Ruminants in Tropical Regions." Univ. of Florida, Gainesville.
McDowell, L. R., Conrad, J. H., and Ellis, G. L. (1984). In "Symposium on Herbivore Nutrition in Sub-Tropics and Tropics—Problems and Prospects" (F. M. C. Gilchrist and R. I. Mackie, eds.), pp. 67–88. The Science Press, Craighall, South Africa.
Milford, R., and Haydock, K. P. H. (1965). *Aust. J. Exp. Agric. Anim. Husb.* **5,** 13–17.
Miller, W. J. (1979). "Dairy Cattle Feeding and Nutrition." Academic Press, New York.
National Research Council (NRC) (1975). "Nutrient Requirements of Sheep," 5th rev. ed. Nutrient Requirements of Domestic Animals, No. 5. Natl. Acad. Sci., Washington, D.C.
National Research Council (NRC) (1978). "Nutrient Requirements of Dairy Cattle," 5th rev. ed. Nutrient Requirements of Domestic Animals, No. 3. Natl. Acad. Sci., Washington, D.C.
National Research Council (NRC) (1980). "Mineral Tolerance of Domestic Animals." Natl. Acad. Sci., Washington, D.C.
National Research Council (NRC) (1981). "Nutrient Requirements of Goats." Nutrient Requirements of Domestic Animals, No. 15. Natl. Acad. Sci., Washington, D.C.
National Research Council (NRC) (1984). "Nutrient Requirements of Beef Cattle." 6th rev. ed. Nutrient Requirements of Domestic Animals, No. 4. Natl. Acad. Sci., Washington, D.C.
Nitis, I. M. (1980). In "Proceedings of the International Workshop on Studies on Feeds and Feeding of Livestock and Poultry—Feed Composition Data Documentation and Feeding Systems in the APHCA Region" (L. C. Kearl and L. E. Harris, eds), pp. 100–107. Int. Feedstuffs Institute, Utah Agric. Exp. Stn., Utah State Univ., Logan.
Payne, W. J. A. (1966). *Nutr. Abstr. Rev.* **36**(3), 653–670.
Paynter, D. I., Anderson, J. W., and McDonald, J. W. (1979). *Aust. J. Agric. Res.* **30,** 703–704.
Perry, T. W. (1980). "Beef Cattle Feeding and Nutrition," pp. 329–332. Academic Press, New York.
Phillips, R. W. (1956). "Recent Developments Affecting Livestock Production in the Americas." FAO Agriculture Development Paper No. 55, pp. 83–98. Food and Agriculture Organization, Rome.
Pitman, W. D., and Pate, F. M. (1984). In "Florida Beef Cattle Research Report," pp. 74–76. Dep. Animal Science, Univ. of Florida, Gainesville.
Rook, J. A. F., and Storry, E. (1962). *Nutr. Abstr. Rev.* **32**(41), 1055–1077.

Smith, A., and Pearson, A. (1984). *Draught Anim. News.* **2,** 1–14.
Smith, C. A. (1962). *J. Agric. Sci.* **58,** 173–178.
Suttle, N. F. (1979). *Br. J. Nutr.* **42,** 89–96.
Sykes, A. R., and Coop, R. L. (1976). *J. Agric. Sci.* **86,** 507–515.
Underwood, E. J. (1981). "The Mineral Nutrition of Livestock." Commonwealth Agricultural Bureaux, London.
Wiener, G., and Field, A. C. (1969). *J. Comp. Pathol.* **79,** 7–14.
Wiener, G., Field, A. C., and Smith, C. (1977). *Vet. Rec.* **101,** 424–425.

3

Water Requirements for Grazing Ruminants and Water as a Source of Minerals

RAY L. SHIRLEY
Department of Animal Science
University of Florida
Gainesville, Florida

I.	Introduction.	37
II.	Water Turnover in Ruminants.	39
III.	Voluntary Intake of Water.	40
IV.	Effect of Dry Matter Intake on Water Consumption.	40
V.	Effect of Water Deprivation on Feed Intake and Utilization.	42
VI.	Effect of Temperature on Water Intake and Digestibility of Nutrients.	44
VII.	Effect of Salinity on Water Consumption.	45
VIII.	Water and Electrolyte Metabolism.	48
IX.	Nutrient Elements in Water.	49
X.	Toxic Elements and Substances in Drinking Water.	50
XI.	Summary.	54
	References.	55

I. INTRODUCTION

Small streams, rivers, lakes, ponds, springs, and wells are common sources of drinking water for ruminants (Figs. 3.1 and 3.2). Due to the universal solvent action of water, many chemical elements and compounds are found in solution as ions, molecules, and radicals.

The nutritional or toxic effects of an element often vary with its ionic form. *Total dissolved solids* refers to the concentration of all dissolved constituents in water. *Salinity* is a synonymous term when applied to total

Fig. 3.1. Small streams and rivers as a source of drinking water for cattle.

Fig. 3.2. Wildebeest drinking water from a pond in South Africa. (Courtesy of L. R. McDowell, University of Florida, Gainesville, Florida.)

ionic concentrations of fresh water. *Hardness* is dependent on the concentration of divalent ions, generally of calcium (Ca) and magnesium (Mg). Very saline water (water with high levels of NaCl) may be *soft* water.

All minerals essential as dietary nutrients for livestock are present in water (Shirley, 1970). It is generally believed that nutrient elements in water are available to animals to about the same extent that they are in feedstuffs or salt mixtures. The radioactive salts of phosphorus (P) and Ca when dissolved in water and administered to steers in a drench were absorbed and deposited in the tissues of steers and excreted at levels equivalent to those of the isotopes incorporated into forages from fertilizer (Shirley *et al.*, 1951, 1957).

Concentrations of mineral nutrients in drinking water for ruminants are generally quite inadequate for meeting dietary requirements. Research indicates that salt should be made available on an ad libitum basis for livestock. Deprivation of water, even to a moderate extent, generally results in a decrease in feed intake and loss of body weight. Cattle prefer to drink several times daily, and if water consumption is limited to once per day or once per two days, cattle will eat less feed and grow more slowly.

II. WATER TURNOVER IN RUMINANTS

Water turnover in ruminants is a factor in their requirement for water. Anand and Parker (1966) found with a tritiated water technique that sheep contain $54.6 \pm 1.8\%$ water on the body weight basis, and the water has a mere biological half-life of 5.4 ± 0.4 days. There is a large content and turnover of water in the alimentary tract of ruminants. Salivary glands provide much of this water and may provide 6–16 liters/day in Merino sheep (Denton, 1956). This indicates that a large volume of water in the alimentary tract is recirculated in the blood.

Sheep are different from cattle in that sheep do not have a large turnover of water in proportion to body size. Renal and fecal water loss form a greater part of the water turnover in cattle than in sheep. MacFarlane *et al.* (1963) observed that water turnover by grazing cattle, Merino sheep, and camels in arid tropics with a mean maximum temperature of 41°C was 148, 110, and 61 ml per kilogram of body weight per 24 hr, respectively. Sheep have feces that are relatively dry compared to that of cattle, and much of the water lost in sheep is by respiration and sweating.

The half-life of water (time in which approximately one-half of the water in the animal's body needs to be replaced) in cattle was reported to be 3.5 ± 0.21 days (Black *et al.*, 1964) with essentially no difference between lactating and nonlactating cows. It was postulated that during lactation a parallel change in pool size and water flux minimizes any change in half-life of body water. Season of the year and ambient temperature had little influence on turnover of water in cattle.

III. VOLUNTARY INTAKE OF WATER

The drinking habits of Friesian cows feeding on pasture on two farms in southwestern England during June, July, and August were observed by Thomas (1971). Voluntary intake of water ranged from 4–14 gal per cow per day, at an average drinking rate of 3.5 gal/min. Peak intake of water occurred between 18:00 and 20:00 hr. Cows normally drank 3–4 times per day. It was concluded that factors which affect voluntary intake of water included milk yield, dry matter (DM) of feed, air temperature, supply of water at peak demand periods, trough refill rate, and cleanliness of troughs. It was indicated that for a 100-cow herd a flow rate of 140 gal of water per hour should be provided in troughs where as many as 6 cows can drink at once.

The grazing behavior and free-water intake of Zebu heifers in Uganda was observed by Wilson (1961). The heifers limited their drinking of water from 2 hr after sunrise to 3 hr before sunset. On the average, the heifers drank 10.1 kg of water during the transitional period and 20.4 kg/day in the dry season when weighing approximately 198.1 kg in body weight.

IV. EFFECT OF DRY MATTER INTAKE ON WATER CONSUMPTION

Many studies have demonstrated the amount of DM in the feed intake markedly affects water consumption of animals. In case of water deprivation, the amount of water intake will similarly affect the amount of ad libitum DM intake. Leitch and Thompson (1944) reviewed early studies on water intake of farm animals and concluded that water intake was quantitatively related to feed intake of ruminants. Brown (1966) concluded from a review of the literature that, in most instances, reducing feed intake of ruminants generally increased digestibility of nutrients. An increase in digestion of nutrients by restriction of feed plus a further increase in digestion when water also was restricted was observed by Thornton and Yates (1968).

Sheep were compared on good pasture when provided water ad libitum, one-half ad libitum, and no free water in two experiments (Calder et al., 1964). Ewes with single lambs were used at two stocking rates one year and weaned lambs the following year. Increasing the stocking rate decreased weight gain by ewes and lambs more than lack of drinking water did. Lactating ewes on adequate pasture had greater weight loss from lack of water than occurred with lambs. In another trial utilizing a variety of diets with 111 individual digestibility determinations, it was observed that sheep consumed a total-water-to-DM ratio of 2.5–1 (Calder et al. 1964). When silage was fed, the ratio increased to 3.3–1, and with frozen pasture herbage, the water-to-DM intake ratio was 4.9–1. There was positive correlation between DM intake ratio and both free water and total water consumption in the various diets with the exception of frozen herbage, which indicated that the frozen herbage had more water than was required by the wethers.

Forbes (1968) conducted several trials with sheep evaluating the effect of dry matter intake, silage compared to hay in diets, pregnancy, and number of lambs per ewe on water intake. Total water intake was closely correlated to dry matter intake and was higher with silage than with dried grass. Total water intake was higher with both dried grass and silage than with long hay. When ewes were fed silage or hay from the ninth to the nineteenth week of pregnancy, total water intake per unit DM intake with each diet doubled during the period. Both twin-bearing and single-lamb-bearing ewes during the last few weeks of pregnancy consumed more total water relative to DM than nonpregnant ewes did. During the first 4 weeks of lactation, total water intake per unit of DM intake was greater with nursing ewes than that of nonpregnant ewes, not counting the water in milk.

Owen and Miller (1968) fed Friesian heifers four diets from 2 months before calving to the end of the first lactation. Two treatments involved ad libitum feeding of a complete diet (barley with supplements) with either ground hay or ground barley straw (25% of diet). In the other two treatments, the hay or straw was fed in the long form ad libitum. The intake of both feed and water showed a consistent pattern in all treatments. Water intake was more closely associated with DM intake than with milk production. Feed intake was lowest around calving time, followed by a sharp rise to a peak in the fourth month of lactation and then a gradual but accelerating decline. A similar pattern of water intake occurred, but the fluctuation was greater and reached a peak in intake during the third month of lactation.

A statistical analysis of 219 adult nonpregnant dry cows fed 71 different diets in metabolism stalls was conducted by Paquay et al. (1970). Dry matter intake was correlated with water intake. Forages with high moisture content enhanced total water intake. Total water intake also was directly related to dietary nitrogen (N), crude fiber, pentosans, fat, Ca, P, Mg, potassium (K), sodium (Na), and chlorine (Cl) intakes. Fecal water losses were positively correlated with dry matter intake and fecal DM. Urinary water was closely related to the amount of absorbed water and DM in the diet. The amount of urinary water may affect the utilization of digestible N and K.

The individual water intakes of 16 (4 groups of 4) lactating dairy cows, measured on 7 consecutive days, were significantly correlated with DM intakes (4.6–14.4 kg/day) and milk yields (13.7–30 kg/day) (Little and Shaw, 1978). A multiple-regression analysis of the data gave the following equation: water intake (kg/day) = 2.15 ($\pm$.42) x DM intake (kg/day) + 0.73($\pm$20) x milk yield (kg/day) + 12.3 ($\pm$5.6).

Reviewing the data, Winchester and Morris (1956) concluded that water intake by cattle is a function of DM consumption and ambient temperature. They graphically plotted water intake versus temperature and obtained curves that remained horizontal between about -12 and $4.4°C$, and then rose with rising temperature at an accelerating rate to $37.8°C$. Data for European cattle (*Bos taurus*) gave a different curve from that for Indian cattle (*Bos indicus*) due to a generally greater water intake by European cattle.

Male Holstein calves 14 days of age were divided into groups and treated with ad libitum feedings of a commercial milk replacer mixed with water to contain 5, 10, 15, 20, or 25% dry matter by Pettyjohn *et al.* (1963). Feed intake increased and dry matter intake decreased as water was increased in diets. Calves fed the 15% DM diet had the best gains and better or equal feed efficiency. Dry matter and protein digestibilities were markedly reduced at higher DM levels. Less scours occurred with the 5 and 10% DM concentrations. Average total water intake per day was greater with the lower DM concentrations.

Guernsey cows fed indoors on cut grass (cut twice per day) were observed to consume 6.7, 2.5, and 5.5. gal of water daily with grass that contained 24.7, 18.0, and 26.0% dry matter (Halley and Dougall, 1962). The total water consumed (i.e., drinking water plus water in the grass) was almost constant at 14 gal per cow per day.

V. EFFECT OF WATER DEPRIVATION ON FEED INTAKE AND UTILIZATION

Availability of drinking water has been observed to markedly affect feed intake, while DM intake has been reported to affect the intake of water. Sheep were observed before, during, and after a period of 4–5 days without water other than that present in the air-dry feed offered (Gordon, 1965). No observed changes took place during 1 or 2 days of water deprivation, but by the fourth day there was a 30% decrease in time actually spent chewing, a 34% decrease in the number of boli regurgitated, and a 46% decrease in feed intake. Watering Hereford heifers once per day decreased tap water intake by about 10% of ad libitum intake, and with watering once per 2 days the intake was decreased about 30% (Weeth et al., 1968).

A trial was conducted by Thickett et al. (1981) that compared the voluntary water intake of Friesian bull calves fed at different nutrient intake levels by varying the type and strength of the milk replacer given until 5 weeks of age. All calves received 4 liters per day of either cow's milk, milk replacer with 100 g of fat per kilogram, milk replacer with 170 g of fat per kilogram, or milk replacer with 200 g of fat per kilogram. Voluntary intakes of water were highest with calves fed on cow's milk and lowest with calves on the milk replacer with 100 g of fat per kilogram. For each liter of water consumed per day, there was extra dry feed intake of 0.082 kg and an increase in liveweight gain of 0.056 kg.

Balch et al. (1953) fed cows 8.4 kg of alfalfa hay daily and allowed 60% of the amount of water drunk during the control period. During restriction of water, some cows refused part of the hay and all lost weight, urine volume fell to 66.4% and fecal water to 67.3% of that during the control period, but no change occurred in nitrogen balance. During water restriction, digestible DM and crude fiber rose slightly, but, due to decreased hay intake, this did not affect the intake of apparently digestible energy, and no changes occurred in the crude protein, ether extract, or nitrogen-free extract digestibilities. It was concluded that no advantage could be gained by restricting intake of water in cattle. Larsen et al. (1917) found that the digestibilities of crude fiber and nitrogen-free extract were increased when cows were restricted to 50% of normal water intake. Due to decreased feed intake during restriction of water, the cows averaged a 45-kg loss in body weight.

Utley et al. (1970) worked with three groups of steers fed high-concentrate diets in a metabolism trial. Each steer was offered a full feed of ground earcorn–urea diet and either (1) water free choice (FC), (2) 80% FC, or (3) 60% FC during the trials. Restriction of water to 60% FC resulted in a reduction ($p<0.05$) in voluntary feed intake, but nitrogen retention tended to increase when water was restricted and was negatively correlated ($p<0.01$) to total nitrogen excretion in the urine. Elevated plasma urea–nitrogen concentration in camels was reported when water intake was restricted (Schmidt-Nielsen et al., 1957).

Total feces and urine were collected from Hereford heifers during 4 days of water deprivation followed by 4 days of ad libitum drinking of water (Weeth et al., 1967). Feed intake decreased by 50% of each preceding day's intake during water deprivation. The heifers lost 16% in body weight during water deprivation, but the weight was recovered during 4 days of ad libitum drinking. Kidney responses to 4 days of water deprivation were a 72% decrease in urine output and a 46% decrease in clearance of creatinine. Little et al. (1978) deprived lactating dairy cows of 40% of their ad libitum water intake during a 3-week period. The deprived cows consumed significantly less hay than control cows (2.93 versus 5.74 kg of dry matter per day) but equivalent amounts of concentrate (7.35 versus 7.71 kg of DM per day). An average of 116 and 143 MJ/day of metabolizable energy were consumed by the deprived and control cows, respectively. The deprived cows had a decrease in milk yield of 16% which was apparent within 24 hr. However, there were no further decreases in volume. The deprived cows had a significant decrease in body weight during the first week but had no further decrease until rehydration, when weight increased to a mean of 14 kg above the weight before deprivation.

VI. EFFECT OF TEMPERATURE ON WATER INTAKE AND DIGESTIBILITY OF NUTRIENTS

Ambient temperature has a marked effect on drinking water intake. Bailey et al. (1962) working with sheep observed that a reduction in environmental temperature from 15 to $-12°C$ caused decreases in the temperatures of the rumen, rectum, and subcutaneous tissues and decreased intake of water from about 1600–800 ml/day. However, at $-12°C$ varying the temperature of drinking water from 0, 10, 20, and 30°C had no effect on water intake.

During three summers, field studies were made on Merino wethers deprived of water while exposed to sun and to maximum air temperatures ranging from 29–42°C at latitude 21 degrees south (MacFarlane et al., 1961). Evaporative cooling determined rate and extent of water and electrolyte changes in the animal's body and produced a different pattern each year. Control of body temperature failed when 31% of body weight was lost by the end of 10 days without water. In hotter weather, 5 days without water caused a 25% loss of body weight and, in some sheep, irreversible circulatory failure. It was concluded that survival in the sun without water depends on insulation, water reserves in the rumen and extracellular fluid, ability to adjust electrolyte concentration, and ability to maintain circulation with decreased volume of blood plasma.

Ittner et al. (1951) compared the effects of the intake of cooled water in a hot climate on two groups of Hereford and one group of Brahman steers. The mean air temperature during the test was 29.7°C (with a mean maximum of 38.1°C and mean minimum of 21.2°C) with an extreme range of 12.2–47.7°C. The weight gains for Herefords drinking water of 18.3°C, Herefords drinking water of 31.2°C, and Brahmans drinking water of 31.2°C were 0.66, 0.49, and 0.66 kg per head per day, respectively. The diet was 75% alfalfa and 25% barley hay. The daily water intake per head averaged 15.4, 16.6, and 10.0 gal, respectively.

Digestion trials with Holstein cows were conducted to evaluate the effect of 1.1, 13.9, 26.7, and 39.4°C water at an average ambient temperature of 11.6°C (trial 1) and −2.8°C (trial 2) by Cunningham et al. (1964). Ruminal temperature depression depended on amount and temperature of ingested water. But no significant differences were found in the digestibilities of DM, energy, and crude protein in the diets in either trial.

McDonald and Bell (1958) observed that lactating Holstein cows increased their free water consumption as ambient temperature decreased from 3.3 to −17.8°C. Free water intake varied from day to day depending on stage of lactation, body size, and feed intake. Ragsdale et al. (1949) in Missouri reported a slight increase in water intake with decreasing temperature and associated it with a greater hay intake with colder weather.

VII. EFFECT OF SALINITY ON WATER CONSUMPTION

The water requirements of ruminants are influenced by the degree of salinity as well as by specific ions and combination of ions in some cases. The effect of NaCl and Na_2SO_4 in the drinking water of sheep was studied by Peirce (1960). Six groups of sheep were fed in pens for 15 months on a diet of chaffed alfalfa and wheat hays. One group was offered rainwater and another group 1.3% NaCl, whereas the others were offered one of the following mixtures of NaCl and Na_2SO_4(%) : 1.22 + 0.10, 1.14 + 0.20, 1.05 + 0.30, and 0.89 + 0.50. Water intake with 1.30% NaCl alone or with 0.10 or 0.50% Na_2SO_4 with NaCl was greater than that of rainwater. The increase was 30–60% greater in the hottest months. Sulfate was higher in the blood plasma of sheep that consumed either 0.2 or 0.5% Na_2SO_4, but the saline drinking water had no effect on concentrations of Na, K, Ca, Mg, or Cl in the blood plasma or adverse effects on the general health, feed intake, weight gains, or wool production of the sheep. Peirce (1959a) reported that the principal effect of water containing high concentrations of NaCl (0.7%) and $MgCl_2$ (0.5%) was a decrease in feed intake.

Peirce (1962) conducted a study with six groups of sheep fed chaffed alfalfa and wheat hays but offered drinking water with various levels of NaCl and $CaCl_2$. The intake of all saline waters was higher than that of rainwater, varying from 100% above for 1.3% NaCl to 20% above for 0.30% $CaCl_2$. The mean daily intakes of water by sheep offered rainwater containing 1.3% NaCl, 1.24% NaCl + 0.05% $CaCl_2$, 1.19% NaCl + 0.10% $CaCl_2$, 1.09% NaCl + 0.20% $CaCl_2$, and 0.98% NaCl + 0.30% $CaCl_2$ were 2.6, 5.2, 4.6, 4.4, 3.8, and 3.1 liters, respectively. The intakes were 45–60% greater in the hottest months than during the coldest months. None of these saline waters had any adverse effect on the general health, feed intake, weight gains, or wool production of the sheep. Heller (1933) provided three cows with drinking water that contained 1.0 or 1.5% $CaCl_2$ and found they drank less water than those cows consuming city water, and they ceased to drink at all after a few days.

Weeth and Haverland (1961) offered drinking water to heifers that contained 1.25, 1.50, or 1.75% NaCl in a winter changeover trial during 30-day periods in Nevada. During summer, a similar trial was conducted using 0, 1.0, or 1.2% NaCl in drinking water. Consumption of water was 24.2 and 42.4% lower when the water contained 1.5 or 1.75% NaCl than when it contained 1.25% NaCl during winter. The 1.75% NaCl level caused anorexia, decreased water intake, and weight gains. During summer, the heifers were adversely affected by water containing 1.2% NaCl and had symptoms of dehydration. Water consumption was increased by 46.6 and 69.0% by the addition of 1.0 and 1.2% NaCl, respectively.

The effects of water loading, NaCl loading, and 4 days of total fasting on kidney function were studied with Hereford heifers by Weeth and Lesperance (1965). Peak urine flow occurred 2–4 hr following drenching, but water diuresis did not affect the excretion of urea N or total N in the urine. The 1.5% NaCl drinking water increased the ratio of urinary excretion to water intake from 0.31 to 0.82. Sodium excretion in the urine increased approximately tenfold with the 1.5% NaCl drinking water. Urine volume decreased from 2.47 liters per day prefast to 0.25 liters per day per 45.5 kg prefast of body weight after 4 days of fasting.

Heifers (weighing 382 kg initial weight) were provided drinking water containing 0, 1, or 2% added NaCl for 30 days during winter in a changeover trial by Weeth *et al.* (1960). The 1.0% NaCl caused a 52.8% increase in water intake and a decrease in blood urea but was not deleterious to growing heifers over the 30-day period. However, the 2.0% NaCl was toxic, causing severe anorexia, weight loss, and anhydremia. Rectal temperature was lowered and the animals were lethargic.

The National Research Council (NRC, 1974) provided a guide for the use of saline waters for livestock (Table 3.1). Their guideline indicates

TABLE 3.1

A Guide to the Use of Saline Waters for Livestock and Poultry[a]

Total soluble salt content of waters (mg/liter)	Comment
Less than 1000	These waters have a relatively low level of salinity and should present no serious burden to any class of livestock or poultry.
1000–2999	These waters should be satisfactory for all classes of livestock and poultry. They may cause temporary and mild diarrhea in livestock not accustomed to them or watery droppings in poultry (especially at the higher levels) but should not affect health or performance.
3000–4999	These waters should be satisfactory for livestock, although they might very possibly cause temporary diarrhea or be refused at first by animals not accustomed to them. They are poor waters for poultry, often causing watery feces and (at the higher levels of salinity) increased mortality and decreased growth, especially in turkeys.
5000–6999	These waters can be used with reasonable safety for dairy and beef cattle, sheep, swine, and horses. It may be well for pregnant or lactating animals to avoid the use of those waters approaching the higher levels of salinity. They are not acceptable waters for poultry, almost always causing some type of problem, especially near the upper limit of salinity, where reduced growth and production or increased mortality will probably occur.
7000–10,000	These waters are unfit for poultry and probably for swine. Considerable risk may exist in using them for pregnant or lactating cows, horses, sheep, the young of these species, or for any animals subjected to heavy heat stress or water loss. In general, their use should be avoided, although older ruminants, horses, and even poultry and swine may subsist on them for long periods of time under conditions of low stress.
More than 10,000	The risks with these highly saline waters are so great that they cannot be recommended for use under any conditions.

[a] From NRC (1974).

that 1000 mg per liter of total soluble salts in drinking water should present no serious burden to any class of livestock. Levels of 1000 to 2999 mg/liter should be satisfactory, though these levels may cause temporary and mild diarrhea in livestock not accustomed to them. Levels of 3000–4999 mg/liter should be satisfactory, though they may cause temporary diarrhea or be refused at first by animals not accustomed to them. Waters that contain 5000–6999 mg/liter can be used with reasonable safety by nonpregnant or lactating beef and dairy cattle and sheep. Drinking water containing 7000–10,000 mg per liter of dissolved solids may involve considerable risk if utilized by pregnant or lactating cattle and sheep; older ruminants may subsist on them for long periods of time. Water with more than 10,000 mg per liter of total dissolved solids cannot be recommended for use under any conditions for ruminants.

Hardness of water is expressed in terms of concentrations of divalent ions, such as Ca and Mg. Graf and Holdaway (1952) compared cows provided water with a total hardness of 290 ppm (about equally divided as to Ca and Mg ions) with cows provided with the same source of water after softening by the zeolite process to 0.0 ppm hardness. Over a period of 57 days, there were essentially no differences between the groups in water and feed intakes or in the production of 4% fat-corrected milk per day per cow.

Over 181 days, cows averaged 8.0 kg of milk daily when consuming well water averaging 190 ppm total hardness compared to 9.0 kg of milk daily when drinking water from the same source that was zeolite treated to 0.0 ppm hardness (Allen *et al.*, 1958). It was concluded that overall softening of water had no measurable effect on milk production, as the lactation period was shorter for the cows provided the softened water. Blosser and Soni (1957) provided dairy cows either hard or soft (zeolite-processed) water in a 6-week double-reversed trial and reported that the cows consumed equivalent amounts of water and feed and produced essentially the same amount of milk.

VIII. WATER AND ELECTROLYTE METABOLISM

Since water output is markedly decreased in animals deprived of drinking water, it seems likely that soluble electrolytes would also be altered in concentration in body fluids as well as in fecal, urinary, and sweat excretions. English (1966) conducted an extensive study on water, Na, K, and Cl balances in Suffolk adult sheep. Water was available ad libitum during stages 1 and 3 but was restricted during stage 2. A small positive balance was observed for each electrolyte in stage 1, and a greater positive balance of each electrolyte occurred in stage 3. However, in stage 2, mean Na and Cl balances were negative, but that of K remained positive. Urinary K–Na ratios during stages 1, 2, and 3 were 9.0, 4.5, and 4.0, respectively. Mean fecal K–Na ratios for these stages were 2.0, 2.7, and 15.0, respectively. Water deprivation (stage 2) resulted in urinary N concentration being elevated in excess of 50%. A decrease in urinary volume was much greater than that of fecal water during water deprivation. Volume of evaporative water loss was estimated to closely parallel urinary output.

Little et al. (1978) deprived lactating dairy cows of 40% of their ad libitum water intake for 3 weeks. They observed a 16% decrease in milk yield within 24 hr. The deprived and control cows had 0.53 and 0.44 g of Na and 1.64 and 1.58 g of K per kilogram of milk, respectively. The deprived and control cows had a positive balance of 0.2 and 0.9 g per day of Na and a negative balance of -10 and -13 g per day of K, respectively.

IX. NUTRIENT ELEMENTS IN WATER

Concentration of most mineral elements in surface waters of the United States was accumulated in STORET [Systems for Technical Data (STORET), 1971] for the period 1957–1969, as presented in Table 3.2. The amounts of nutrient elements in drinking water for the various species can be calculated by multiplying the concentration times the volume of water intake per day. Concentrations of mineral nutrients in drinking water for ruminants are generally quite inadequate for dietary requirements. According to the National Research Council (NRC, 1974) average concentrations of NaCl in surface waters in the United States would supply beef

TABLE 3.2

Composition of United States Surface Water, 1957–1969[a]

Substance	Mean	Maximum	Minimum	Determinations
Phosphorus (mg/liter)	0.087	5.0	0.001	1,729
Calcium (mg/liter)	57.1	173.0	11.0	510
Magnesium (mg/liter)	14.3	137.0	8.5	1,143
Sodium (mg/liter)	55.1	7,500.0	0.2	1,801
Potassium (mg/liter)	4.3	370.0	0.06	1,804
Chloride (mg/liter)	478.0	19,000.0	0.0	37,355
Sulfate (mg/liter)	135.9	3,383.0	0.0	30,229
Copper (μg/liter)	13.8	280.0	0.8	1,871
Iron (μg/liter)	43.9	4,600.0	0.10	1,836
Manganese (μg/liter)	29.4	3,230.0	0.20	1,818
Zinc (μg/liter)	51.8	1,183.0	1.0	1,883
Selenium (μg/liter)	0.016	1.0	0.01	234
Iodine[b] (μg/liter)	46.1	336.0	4.0	15
Cobalt[c] (μg/liter)	1.0	5.0	0	720

[a] From NRC (1974).
[b] Dantzman and Breland (1970).
[c] Durum et al. (1971).

cattle with approximately 34% of their daily requirement, lactating dairy cattle with 19%, and fattening lambs and lactating ewes with 6–7%. Research indicates that salt should be made available on an ad libitum basis for ruminants or adequate amounts should be added to their diets.

Calcium at average concentrations in surface waters is present in sufficient quantities to provide approximately 5–8% of the requirements for sheep and lactating dairy cows and 10–16% for beef cattle (NRC, 1974). *Phosphorus* in such drinking water would provide only 1% or less of the daily requirement of ruminants. *Magnesium* at average concentrations could provide 4–11% of the requirements of beef, dairy cattle, and sheep. *Sulfur* (S) at average levels would meet approximately 28% of beef cattle requirements, 20–45% of dairy cattle requirements, and 10–11% of sheep requirements.

Potassium, iron (Fe), *zinc* (Zn), *copper* (Cu), *cobalt* (Co), *manganese* (Mn), and *selenium* (Se) at average concentrations would provide approximately 1, 1, 2, 2, 12, 6, and 1% or less of cattle and sheep daily requirements, respectively (NRC, 1974). It is apparent that for optimum growth and performance ruminants need to be provided with essential mineral elements in their feedstuffs or mineral boxes.

X. TOXIC ELEMENTS AND SUBSTANCES IN DRINKING WATER

Water occasionally contains elements and substances in toxic concentrations (Table 3.3). Generally toxic levels are due to point sources and may occur during flooding or by human error. The amount of potentially toxic substance present in the feedstuff also needs to be taken into account, as the extent of toxicity may depend on the total intake. The NRC (1974, 1980) presented thorough reviews of toxic elements in feedstuffs, drenches, and drinking water for ruminants as well as nonruminants. Limited information is available on experimentally determined toxic levels of various elements and substances in drinking water provided to livestock.

A number of problems are involved in assigning toxic levels to waterborne substances. Elements and substances may vary in toxicity depending on whether they are suspended as part of solids or in solution. Different valences and chemical forms of elements have different levels of toxicity. Also, young and healthy animals may not react in the same way as mature or unthrifty animals. Finally, rate of consumption is a factor to be considered in determining toxicity.

Elements such as Fe, aluminium (Al), beryllium (Be), boron (B), chro-

TABLE 3.3

Recommended Limits of Concentration of Some Potentially Toxic Substances in Drinking Water for Livestock and Poultry[a]

Item	Safe upper limit of concentration (mg/liter)
Arsenic	0.2
Barium	Not established
Cadmium	0.05
Chromium	1.0
Cobalt	1.0
Copper	0.5
Cyanide	Not established
Fluoride	2.0
Iron	Not established
Lead	0.1
Manganese	Not established
Mercury	0.010
Molybdenum	Not established
Nickel	1.0
Nitrate—N	100.0
Nitrite—N	10.0
Salinity	See Table 3.1
Vanadium	0.1
Zinc	25.0

[a] From NRC (1974).

mium (Cr), Co, Cu, iodine (I), Mn, molybdenum (Mo) and Zn are found in water, generally at low levels, in soluble form or are toxic only in excessive concentrations (NRC, 1974). However, Coup and Campbell (1964) observed that 17 mg of Fe per liter in pasture irrigation water caused scouring and decreased milk production and body weight in cows.

This section presents a brief summary of reports on observations for concentrations of elements and substances in drinking water and drenches that have caused toxicity signs in cattle and sheep. The concentrations are expressed as milligrams per liter (ppm) of water or as daily intakes per kilogram of body weight of the animal.

Generally, inorganic *arsenicals* are more toxic than the organic forms, and many generalizations concerning arsenicals have important exceptions (Frost, 1967). Fitch *et al.* (1939) reported that 343 mg per kilogram of arsenic (As) as As_2O_3 in daily intake increased As in milk of cows. *Boron* intake of 150 mg per liter of water as borax decreased feed intake of cattle, while 300 mg/liter resulted in weight loss and lethargy (Green and Weeth,

1977). Plants require B, but it has no known function in animals. *Cadmium* (Cd) in diets fed cattle and sheep has caused decreased feed intake, loss of body weight, and, depending on concentration, death (NRC, 1980). Cadmium in the drinking water (160 mg/liter) resulted in decreased growth, hematocrit, and hemoglobin of rabbits (Stowe *et al.*, 1972), and 5 mg/liter of Cd in water shortened the lifespan of rats (Kanisawa and Schroeder, 1969).

Chromium as chromium oxide has been widely used as a fecal marker in metabolism studies with cattle and sheep due to its low toxicity and lack of solubility in the digestive tract. The chromate and chloride salts are more soluble than oxides of Cr and would be expected to be more toxic. The NRC (1980) suggests a maximum tolerable level of 3000 ppm Cr as the oxide and 1000 ppm as the chloride for livestock. A level of 7.7 ppm as $K_2Cr_2O_7$ in drinking water of rats increased Cr in the tissues (MacKenzie *et al.*, 1958). Keener *et al.* (1949) found that 1–9 mg of *cobalt* as cobaltous sulfate per kilogram of body weight in drinking water of cattle caused hyperchromenia, and 22–44 mg per kilogram of body weight gave loss of appetite, rough hair coat, and incoordination.

Copper in water solution is much more toxic to cattle than copper in dry feed. Chapman *et al.* (1962) gave cattle 12 g of $CuSO_4$ daily in capsules for 12 months without ill effects, but when this amount was in water drenches, two treated animals died within 65 days. When sheep were given 250 mg Cu as $CuSO_4$ per kilogram of body weight daily in a drench for 10 weeks, they had a hemolytic crisis and 80% died (Morgan, 1973). Small intakes of *fluorine* (F) appear to be beneficial to formation of bone and teeth, but high levels will cause mottling of teeth, decreased feed intake, and weight gains. Merriman and Hobbs (1962) reported that 50–1812 mg of F per liter in drinking water with hay diets caused mottling of teeth in heifers. Levels of 100 mg of F per liter of water provided cattle over an 11 month period resulted in decreased feed intake and weight gains (Ramberg *et al.*, 1970). Sheep provided drinking water containing 20 mg of F per liter had marked dental lesions (Peirce, 1952) and decreased wool production (Peirce, 1959a).

Pregnant goats provided a water drench with 50 mg of *lead* (Pb) as lead acetate per kilogram of body weight aborted after 6 days and died after 41 days (Dollahite *et al.*, 1975). Aronson *et al.* (1968) made infusions of calcium–EDTA for treatment of Pb toxicosis in cattle. Intakes of 2.2.–4.4 g of *Mg* per day in drinking water provided sheep over 16 months resulted in decreased feed intake and some diarrhea (Peirce, 1959b). Allison (1930) observed high Mg levels of about 1% in water in parts of Minnesota, North Dakota, South Dakota, and Montana. Cattle drinking these waters could not be fattened for market and exhibited a "run-down" appearance, and many died prematurely.

Excessive *Mn* levels, as high as 1100 ppm, in diets fed various species of animals produced serious health problems in only one of 21 experimental groups (NRC, 1980). However, when Mn was present in water provided rabbits at a level of 24.4 mg per kilogram of body weight per day, there was a loss in body weight and transitory paralysis (Umarji *et al.*, 1969). Methylmercury and other organic *mercury* (Hg) compounds are more toxic than inorganic mercuric compounds. Methylmercury dicyandiamide in diets containing 0.225 mg of Hg per kilogram of body weight of sheep and cattle was toxic (Wright *et al.*, 1973). When mice were provided water with 5 mg Hg per liter as CH_3HgCl for 90 days, they died, but the same amount of Hg in water as $HgCl_2$ throughout their life span resulted in no adverse effects (Schroeder and Mitchener, 1975).

Levels of 50 and 250 ppm of *nickel* (Ni) as nickel carbonate in diets fed lactating dairy cows had no adverse effects (O'Dell *et al.*, 1970). However, 5 mg of Ni in water provided rats over three generations resulted in death of young and runts in the F_1 through F_3 generations (Schroeder and Mitchener, 1971). This level of Ni in the drinking water of mice over their lifetime resulted in more tumors in males than in females (Schroeder *et al.*, 1964). Excess levels of *K* in supplements may cause toxicosis (NRC, 1980). An aqueous drench containing 71 mg of K as KCl per kilogram of body weight given cows resulted in milk fever (Dennis and Harbaugh, 1948), while 105 mg caused death in 10 min (Ward, 1966). When a drench of 820 mg of K as KCl per kilogram of body weight and 1570 mg of transaconitic acid was given yearling cattle, 10–12 animals developed tetany (Bohman *et al.*, 1969).

Selenium poisoning may occur when grazing ruminants eat Se accumulator plants, especially when pasture is limited. Generally, dry feedstuffs and drinking water are marginal or below the dietary requirements of ruminants, and diets may need supplementation of the element. Selenium may be toxic at relatively low levels. Miller and Williams (1940) observed that a drench of 10.1 mg of Se as Na_2SeO_3 per kilogram of body weight of cattle caused anorexia within 2 days and decreased milk production, while 14.1 mg resulted in anorexia, garlic breath, and death in 4 days. Sheep given a drench of 6.4 mg of Se as Na_2SeO_3 per kilogram of body weight developed pulmonary congestion and died in 1–15 days (Gabbedy and Dickson, 1969).

Silicon (Si) appears to aid in early formation of bone and in growth of animals, but excess dietary Si may depress digestibility of forages, slow growth and reproduction, and increase kidney stones in ruminants (NRC, 1980). However, Si toxicosis is not a serious problem with practical farm or ranch conditions. Smith *et al.* (1972) reported that 374 mg of Si as sodium silicate per liter of drinking water provided sheep increased weight gain and feed efficiency in wethers only. This level of Si in drinking water was slightly detrimental to weight gains by females fed three different diets.

Uranium (U) does not appear to be essential for ruminants, and cattle are fairly tolerant of the element. Garner (1963) reported that cows given 4 g of uranyl nitrate per kilogram of body weight became ill, and milk production decreased before the cows returned to normal. While *Zn* is an essential element for ruminants at levels of 40–100 ppm in the diet, levels of 600–1000 ppm in the diet are toxic (NRC, 1980). Smith (1977) found that 20 mg of Zn as zinc sulfate per kilogram of body weight in a drench given to sheep daily over a 6-week period resulted in loss of body weight, and death occurred in 16–35 weeks.

Nitrate is frequently found in ponds, wells, and streams near feedlots and septic tanks. Fine-textured soils release less nitrate than coarse-textured soils. The amount of nitrate in water around feedlots was found to correlate directly with the number of animals and inversely with the depth of wells (Keller and Smith, 1967). Sheep had approximately 5 g of methemoglobin per 100 ml of blood in about 8 hr after being given a drench containing 112 mg of NO_3–N per kilogram of body weight (Emerick *et al.*, 1965). Buchman *et al.* (1968) had one sheep die when animals consumed 99 mg NO_3–N per kilogram of body weight in hay plus a drench.

Nitrite was not correlated in water with livestock or fertilizer use by Smith (1965) except in sand point wells in river floodplains. Nitrite is formed when nitrate in the rumen is reduced and reacts with the hemoglobin in the blood to form methemoglobin resulting in a decrease in the hemoglobin's oxygen carrying capacity.

XI. SUMMARY

Nutrient minerals are present in drinking water for ruminants but at concentrations that are generally quite inadequate for meeting daily dietary requirements. Nutrient minerals need to be provided ad libitum in salt blocks or salt mixes or mixed in diets in proper amounts. Sodium, Cl, Ca, sulfate, bicarbonate, and Mg are ions most commonly present in highly saline waters. In arid and semiarid regions, salinity of drinking water may exceed tolerance of ruminants for maximum growth, lactation, and reproduction. Nitrates and nitrates occasionally occur at toxic levels in drinking waters from point sources of contamination. Other toxic substances, at times, may contaminate drinking water.

Water intake by ruminants has been found to increase with increasing DM intake, increased milk production, pregnancy, and elevated ambient temperature. Cattle generally drink only slightly more water going from −

12 to 4°C, but the intake increases at an accelerating rate with increasing temperature from 4–38°C. A 450-kg steer would drink approximately 28, 49, and 66 liters of water daily at 4, 21, and 32°C, respectively.

Cattle have a greater turnover of water in proportion to body size than do sheep. Renal and fecal water loss is a greater part of the water turnover in cattle. Deprivation of drinking water generally decreases feed intake, which results in loss of body weight. Cattle prefer to drink several times daily and if limited to once per day or once per 2 days will eat less feed and grow more slowly. European cattle (*Bos taurus*) drink more water than Indian cattle (*Bos indicus*).

REFERENCES

Allen, N. N., Ansman, D., Patterson, W. N., and Hays, O. E. (1958). *J. Dairy Sci.* **41,** 688–691.
Allison, G. S. (1930). *Science* **71,** 559–560.
Anand, R. S., and Parker, H. R. (1966). *Am. J. Vet. Res.* **27,** 899–906.
Aronson, A. L., Hammond, P. B., and Strafuss, A. C. (1968). *Toxicol. Appl. Pharmacol.* **12,** 337–349.
Bailey, C. B., Hironaka, R., and Slen, S. B. (1962). *Can. J. Anim. Sci.* **42,** 1–8.
Balch, C. C., Balch, D. A., Johnson, V. W., and Turner, J. (1953). *Br. J. Nutr.* **7,** 212–224.
Black, A. L., Baker, N. F., Bartley, J. C., Chapman, T. E., and Phillips, R. W. (1964). *Science* **144,** 876–878.
Blosser, T. H., and Soni, B. K. (1957). *J. Dairy Sci.* **40,** 1519–1524.
Bohman, V. R., Lesperance, A. L., Harding, G. D., and Grunes, D. L. (1969). *J. Anim. Sci.* **29,** 99–102.
Brown, L. D. (1966). *J. Dairy Sci.* **49,** 223–230.
Buchman, D. T., Shirley, R. L., and Killinger, G. B. (1968). *Proc.—Soil Crop Sci. Soc. Fla.* **28,** 209–215.
Calder, F. W., Nicholson, J. W. G., and Cunningham, H. M. (1964). *Can. J. Anim. Sci.* **44,** 266–271.
Chapman, H. L., Jr., Nelson, S. L., Kidder, R. W., Sippel, W. L., and Kidder, C. W. (1962). *J. Anim. Sci.* **21,** 960–962.
Coup, M. R., and Campbell, A. G. (1964). *N. Z. J. Agric. Res.* **7,** 624–638.
Cunningham, M. D., Martz, F. A., and Merilan, C. P. (1964). *J. Dairy Sci.* **47,** 382–385.
Dantzman, C. L., and Breland, H. L. (1970). *Proc.—Soil Crop Sci. Soc. Fla.* **29,** 18–28.
Dennis, J., and Harbaugh, F. A. (1948). *Am. J. Vet. Res.* **9,** 20–25.
Denton, D. A. (1956). *J. Physiol. (London)* **131,** 516–525.
Dollahite, J. W., Rowe, L. D., and Reagor, J. C. (1975). *Southwest. Vet.* **28,** 40–45.
Durum, W. H., Hem, J. D., and Heidel, S. G. (1971). *Geol. Surv. Circ. (U.S.)* **643,** 1–49.
Emerick, R. J., Embry, L. B., and Seerley, R. W. (1965). *J. Anim. Sci.* **24,** 221–230.
English, P. B. (1966). *Res. Vet. Sci.* **7,** 223–257.
Fitch, L. W. N., Grimmett, R. E. R., and Wall, E. M. (1939). *N. Z. J. Sci. Technol. Sect. A* **21,** 146–149.
Forbes, J. M. (1968). *Br. J. Nutr.* **22,** 33–43.
Frost, D. V. (1967). *Fed. Proc. Fed. Am. Soc. Exp. Biol.* **26,** 194–208.
Gabbedy, B. J., and Dickson, J. (1969). *Aust. Vet. J.* **45,** 470–472.

Garner, R. J. (1963). *Health Phys.* **9**, 597–605.
Gordon, J. A. (1965). *J. Agr. Sci.* **64**, 31–35.
Graf, G. C., and Holdaway, C. W. (1952). *J. Dairy Sci.* **35**, 998–1000.
Green, G. H., and Weeth, H. J. (1977). *J. Anim. Sci.* **46**, 812–818.
Halley, R. J., and Dougall, B. M. (1962). *J. Dairy Res.* **29**, 241–248.
Heller, V. G. (1933). *Bull.—Okla., Agric. Exp. Stn.*, **217**, 1–23.
Ittner, N. R., Kelly, C. F., and Guilbert, H. R. (1951). *J. Anim. Sci.* **10**, 742–751.
Kanisawa, M., and Schroeder, H. A. (1969). *Cancer Res.* **29**, 892–895.
Keener, H. A., Percival, G. P., and Marrow, K. S. (1949). *J. Dairy Sci.* **32**, 527–533.
Keller, W. D., and Smith, G. E. (1967). *Spec. Pap.—Geol. Soc. Am.* **90**, 47–59.
Larsen, C., Hungerford, E. H., and Bailey, D. E. (1917). *S. D. Agric. Exp. Stn., Bull.* **175**, 646–680.
Leitch, J., and Thomson, J. S. (1944). *Nutr. Abstr. Rev.* **14**(2), 197–223.
Little, W., and Shaw, S. R. (1978). *Anim. Prod.* **26**, 225–227.
Little, W., Sanson, B. F., Manston, R., and Allen, W. M. (1978). *Anim. Prod.* **27**, 79–87.
McDonald, M. A., and Bell, J. M. (1958). *Can. J. Anim. Sci.* **38**, 23–32.
MacFarlane, W. V., Morris, R. J. H., Howard, B., McDonald, J., and Budtz-Olsen, O. E. (1961). *Aust. J. Agric. Res.* **12**, 889–912.
MacFarlane, W. V., Morris, R. J. H., and Howard, B. (1963). *Nature (London)* **197**, 270–271.
MacKenzie, R. D., Byerrum, R. U., Decker, C. F., Hoppert, C. A., and Langham, R. F. (1958). *AMA Arch. Ind. Health* **18**, 232–234.
Merriman, G. M., and Hobbs, C. S. (1962). *Tenn. Agric. Exp. Stn., Bull.* **347**, 46.
Miller, W. T., and Williams, K. T. (1940). *J. Agri. Res.* **60**, 163–173.
Morgan, K. T., (1973). *Res. Vet. Sci.* **15**, 88–95.
National Research Council (NRC) (1974). "Nutrient and Toxic Substances in Water for Livestock and Poultry." Natl. Acad. Sci., Washington, D.C.
National Research Council (NRC) (1980). "Mineral Tolerance of Domestic Animals." - Natl. Acad. Sci., Washington, D.C.
O'Dell, G. D., Miller, W. J., King, W. A., Ellers, J. C., and Jurecek, H. (1970). *J. Dairy Sci.* **53**, 1545–1559.
Owen, J. B., and Miller, E. L. (1968). *J. Agri. Sci.* **70**, 223–235.
Paquay, R., De Baere, R., and Lousse, A. (1970). *J. Agri. Sci.* **74**, 423–432.
Peirce, A. W. (1952). *Aust. J. Agric. Res.* **3**, 326–340.
Peirce, A. W. (1959a). *Aust. J. Agric. Res.* **10**, 186–198.
Peirce, A. W. (1959b). *Aust. J. Agric. Res.* **10**, 725–735.
Peirce, A. W. (1960). *Aust. J. Agric. Res.* **11**, 548–556.
Peirce, A. W. (1962). *Aust. J. Agric. Res.* **13**, 479–486.
Pettyjohn, J. D., Everett, J. P., Jr., and Mochire, R. D. (1963). *J. Dairy Sci.* **46**, 710–714.
Ragsdale, A. C., Worstell, D. M., Thompson, H. J., and Brody, S. (1949). *Mo., Agric. Exp. Stn., Bull.* **449**.
Ramberg, C. F., Jr., Phang, J. M., Mayer, G. P., Norberg, A. J., and Kronfeld, D. S. (1970). *J. Nutr.* **100**, 981–989.
Schmidt-Nielsen, K., Schmidt-Nielsen, B., Haupt, T. R., and Jarnum, S. A. (1957). *Am. J. Physiol.* **188**, 103–112.
Schroeder, H. A., and Mitchener, M. (1971). *Arch. Environ. Health* **23**, 102–106.
Schroeder, H. A., and Mitchener, M. (1975). *J. Nutr.* **105**, 452–458.
Schroeder, H. A., Balassa, J. J., and Vinton, Jr., W. H. (1964). *J. Nutr.* **83**, 239–250.
Shirley, R. L. (1970). *Proc.—AFMA Nutr. Counc.* **30**, 23–25.
Shirley, R. L., Davis, G. K., and Neller, J. R. (1951). *J. Anim. Sci.* **10**, 335–336.

Shirley, R. L., Robertson, W. K., McCall, J. T., Neller, J. R., and Davis, G. K. (1957). *J. Fla. Acad. Sci.* **20**, 133-138.
Smith, B. L. (1977). *N.Z. Vet. J.* **25**, 310-312.
Smith, G. E. (1965). *Spec. Rep.—Mo., Agric. Exp. Stn.* **55**, 42-52.
Smith, G. S., Neumann, A. L., Nelson, A. B., and Ray, E. E. (1972). *J. Anim. Sci.* **34**, 839-845.
Stowe, H. D., Wilson, M., and Goyer, R. A. (1972). *Arch. Pathol.* **94**, 389-405.
Systems for Technical Data (STORET) (1971). Water Programs Office, Environmental Protection Agency, Washington, D.C.
Thickett, W. S., Cuthbert, N. H., Brigstocke, T. D. A., Linderman, M. A., and Wilson, P. N. (1981). *Anim. Prod.* **33**, 25-30.
Thomas, T. P. (1971). *Anim. Prod.* **13**, 399-400.
Thornton, R. F., and Yates, N. G. (1968). *Aust. J. Agric. Res.* **19**, 665-672.
Umarji, G. M., Anantanarayanan, K. G. and Bellare, R. A. (1969). *C. R. Seances Soc. Biol. Ses Fil.* **162**, 1725-1728.
Utley, P. R., Bradley, N. W., and Boling, J. A. (1970). *J. Anim. Sci.* **31**, 130-135.
Ward, G. M. (1966). *J. Dairy Sci.* **49**, 268-276.
Weeth, H. J., and Haverland, L. H. (1961). *J. Anim. Sci.* **20**, 518-521.
Weeth, H. J., and Lesperance, A. L. (1965). *J. Anim. Sci.* **24**, 441-447.
Weeth, H. J., Haverland, L. H., and Cassard, D. W. (1960). *J. Anim. Sci.* **19**, 845-851.
Weeth, H. J., Sawhney, D. S., and Lesperance, A. L. (1967). *J. Anim. Sci.* **26**, 418-423.
Weeth, H. J., Lesperance, A. L., and Bohman, V. R. (1968). *J. Anim. Sci.* **27**, 739-750.
Wilson, P. N. (1961). *J. Agric. Sci.* **56**, 351-364.
Winchester, C. F., and Morris, M. J. (1956). *J. Anim. Sci.* **15**, 722-740.
Wright, F. C., Palmer, J. S., and Riner, J. C. (1973). *J. Agric. Food Chem.* **21**, 414-416.

4

Thermal Stress as a Factor Associated with Nutrient Requirements and Interrelationships

R. J. COLLIER AND D. K. BEEDE
Dairy Science Department
University of Florida
Gainesville, Florida

I.	Introduction.	59
II.	Physiological Responses of Ruminants to Thermal Stress.	60
III.	Metabolic and Hormonal Responses to Thermal Stress.	61
IV.	Effects of Thermal Stress on Nutrient Requirements.	62
	A. Feed Consumption.	62
	B. Thermal Environment by Forage Quality Interrelationship.	64
	C. Digestion.	64
	D. Energy Metabolism.	65
	E. Water Balance and Requirement.	66
	F. Mineral Needs during Thermal Stress.	67
	G. Attenuating Acid–Base Imbalance.	68
	H. Protein Nutrition.	69
	I. Vitamin A.	69
	References.	70

I. INTRODUCTION

Thermal stress poses a major economic loss to producers, resulting in reduced growth rates, milk yields, and reproductive performance. This is especially true in grazing situations where opportunities to reduce stress by use of shade structures are not applicable. Effects of thermal stress are mediated via direct and indirect mechanisms (Collier *et al.*, 1982). Direct effects involve alteration of metabolism to accommodate increased heat load. Indirect effects are associated with alteration in feed quality and quantity and populations of disease-causing organisms. Several factors influence degree of stress on animals. These include breed, stage of life cycle, physiological state, and variability among animals. Little progress has been made to date in simultaneously selecting animals for heat tolerance and high productivity. This is due to a basic antagonism existing between heat tolerance and high metabolic rates. Animals adapted to stressful conditions have lower metabolic rates and levels of production. Under conditions of grazing, there are fewer opportunities to protect animals from the environment. This requires that animals be adapted to heat and normally results in a lower level of productivity.

II. PHYSIOLOGICAL RESPONSES OF RUMINANTS TO THERMAL STRESS

Net effect of heat stress is to increase heat loss by evaporation and to decrease heat production by metabolism. This is dictated by the fact that animals have four major routes of heat loss available: radiation, conduction, convection, and evaporation. The first three require a thermal gradient. Therefore, when effective ambient temperature approaches body temperature, evaporation remains the only available route of heat loss. Fortunately, relative humidity declines as air temperature rises due to expansion of air. This increases the vapor pressure gradient needed for evaporation and increases the efficiency of evaporation. However, ruminants do not sweat as heavily as humans or horses and must rely on respiratory heat loss in addition to sweating. This has negative effects on production, as increasing respiration rates necessitate reduction in feed intake and rumination (Collier *et al.*, 1981).

In addition to increased respiration rates, ruminants often must store heat, as they are unable to dissipate the entire heat load absorbed. Fig. 4.1 illustrates the rectal temperature patterns of Jersey cows with and without access to shade. Clearly, animals denied shade store heat during daylight hours which is dissipated at night. In a 500-kg animal, a 1°C rise in body temperature represents about 410 kcal of stored heat to dissipate. This excess heat has negative effects on metabolism, since the animal will tend to lower its own metabolic heat production to accommodate the extra heat load. Thermal stress thus alters metabolism of ruminants, requiring associated changes in the endocrine system which regulate metabolism.

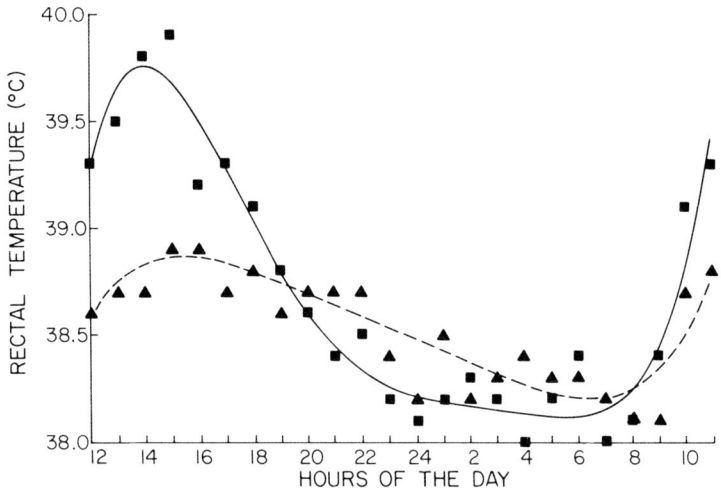

Fig. 4.1. Effect of shade (▲) or no-shade (■) environment on rectal temperatures of Jersey cows over a 24-hr period.

III. METABOLIC AND HORMONAL RESPONSES TO THERMAL STRESS

The endocrine system is affected markedly by thermal stress. Among the hormones known to be associated with adaptation to thermal stress are antidiuretic hormone (ADH), prolactin, growth hormone, thyroxine, glucocorticoids, and aldosterone.

Increased sweating and evaporative water loss at the lungs are reflected in increased antidiuretic hormone concentrations in plasma (El-Nouty *et al.*, 1980). Likewise, increased electrolyte turnover is associated with altered aldosterone concentrations. Cattle produce sweat which contains high amounts of potassium (K) but lower amounts of sodium (Na). During periods of intense sweating, the increased K loss at the skin requires an increase in Na excretion at the kidney. This is reflected in lower aldosterone concentrations in blood. Thus, thermal stress increases turnover of both Na and K.

Prolactin concentrations are elevated during thermal stress (Collier *et al.*, 1982). The role of prolactin in adaptation to thermal stress is not well defined. However, increasing dietary K from 0.66 or 1.08 to 1.64% markedly reduced plasma prolactin concentrations in heat-stressed cattle (Collier *et al.*, 1982).

As pointed out earlier, chronic thermal stress results in lower metabolic rate. This is reflected in lower concentrations of growth hormone, thyroxine, and glucocorticoids in blood (Mitra *et al.*, 1972; Van Jonack and Johnson, 1975; Collier *et al.*, 1982). Presently, there is no evidence that increasing blood concentrations of these hormones via exogenous hormone treatment is beneficial to animal performance.

IV. EFFECTS OF THERMAL STRESS ON NUTRIENT REQUIREMENTS

Major effects of thermal stress on ruminants relate to compensation strategies aimed at maintaining body temperature. These physiological responses generally do not optimize productive functions and may alter the animal's requirements for nutrients. Though most of the knowledge on this topic was not acquired through research with the grazing ruminant, a review of existing information would serve as a basis for improved husbandry practices and future research. Included in this section are discussions of effects of thermal stress on: (1) feed consumption, (2) environment by forage quality interrelationship, (3) digestion, and (4) metabolism and requirements of specific nutrients.

A. Feed Consumption

Reduction in voluntary feed intake near or above the upper critical temperature of the animal is widely accepted as a major negative influence on productivity. Fig. 4.2 depicts the general response of feed consumption to a range of constant environmental chamber temperatures. Dry matter intake begins to decline markedly as temperatures exceed about 25–27°C. However, one should note that other environmental factors such as wind velocity, humidity, and radiation also directly affect homeothermy under natural conditions and, thus, likely are interrelated with ambient temperature in affecting feed consumption. The environmental temperature at which feed consumption begins to decline is influenced by diet composition. By the nature of its diet, the grazing ruminant would appear to be more deleteriously affected than intensively managed ruminants are. Reduction in feed intake is due mainly to reduced forage consumption. Using the lactating dairy cow as example, the National Research Council (NRC, 1981) suggested that the greater the proportion of roughage in the ration, the greater and the more rapid the reduction in dry matter consumption would be as environmental temperatures increase. In

4. Thermal Stress and Nutrient Requirements

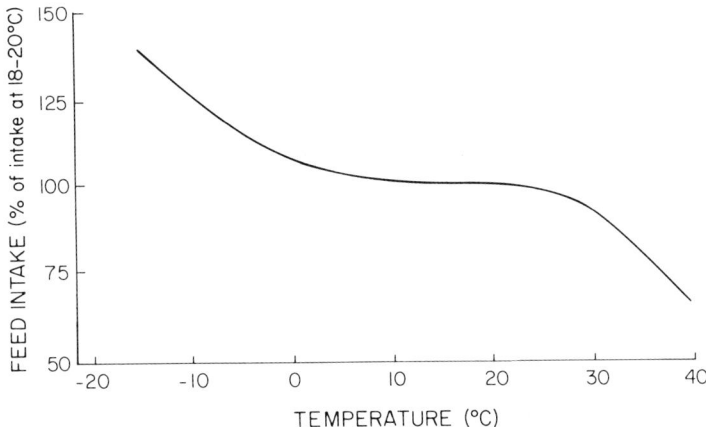

Fig. 4.2. Effect of environmental temperatures on feed consumption of cattle maintained in environmental chambers. (Modified from NRC, 1981.)

general, the less digestible the ration fed to a heat-stressed animal, the more will be the rate and extent of reduction in consumption; this holds whether comparing two forages with vastly different qualities or comparing higher- to lower-concentrate diets, if they vary appreciably in digestibility.

Many physiological responses to thermal stress are strategies for maintaining body core temperature. Reducing dry matter intake and, thus, heat generated during ruminal fermentation and body metabolism aids in maintaining heat balance. Additionally, elevated respiration rates and water intake, resulting from increased environmental temperatures, lead to concomitant reductions in feed dry matter consumption (Beede *et al.*, 1983; Roman-Ponce *et al.*, 1977). An associated effect is reduced gut motility and rumination that, along with increased water intake, lead to gut fill. Rates of rumen contractions are reduced at high environmental temperatures (Attebery and Johnson, 1969). Also, reduced rates of passage in steers fed forage diets during thermal stress increased gut fill, probably depressing appetite (Warren *et al.*, 1974). Another result may be a direct negative effect of elevated temperatures on the appetite center of the hypothalamus (Baile and Forbes, 1974).

Reduced feed intake results in smaller quantities of essential nutrients and metabolizable energy being consumed. This consequence may be deleterious, particularly to performance of grazing ruminants, unless some alternate approach (e.g., supplementation) can be employed to provide required nutrients and energy.

B. Thermal Environment by Forage Quality Interrelationship

We have considered direct effects of thermal stress on feed intake. However, with special reference to the grazing ruminant, one should not ignore the impact of high temperatures on changes in forage quality. This aspect has been reviewed by Mannetje (1984). High environmental temperatures result in rapid maturing of forages and an increase in cell wall content (Van Soest, 1981). This occurs in both temperate (C_3 pathway of carbohydrate synthesis) and tropical (C_4) grasses but is more pronounced in tropical species. Legumes are affected similarly. Increasing cell wall content is inversely related to digestibility (NRC, 1981). Also, rate of plant maturation increases with increasing environmental temperatures. For the grazing ruminant, the net result is reduced consumption (because of direct effects of heat stress) of poorer-quality forages, further reducing absolute intake of digestible energy and required nutrients. The thermal environment by forage quality interaction complicates making estimates of effects of thermal stress on intake and productivity of grazing ruminants.

C. Digestion

Separate from any effects of warm climates on plant quality or feed consumption, there appear to be influences of thermal stress on digestibility of ingested feed. Changes in the dynamic characteristics of digestion and neuroendocrine factors affect digestibility. A number of studies assessing effects of increasing environmental temperatures on digestibility have been summarized elsewhere (NRC, 1981). In general, in more temperate regions, as environmental temperatures rise, digestibility of roughages by cattle increases (Lippke, 1975; Colditz and Kellaway, 1972; McDowell et al., 1969). However, in sheep experiencing severe tropical thermal stress, dry matter digestibility was depressed (Bhattacharya and Hussain, 1974). Differences in responsiveness of sheep and cattle to thermal stress may exist, as Lippke (1975) reported. There were significant increases in dry matter and fiber component digestibilities of alfalfa pellets fed to steers housed at 32°C compared to 21°C, but there were no environmental temperature effects on digestibility among wethers.

Classically, it is surmised that a variety of factors affect digestibility, e.g., rate of feed consumption, feed quality, nutrient composition, rates of passage of digesta, and volumes of rumen and postruminal digestive organs (Ellis et al., 1984). All likely are influenced by environmental temperatures. Lower feed consumption and feed quality, as previously discussed, are affected by high environmental temperatures and could alter digestibility, with reduced intake increasing and poorer feed quality decreasing digestibility. However, increases in digestibility are not due solely to lower rates of feed intake, because when dry matter intake was equalized by sham feeding through ruminal cannulae, digestibilities were higher still among heat-stressed cattle than those in a thermoneutral environment (Lippke, 1975; Warren et al., 1974). Some reports indicate that heat stress may alter digestibility by causing transient or longer-lasting changes in rates of passage and digestive tract volume (Warren et al., 1974; Schneider et al., 1984a). In general, rates of passage of ingesta are slower (and ruminal mean retention times greater) and ruminal volumes greater, allowing for greater time to digest potentially digestible feed. These alterations in digestive function would be helpful, particularly for grazing ruminants, to more completely digest ingested feed. However, this advantage is offset largely by lower feed intake, resulting in lesser amounts of total nutrients being available to the heat-stressed animal. Another important consideration here, which has received sparse evaluation, is the fact that there may be fewer available nutrients and energy actually absorbed. Von Englehardt and Hales (1977) noted that blood flow to and from stomach compartments of sheep was reduced 15–30% due to heat stress. Thus, even if digested and readied for absorption, a reduced proportion of nutrients and energy may be absorbed and made available to the animal for maintenance and productive functions.

Hypothyroidism is known to result in reduced gut motility (Levin, 1969). Miller et al. (1974) noted that dietary provision of thyroprotein to cattle with damaged thyroid glands enhanced rate of passage of digesta. Thermal stress also is associated with reduced thyroid activity (de Andrade et al., 1977; Gale, 1973; Johnson, 1976). This suggests an association between reduced gut motility, rate of passage during thermal stress, and hormonal influence of the thyroid gland. More evidence is needed to define relationships between the endocrine system and digestive function, such as the potential temporal nature of such events, before attempts are made to modulate these factors to enhance digestibility and animal performance.

D. Energy Metabolism

In grazing ruminants, the vast majority of metabolizable energy available to the animal are volatile fatty acids (VFA) from ruminal fermentation (Annison and Armstrong, 1970). Thermal stress reduces the quantity of VFA produced in the rumen. Lower apparent production of VFA likely is related to reduced feed consumption (Gengler *et al.*, 1970). McDowell (1972) also reported a reduction in ruminal concentrations of acetate and propionate in heat-stressed cattle. Somehow, reduction in apparent VFA production is not exclusively a result of decreased feed consumption, because production was restored only partially when orts of heat-stressed cattle were force-fed through the ruminal cannulae to thermoneutral intake levels (Kelly *et al.*, 1967). Estimations of actual ruminal VFA production rates using isotopically labeled metabolites (Bergman *et al.*, 1965) have not been made in heat-stressed cattle. This would provide further information about the true energy economy of the heat-stressed grazing ruminant.

Although digestibility of dietary energy and fiber are enhanced in a hotter environment, efficiency of utilization of energy is reduced (NRC, 1981; McDowell, 1972). This is due to a higher maintenance requirement of heat-stressed animals resulting from elevated body metabolism and activity to alleviate excess heat load, and lower production caused partly by reduced intake of energy and required nutrients. For example, accelerated panting may increase maintenance requirement 7–25%, depending on its intensity (NRC, 1981).

E. Water Balance and Requirement

Water unequivocally is one of the most important nutrients for the heat-stressed grazing ruminant. Surprisingly little is known about actual requirements for water under these conditions. Winchester and Morris (1956) and NRC (1981) have summarized available information. Water requirements of grazing livestock are supplied from: (1) metabolic water, derived from tissue oxidation of organic substrates, (2) water contained in ingested feed, and (3) drinking water. The last is most significant quantitatively in meeting the needs of the heat-stressed ruminant. Water in feed is highly variable depending on season, rainfall, and time of day grazing occurs.

Increased water consumption is a major response to thermal stress (McDowell, 1972; Johnson and Yeck, 1964). Consumed water exerts effects on animal comfort by direct cooling in the reticulorumen (Bianca, 1964) and by serving as the primary vehicle for heat transfer and dissipation through sweating and panting. McDowell (1972) illustrated how dramatically total body-water balance was affected during thermal stress. In climate chambers (18 versus 30°C) drinking water consumption of lactating cows increased 29% at the warmer temperature, whereas fecal water loss declined 33%, but loss of water via urine, skin surface, and respiratory evaporation increased 15, 59, and 50%, respectively. Comparable results were noted for nonlactating cattle, though water loss via sweating was considerably greater [176% increase at 30 compared to 20°C (McDowell and Weldy, 1967)]. From these chamber studies and using a factorial-type estimation, it appears that water needs may rise 1.2- to 2-fold compared to requirements in thermoneutrality. Under natural conditions, such as grazing, water loss may be even greater because of potential for greater natural ventilation and sweating.

Numerous factors, such as level of feed intake and physical form of diet, physiological state, species and breed of animal, and quality, accessibility, and temperature of the water, may influence intake during thermal stress (NRC, 1981). For grazing ruminants, quantity of drinking water consumed is related inversely to water content of forages. At maximum daily environmental temperatures ranging from 13 to 28°C, water consumption of grazing cattle was correlated positively with maximum temperatures and daily hours of sunlight but negatively correlated with forage moisture content, rainfall, and relative humidity (Cowan et al., 1978).

Under field conditions, water intake seems to increase rather rapidly above 27°C ambient temperature (NRC, 1981). However, after an acclimation period of perhaps several weeks, water consumption may become relatively constant, nearer intake levels found at thermoneutrality. Because of the circadian patterns of ambient temperature, animals rarely are exposed to temperatures above 27°C through extended periods. The key husbandry practice for grazing ruminants in warm climates is to provide an abundant clean source of drinking water in reasonable proximity to the grazing area and shade.

F. Mineral Needs during Thermal Stress

Mineral requirements of grazing ruminants are discussed elsewhere in this volume. Intent here is to address only the effect of heat stress on needs for potassium (K) and sodium (Na). Specific requirements under grazing conditions are difficult to define because of differences in physiological status of the animal and potential interrelationships of various minerals.

Since animals reduce their voluntary intake during thermal stress, it is logical that mineral intake may be less than optimal. Also, nutritional–physiological ramifications of heat stress change macromineral needs (Schneider et al., 1984b; Beede et al., 1983; Collier et al., 1982). For example, in climate rooms, increased sweating during hyperthermia also increased loss of K in skin secretions (Singh and Newton, 1978, Jenkinson and Mabon, 1973; Johnson, 1970). This recently was characterized further under natural subtropical conditions, where a 5-fold increase in relative K loss from the skin was measured in shaded compared to unshaded animals during peak thermal stress (Beede et al., 1983). Additionally, Jenkinson and Mabon (1973) also noted marked increases in rate of loss of Na, magnesium (Mg), calcium (Ca), and chloride (Cl), but not phosphorus (P), and there were significant correlations of these losses with sweating rate. Significance of this to the grazing ruminant has not been studied. However, for lactating cows fed complete mixed diets, supplementation of K and Na above current recommendations (NRC, 1978) during heat stress resulted in 3–9% increases in milk yield (Beede et al., 1985). El-Nouty et al. (1980) reported the relationship among thermal stress, plasma aldosterone concentration, and urine electrolyte excretion. With prolonged exposure to 35°C, plasma aldosterone concentrations of nonlactating Holstein cows were 40% lower than at 20°C. Also, heat stress caused reductions of serum and urinary K concentrations. It was suggested that the fall in serum K depressed aldosterone secretion, possibly reducing urinary K losses. Urinary Na excretion also increased. Therefore, under heat stress, cattle increased sodium excretion while decreasing K losses. These physiological events are consistent with the premise of increased K and Na requirements during heat stress.

Mallonee et al. (1982) reported significant interactions of dietary Na and K on milk yield and feed intake of lactating cattle. When total dietary K was 1.6% of dietary matter, milk yield was greater with 0.7% Na than with 0.46% Na. This general concept may have application for grazing ruminants, when forages contain higher quantities of K, particularly with fertilization. Turner (1981) suggested also that the K–Na ratio in pasture herbage may be an important factor in the etiology of bloat on New Zealand dairy farms.

Direct effects of heat stress on the macromineral requirements of grazing ruminants has not been characterized under controlled experimental conditions. However, most natural feedstuffs are relatively low in Na, and mineral supplements are required to fulfill the animal's needs. Interactions of Na and K, and perhaps other elements, deserve attention. McDowell et al. (1983) suggested that K deficiency might occur in ruminants grazing mature or leached forages. Additionally, if urea provided supplemental nitrogen instead of plant protein supplements, adequate K would not be concomitantly supplied.

G. Attenuating Acid–Base Imbalance

Dale and Brody (1954) first characterized alteration in acid–base balance during thermal stress in cattle. This condition was reviewed further by Collier et al. (1982). Ramifications may include blood acid–base imbalance plus a decrease in the salivary bicarbonate pool available for ruminal buffering. Schneider et al. (1984b) showed enhanced lactational performance of heat-stressed lactating cows fed high-concentrate diets by providing 0.85% dietary sodium bicarbonate, presumably buffering the rumen and maintaining a higher ruminal pH. Additionally, it appeared that during heat stress lactating dairy cows were extremely resilient to challenges to blood acid–base balance by dietary or environmental stressors. It is not known whether thermally induced changes in acid–base homeostasis occur in grazing ruminants or, if so, whether they are of any consequence on performance.

H. Protein Nutrition

It is not known if thermal stress affects the protein requirement of the grazing ruminant. However, metabolism studies showed that acutely heat-stressed cattle were in negative nitrogen balance (Kamal and Johnson, 1970) largely because of reduced ration consumption. In management systems where protein and energy supplementation are not routinely practiced, additional quantities could be provided only with higher-quality forages (Mannetje, 1984). Effectiveness of this approach is dependent obviously on environmental influences on forage quality. Leng (1984) outlined practices for supplementation of sheep and cattle on dry or green pastures with protein meals or molasses. If major nutritional constraints of the pastures and animals are known, then these approaches would appear theoretically sound.

In intensive management systems with naturally heat-stressed feedlot cattle and sheep, Ames *et al.* (1980) observed that the efficiency of dietary protein utilization above maintenance was improved (by a 5–15% higher protein efficiency ratio) by adjusting protein consumption to expected decline in growth rate resulting from heat stress. This idea has not been tested with grazing ruminants in warm climates. However, perhaps it should be, since high-quality protein is often scarce, and more efficient use of available resources is paramount.

I. Vitamin A

Effects of thermal stress on vitamin needs of the grazing ruminant have not been evaluated extensively. However, Page *et al.* (1959) noted that short-term thermal stress caused a 30% decline in hepatic vitamin A stores of steers. Reduction of liver vitamin A content occurred whether animals were in sufficient vitamin A status before being subjected to thermal stress or whether they were first depleted by feeding a vitamin A–deficient diet for 105 days before thermal stress was imposed. Potential impact of this on reproductive performance, epithelial cell function, and general health of grazing animals in warm climates has not been explored. Direct effects of heat stress on requirements of other vitamins and resulting potential changes in intermediary metabolism have not been characterized. However, with reduction in feed consumption associated with thermal stress, supplementation of vitamins to amounts prescribed for normothermic animals, at least, would seem warranted. Further assessment will be required to ascertain if thermal stress increases requirements for various vitamins and dictates additional supplementation above normal recommendations for grazing ruminants.

REFERENCES

Ames, D. R., Brink, D. R., and Willms, C. L. (1980). *J. Anim. Sci.* **50**, 1–6.
Annison, E. F., and Armstrong, D. G. (1970). *In* "Physiology of Digestion and Metabolism in the Ruminant" (A. T. Phillipson, ed.), pp. 422–437. Oriel Press, Newcastle Upon Tyne, England.
Attebery, J. T., and Johnson, H. D. (1969). *J. Anim. Sci.* **29**, 734–737.
Baile, C. A., and Forbes, J. M. (1974). *Physiol. Rev.* **54**, 160–214.
Beede, D. K., Mallonee, P. G., Schneider, P. L., Wilcox, C. J., and Collier, R. J. (1983). *S. Afr. J. Anim. Sci.* **13**, 198–200.
Beede, D. K., Collier, R. J., Wilcox, C. J., and Thatcher, W. W. (1985). *In* "Milk Production in Developing Countries" (A. J. Smith, ed.). Univ. of Edinburgh, Edinburgh (in press).
Bergman, E. N., Reid, R. S., Murray, M. G., Brockway, J. M., and Whitelow, F. G. (1965). *Biochem. J.* **97**, 53–61.
Bhattacharya, A. N., and Hussain, F. (1974). *J. Anim. Sci.* **88**, 877–886.
Bianca, W. (1964). *Res. Vet. Sci.* **5**, 75–80.
Colditz, P. J., and Kellaway, R. C. (1972). *Aust. J. Agric. Res.* **23**, 717–724.
Collier, R. J., Eley, R. M., Sharma, A. K., Pereira, R. M., and Buffington, D. E. (1981). *J. Dairy Sci.* **64**, 844–849.
Collier, R. J., Beede, D. K., Thatcher, W. W., Israel, L. A., and Wilcox, C. J. (1982). *J. Dairy Sci.* **65**, 2213–2227.
Cowan, R. T., Shackel, D., and Davison, T. M. (1978). *Aust. J. Agric. Anim. Husb.* **18**, 190–197.
Dale, H. E., and Brody, S. (1954). *Res. Bull.—Mo., Agric. Exp. Stn.* **562**, 12.
de Andrade, A. N., Rogler, J. C., Featherstone, W. R., and Allison, C. W. (1977). *Poult. Sci.* **56**, 1178.
Ellis, W. C., Matis, J. H., Pond, K. R., Lascano, C. E., and Telford, J. P. (1984). *In* "Symposium on Herbivore Nutrition in the Sub-tropics and Tropics—Problems and Prospects" (F. M. C. Gilchrist and R. I. Mackie, eds.), pp. 269–293. The Science Press, Craighall, South Africa.
El-Nouty, F. D., Elbanna, I. M., Davis, T. P., and Johnson, H. D. (1980). *J. Appl. Physiol. Respir. Environ. Exercise Physiol.* **48**, 249–255.
Gale, C. C. (1973). *Annu. Rev. Physiol.* **35**, 391–402.
Gengler, W. R., Martz, F. A., Johnson, H. D., Krause, G. F., and Hahn, L. (1970). *J. Dairy Sci.* **53**, 434–437.
Jenkinson, D. M., and Mabon, R. M. (1973). *Br. Vet. J.* **129**, 282–295.

Johnson, H. D. (ed.) (1976). "Progress in Biometerology," Vol. 1, Part 1, Swets and Zeithlinger, Amsterdam.
Johnson, H. D., and Yeck, R. G. (1964). *Res. Bull.—Mo., Agric. Exp. Stn.* **865**, 1–31.
Johnson, K. G. (1970). *J. Agric. Sci.* **75**, 397–402.
Kamal, T. H., and Johnson, H. D. (1970). *J. Dairy Sci.* **53**, 1734–1738.
Kelly, R. O., Martz, F. A., and Johnson, H. D. (1967). *J. Dairy Sci.* **50**, 531–533.
Leng, R. A. (1984). *In* "Symposium on Herbivore Nutrition in the Subtropics and Tropics—Problems and Prospects" (F. M. C. Gilchrist and R. I. Mackie, eds.), pp. 129–144. The Science Press, Craighall, South Africa.
Levin, R. J. (1969). *J. Endocrinol.* **45**, 315–335.
Lippke, H. (1975). *J. Dairy Sci.* **58**, 1860–1864.
McDowell, L. R., Conrad, J. H., Ellis, G. L., and Loosli, J. K. (1983). "Minerals for Grazing Ruminants in Tropical Regions." Dept. Anim. Sci., Univ. of Florida, Gainesville.
McDowell, R. E. (1972). "Improvement of Livestock Production in Warm Climates." Freeman, San Francisco, California.
McDowell, R. E., and Weldy, J. R. (1967). *Biometerology* **2**, 414–421.
McDowell, R. E., Moody, E. G., Van Soest, P. J., Lehmann, R. P., and Ford, G. L. (1969). *J. Dairy Sci.* **52**, 188–194.
Mallonee, P. G., Beede, D. K., and Wilcox, C. J. (1982). *J. Dairy Sci.* **65**(Suppl. 1), 212.
Mannetje, L. 't. (1984). *In* "Herbivore Nutrition in the Subtropics and Tropics" (F. M. C. Gilchrist and R. I. Mackie, eds.), pp. 51–66. The Science Press, Craighall, South Africa.
Miller, J. K., Swanson, E. W., Lyke, W. A., Moss, B. R., and Byrne, W. F. (1974). *J. Dairy Sci.* **57**, 193–197.
Mitra, R., Christinson, G. I., and Johnson, H. D. (1972). *J. Anim. Sci.* **34**, 776–779.
National Research Council (NRC) (1978). "Nutrient Requirements of Dairy Cattle," 5th rev. ed. Nutrient Requirements of Domestic Animals, No. 3. Nat'l. Acad. Sci., Washington, D.C.
National Research Council (NRC) (1981). "Effect of Environment on Nutrient Requirements of Domestic Animals." Nat. Acad. Sci., Washington, D.C.
Page, H. M., Erwin, E. S., and Nelms, G. E. (1959). *Am. J. Physiol.* **196**, 917–918.
Roman-Ponce, H., Thatcher, W. W., Buffington, D. E., Wilcox, C. J., and Van Horn, H. H. (1977). *J. Dairy Sci.* **60**, 424–430.
Schneider, P. L., Beede, D. K., Hirchert, E. M., and Wilcox, C. J. (1984a). *J. Dairy Sci.* **67**(1), 120.
Schneider, P. L., Beede, D. K., Wilcox, C. J., and Collier, R. J. (1984b). *J. Dairy Sci.* **67**, 2546.
Singh, S. P., and Newton, W. M. (1978). *Am. J. Vet. Res.* **39**, 799–805.
Turner, M. A. (1981). *Vet Sci. Commun.* **5**, 159–162.
Van Jonack, W. J., and Johnson, H. D. (1975). *J. Dairy Sci.* **58**, 507–511.
Van Soest, P. J. (1981). "Nutritional Ecology of the Ruminant." O and B Books, Corvallis, Oregon.
Von Engelhardt, W., and Hales, J. R. S. (1977). *Am. J. Physiol.* **232**, E53–E56.
Warren, W. P., Martz, F. A., Asay, K. H., Hilderbrand, E. S., Payne, C. G., and Vogt, J. R. (1974). *J. Anim. Sci.* **39**, 93–96.
Winchester, C. F., and Morris, M. J. (1956). *J. Anim. Sci.* **15**, 722–744.

5

Forages for Grazing Systems in Warm Climates

PAUL MISLEVY

Department of Agronomy
University of Florida
Agricultural Research Center
Ona, Florida

I.	Introduction.	73
II.	Selected Warm-Climate Perennial Grass Types.	75
III.	Selected Warm-Climate Perennial Legumes.	91
	References.	101

I. INTRODUCTION

The tropical and subtropical region of the world can be described as that area between 30° N latitude and 30° S latitude. This chapter will consider this geographic area as the "warm-climate" (tropical) region, based on the predominate forage species used for grazing. More than 3000 grass species and over 1000 legume species (McDowell, 1972) may provide forage for herbivores, but the predominate warm-climate forages may number only 20–30 species. Forages grown in the warm climate utilize about 29% more land area than forages from the temperate areas; however, this warm-climate region only produces 36 and 19% of the worlds meat and milk, respectively [Food and Agriculture Organization of the United Nations, (FAO) 1980].

Forage grasses grown in the warm climate generally contain the C_4 carbon pathway of carbohydrate synthesis. These plants basically produce more dry matter per unit leaf area, thus having a greater water use efficiency (Downes, 1969). More dry matter is produced per unit of water because the C_4 plants transpire less water and support a larger shoot–root ratio than C_3 plants (Boote, 1976). Warm-climate plants have a distinct advantage over temperate (C_3) plants in areas of high irradiation, high temperature, and low moisture (Loomis et al., 1971; Ludlow, 1976).

In recent years, there has been a growing realization that the warm-climate forages are different from the temperate forages. Plant species and environmental factors appear to be responsible for differences in warm-climate forage quality. Some of these factors include neutral detergent solubles (NDS), neutral detergent fiber (NDF), plant morphology (e.g., leaf–stem ratio and tiller density), and plant anatomy (low proportion of mesophyll and high concentrations of epidermal, vascular, and sclerenchyma cells) (Dirven, 1977; Minson and Wilson, 1980; Wilson, 1982; and McLeod and Minson, 1974). Legumes have been found to contain more NDS, which are highly digestible, than do grasses at similar stages of maturity. The NDF of cell walls in legumes have a much lower hemicellulose and a higher lignin content (percentage of cell wall) than do grasses. Temperate grasses are generally higher in NDS and lower in lignin, crude fiber (CF), NDF, acid detergent fiber (ADF), and cellulose content than tropical grasses at similar stages of development. The temperate annual grasses are also higher in NDS and lower in NDF than temperate perennial grasses. Tropical annual grasses follow a similar pattern to temperate grasses, being higher in NDS and lower in NDF than tropical perennial grasses.

Tropical grasses generally have a lower leaf–stem ratio; consequently, more rapidly maturing stem material is produced, resulting in lower digestible forage and generally lower intake. Since tropical forages are generally higher yielding than temperate entries, a higher demand for plant nutrients exists, which could result in a dilution effect if mineral concentration in the soil is low.

Temperature plays an important role in *in vitro* dry matter digestibility (IVDMD) of temperate and tropical forages (Wilson and Ford, 1971). Temperate grasses appear inherently higher in IVDMD, than tropical grasses at various temperatures. However, tropical grasses grow faster at all but low (15°C daytime to 10°C nighttime) temperature regimes. Increased temperature has a tendency to increase structural carbohydrates with little effect on nonstructural carbohydrates for both temperate and warm climate forages; however, the increase in structural carbohydrates is always higher for the tropical plants. For every degree Celsius increase in optimum temperature, the digestibility of tropical grasses decreases by 0.6 to 1.0 percentage units. Studies conducted by Milford and Minson (1968) and Minson (1967) indicate that the overall lower digestibility of tropical grasses is not caused by poor plant nutrition. They conclude that there is a consistent difference between dry matter digestibilities of temperate and tropical grasses that is not associated with stage of growth or fertilizer treatment.

Comparisons conducted by Laksesvela and Said (1978) revealed temperate grasses are about 20–35% higher in digestible crude protein (DCP) and net energy (NE) than warm climate grasses.

Forage digestibility of water-stressed plants was either similar to or higher than that of plants receiving adequate moisture. Water-stressed plants generally have a higher leaf–stem ratio, resulting in higher digestibility of both leaves and stems (Wilson, 1981).

Perhaps the world's greatest potential for forage, and hence for ruminant livestock production, lies in the humid tropics (comprising 5 billion hectares of land) with year-round warm temperature, year-long growing season, and high rainfall. These environmental factors are responsible for high, dry matter yields of most warm-climate forages.

One of the most important factors when selecting forages to be used in a grazing system is management. That is, the individual must be well versed in all options pertaining to species selection and variables as they relate to each plant entry. The performance of a species on a specific soil type, time of year for maximum forage production, need for irrigation, and day-to-day management varies, as related to a specific forage entry. The type of grazing system most economical, grazing frequency, fertility requirement, species persistence, and advantages and disadvantages of each forage will be discussed for the most common warm-climate forages.

II. SELECTED WARM-CLIMATE PERENNIAL GRASS TYPES

1. Gambagrass (*Andropogon gayanus*)
   Common synonyms: Gambagrass, Sadabahar.
   Common cultivars: Gambagrass has three recognized subspecies: (1) var. *gayanus*—joints and pedicels ciliate on one margin; (2) var. *squamulatus*—joints and pedicels ciliate on two margins, moderately vigorous plant; (3) var. *bisquamulatus*—joints and pedicles ciliate on two margins, large, vigorous plant.
   Season of growth: Warm-season grass, generally found where average minimum winter day temperature does not drop below 4.4°C (Bowden, 1964).
   Soil condition: Plants tolerate a wide range of soil moisture from drought conditions to seasonal flooding. Some types tolerate 6

months of drought, remain green, and even produce some growth. Plants perform poorly on heavy clay soils.

Establishment: Vegetative plantings of rooted tillers or seeding into well-prepared seedbed and covering with 1–3 cm soil.

Growth habit: Perennial bunch grass forming a crown up to 1 m in diameter.

Management and utilization: Seed yields are poor, with 1 ha providing enough seed to establish 2–4 ha. Present management of native lands requires burning *A. gayanus* near the end of each dry season followed by grazing of developing tillers at the start of the rainy season. Dry matter (DM) yields range from 10 (plants cut 12 times per year at 60 cm) to 14 Mg ha^{-1}, with higher yields obtained when plants were cut 7 times per year at an average height of 150 cm. Plants respond to nitrogen (N) in the order of 20–25 kg ha^{-1} DM and 2 kg ha^{-1} crude protein (CP) for each kilogram of applied N to about 50 kg ha^{-1}. This grass tends to grow best in association with Centro and Stylo. Forage CP content tends to be lower than for most perennial grasses, ranging from 10.1% at a 4-week cutting frequency to 4.8% at full flower (Bowden, 1963). Seed production increases linearly from 38 to 75 kg ha^{-1} as applied N increases from 56 to 224 kg ha^{-1}.

Advantages:
  a. Resistant to grass fires.
  b. Var. *gayanus* will tolerate seasonal flooded conditions to a water depth of 2 m.
  c. Drought tolerant and remains green, with some production during the drought season.
  d. Produces high DM yields.

Disadvantages:
  a. Seed yield is poor on a commercial basis.
  b. Silage results in poor quality partly due to slow fermentation.
  c. Plant CP tends to be lower than for most other tropical grasses.

2. Signalgrass (*Brachiaria brizantha*)
   Paragrass (*B. mutica*)
   Ruzigrass (*B. ruziziensis*)

   Common synonyms: Signalgrass, Palisadegrass; Paragrass, Mauritiusgrass, Malohillo, Angolagrass, Capim angola, Egipto, Amirable, Penhalongagrass; Ruzigrass, Congo signalgrass, Congograss, Kennedy Ruzi.

   Common cultivars: Only the *B. ruziziensis* has cv. Kennedy.

   Season of growth: Warm season with soil moisture.

Soil condition: *B. mutica* prefers low-lying, seasonally flooded areas, withstands waterlogging and long-term flooding, but will not tolerate arid or semiarid areas. *B. ruziziensis* prefers moist non-waterlogged soil.

Establishment: *B. brizantha* is generally propagated by seed, *B. mutica* by cuttings or pieces of creeping shoots spaced about 1 m apart, and *B. ruziziensis* by either seed or vegetative rooted cuttings.

Growth habit: *B. brizantha*—tufted perennial with erect–suberect stems. *B. mutica* and *B. ruziziensis*—perennial with flowering stems 1–2 m high, ascending from long, many–noded prostrate shoots, freely rooting at the nodes and forming dense cover.

Irrigation: *B. mutica* responds to irrigation in areas where temperature is favorable, and rainfall is less than that received in humid tropics.

Management and utilization: Paragrass persists best when harvested or grazed at 45–60 cm, back to a 7-cm stubble. Some reports indicate that low cutting or grazing *B. mutica* swards is favored. Its popularity persists because of ease of propagation, competitive vigor, high yields, and quality. Reports indicate 1 kg of applied N increases DM by 21 to 47 kg. Clipping paragrass every 6–8 weeks appears most desirable. Crude protein in forage ranges from 2.8 to 16%, with digestibility of CP ranging from 54 to 62% in leaves and from 72 to 77% in stems. Since there is a high leaf–stem ratio, leaves contain about twice as much CP as the stems. Poor results have been obtained from ensiling paragrass due to a high pH; also, silage fed to cattle results in low intake and low milk production. Milk and beef production is high from fresh paragrass, especially if well fertilized. *B. brizantha* in general is inferior to other cultivated species of *Brachiaria*, with CP digestibility of 36–40%. *B. ruziziensis* forage has high digestibile CP averaging 65–70%, high fresh herbage yields (133–145 Mg ha^{-1}), and high liveweight gains (Hunkar, 1969).

Advantages:
  a. Prefers wet waterlogged soils.
  b. Grows well in combination with kudzu.
  c. Popular due to ease of propagation, vigorous high yield and quality.
  d. *B. ruziziensis* is highly productive with Stylo during the dry season.

Disadvantages:
  a. Will not tolerate dry soil.

b. Gives poor results when ensiled.
c. *B. brizantha* has low CP digestibility.
3. Rhodesgrass (*Chloris gayana*)
Common synonyms: Rhodesgrass, Pasto rodes.
Common cultivars: Pioneer (diploid), Katambora (diploid), Samford (tetraploid), Callide (tetraploid) identical with Mpwapwa, Giant, and possibly Kongwa, Masaba or Endebess, Mbarara (tetraploid), Nzoia (diploid), Pokot (tetraploid).
Season of growth: Warm season, expresses growth from 10 to 50°C but 35°C is optimum for photosynthesis. Rhodesgrass is tolerant to cold, having 97% survival at −9°C (Jones, 1969).
Soil condition: Grows under various soil conditions except heavy clay or soils of high acidity, tolerates high-akaline soils, develops well on volcanic ash, and has high salt tolerance. Plants grow well on high sodium (Na) soils but not in the form of NaCl or $NaNO_3$. Good establishment on saline–sodic soils. Withstands flooding up to 10–15 days.
Establishment: From vegetative plant parts or from seed. Seed should be spread on a well-prepared seed bed, covered very shallow, and rolled. Seeds do not germinate below a soil depth of 1.5–2.5 cm. Seeding rate is 0.5–1.0 kg of pure live seed per hectare.
Growth habit: Stoloniferous or occasionally tufted perennial, with erect or geniculated ascending stems 0.5–2 m high. A pioneer grass in abandoned cultivated fields, it occurs naturally in most tropical and subtropical plains up to 2000 m in elevation.
Irrigation: Responds well to irrigation, even though it is drought tolerant. Grown under irrigation in arid and semiarid countries outside the tropics, e.g., the southern United States of America.
Management and utilization: Good seed production, forage yields reasonable to high but seldom approaches most other tropical grasses, with average forage quality. Productivity of the sward lasts only 3–5 years. Grazing or cutting for hay can begin when grass reaches 50 cm in height. Allowing plants to attain taller heights before harvesting or grazing results in decreased forage quality. Generally responds to phosphorus (P) fertilizer, especially when followed by N application. Grows well in association with Centro, Stylo, Alfalfa, and white clover. Plants are affected by *Helminthosporium,* which causes dieback of leaves and shoot bases and by *Fusarium graminium,* which attacks spikelets and causes seed loss, and to a lesser extent by Rhodesgrass scale

(*Antonina graminis*), which also parasitizes other grasses. Maximum DM yields are generally obtained the second year of growth. This is one of the best tropical grasses for hay because it produces even stands. Generally poor results are obtained from silage making because of poor packing and low lactic acid content. Plants are generally low in potassium (K) and magnesium (Mg) and high in Na. Palatability of 4- to 5-week-old grass is good but falls rapidly with plant age and is lower than that of many other tropical grasses.

Advantages:
- a. Warm-season grass which expresses some vegetative growth at 10°C and tolerates temperatures to −9°C.
- b. Tolerates high salt and alkaline soils.
- c. Expresses considerable drought tolerance.
- d. Excellent upright grass for hay.

Disadvantages:
- a. Generally produces lower yields than most other tropical grasses.
- b. Short-lived perennial grass lasting 3–5 years.
- c. Seriously affected by *Helminthosporium* leaf spot, causing dieback of leaves and shoot bases.
- d. Generally produces low-quality silage because of poor packing and low lactic acid content.
- e. Forage quality drops rapidly after 5 weeks of growth.

4. Stargrass (*Cynodon nlemfuensis* and *C. aethiopicus*)

   Bermudagrass (*C. dactylon* var. *dactylon*) may be a weed species but is important as a pasture grass: var. *aridus*—large robust, more arid plant, important as a pasture species; var. *elegans*—low growing turf species; var. *coursii*—nonrhizomatous.

   Common synonyms: Stargrass and giant stargrass

   Bermudagrass, Dhub, Hariali; Dog's toothgrass, stargrass, Chiendent, Zacate Bermuda, Pasto Bermuda, Pasto Argentina.

   Common cultivars: Clonal stargrass—Ona, McCaleb stargrass, stargrass No. 2, Muguga stargrass, Cynodon IB-8 (Ibadan 8). Clonal bermudagrass—Alicia, Callie.

   Hybrid bermudagrass—Coastal, Midland, Suwannee, Coastcross-1.

   Season of growth: Stargrass and bermudagrass are warm-season perennials, with maximum photosynthetic activity at about 37°C, which is typical for grass species with C_4 pathway of photosynthesis. Plants tolerate several months of drought but produce little or no forage.

   Soil condition: Both prefer moist, well-drained soils ranging from sands to clays. Bermudagrass tolerates considerable salinity in irrigation water and withstands short periods of flooding.

Establishment: Established vegetatively using pieces of rhizomes, stolons, or stem pieces. When placed in a moist, firm seed bed, nodes germinate in about 7–10 days. This is accomplished by distributing freshly harvested planting material on cultivated soil, covering by disking to a 10- to 12-cm depth, followed by an extremely firm packing.

Growth habit: Generally stargrass is strongly stoloniferous without rhizomes and occurs mainly in the tropics. Bermudagrass has stolons and rhizomes and generally occurs in subtropics. If N rate is low an open sward will result.

Irrigation: Responds to irrigation under dry conditions if adequate temperature, when adequate fertility is available.

Management and utilization: Both perform best (in yield and quality) when a 5-week grazing frequency is allowed. Plants can be grazed or clipped more frequently during rainy season and less frequently during critical stress periods. Plants should not routinely be clipped or grazed below a 10-cm stubble, since continuous close grazing could jeopardize species persistence. Both are quite competitive and may be planted on land contaminated with other grasses if thorough site preparation and intensive planting practices are used. Some *Cynodon* spp. contain hydrocyanic acid potential (HCN-p) with some types very high (Ona and McCaleb) and some low (Callie and Alicia). Palatability decreases rapidly after a frost. Broadleaf weed control can be accomplished by applying 1.2 kg ha^{-1} 2,4-D Amine + 0.5 kg ha^{-1} Dicamba. Little difference in CP observed between varieties, averaging 16, 15, 12, 11, and 8% for grazing frequency of 2, 3, 4, 5, and 7 weeks, respectively. *C. dactylon* is a common weed species, will grow on alkaline and saline soils, and tolerates drought. Shading reduces DM yields considerably. Responds well to N and other plant elements. One kilogram of applied N usually increases DM yields 30–35 kg. *C. dactylon* supports a fungus on dead leaves causing a hypersensitivity to sunlight in cattle grazing on this grass.

Advantages:
  a. Produces high yields of good-quality forage when a 5-week rest period is allowed.
  b. Competes well with other grasses and may be planted on land where grassy weeds are a problem.
  c. Selected varieties like Callie and Alicia are low in HCN-p.

d. When managed properly, persist for many years under an intensive grazing system.

Disadvantages:
a. Generally, ensiling has met with little success. It is difficult to compress, and a pH of 5.0 results in a nonlactic acid-type fermentation.
b. Requires high levels of fertilizer, averaging about 25% higher than that required for *Digitaria* spp.
c. Selected cultivars are high in HCN-p and low in digestibility.
d. Palatability and quality drops rapidly after a 5-week rest period.

5. Pangola (*Digitaria decumbens*)

Common synonyms: Pangolagrass, Digitgrass, Pongolagrass.

Common cultivars: Pangolagrass.

Season of growth: Warm-season perennial. Produces forage when mean temperature is 10°C or above, with maximum shoot growth at 43°C.

Soil condition: Performs well on moist soil; however, will not tolerate long periods of flooding; can withstand drought but produces little forage under drought stress conditions. Grows on various soil types from acid (pH 4.3) to alkaline (pH 8.0); tolerates considerable sodium chloride.

Establishment: Mainly from stem cuttings, generally no viable seed produced. Weed control accomplished by applying 1.2 kg ha^{-1} 2,4-D Amine + 0.5 kg ha^{-1} Dicamba about 2 weeks after vegetative planting, when developing tillers are 2–5 cm tall.

Growth habit: Strongly stoloniferous grass rooting from nodes.

Irrigation: Irrigation applied during moisture stress results in moderate yield increases, provided favorable temperature is available.

Management and utilization: Should be harvested or grazed at 5-week intervals or when plants attain a height of 30–45 cm to obtain high quality. A successful rule of thumb is to cut half and leave half above ground forage by weight. Therefore, most species would be grazed or cut much lower than half the plant height, allowing a grass stubble of 10 cm. Broadleaf weeds can be controlled by close mowing or applying 1.2 kg ha^{-1} 2,4-D Amine + 0.5 kg ha^{-1} Dicamba. Pangola should never be planted on land contaminated with bermudagrass. Pangola forage decreases slowly in quality over time while remaining high in palatability. This grass can be made into excellent high-quality hay. Performs well with 112 kg per hectare of N per year alone, or even better yields can be obtained when grown in association with a compatible legume. Generally an increase in DM production is obtained as harvest frequency decreases; however, forage quantity consumed by the cattle and forage quality also decrease. Unlike most other tropical grasses, pangola tends to respond positively to Na. This grass is sensitive to copper (Cu) deficiency and consequently responds to Cu and, to a lesser extent, other micronutrients. Pangola grows well in association with Greenleaf *desmodium,* white clover, and Siratro when no N is applied to the mixture. The yellow sugarcane aphid (*Sipha flava*), spittlebug (*Prosapia bicincta*), army worms and loopers (*Spodoptera* and *Mocis*), and stunting virus can be serious pests on pangola. The stunt virus transmitted by an aphid causes pangola to recover slowly after defoliation, producing no stolons, and producing only dwarf leaves crowded on stems of compacted internodes. The virus appears to be most prevelant in the Caribbean and surrounding areas. The University of Florida has developed a variety called Transvala, which is resistant to stunt virus.

Advantages:
  a. Generally high in forage quality even after 5 weeks of growth but slowly drops in quality with age.
  b. High palatability even after forage matures and drops in quality.
  c. Excellent species for hay.
  d. Does not contain HCN-p.

Disadvantages:
  a. Will not successfully compete over a long period of time with common bermudagrass and should not be planted on land contaminated with this grass.
  b. Generally damaged by spittlebug, yellow sugarcane aphid, army worms (along with many other tropical grasses), and stunt virus.

6. Jaragua (*Hyparrhenia rufa*)

Common synonyms: Jaragua (Yaragua, Faragua), Yayale, Veyale.

Common cultivars: None.

Season of growth: Warm-season perennial; however, will grow well at elevations of 1800 m.

Soil condition: Requires relative dry soil areas.

Establishment: Seed or vegetative cuttings, but is more difficult to establish than are many other grasses. Seed contains no postharvest dormancy.

Growth habit: Bunch-type, upright plant forming an intermediate-type sward.

Irrigation: Grass responds well to irrigation, with research indicating a 3- to 4-fold herbage yield increase.

Management and utilization: Will not tolerate flooding but withstands burning; however, seed lying on the ground will not survive fires. To maintain high-quality swards the grass should be grazed frequently or continuously; however, a 30 cm stubble should remain to improve persistence. Research indicates good persistence when grown with a legume and when rotational grazing (grazing 7 and resting 14 days) is used. Jaragua responds to N, P, sulfur (S), but not K. Crude protein content ranges from 1 to 15%, depending on the physiological stage at which plants are harvested. Grass generally produces low-quality silage. Animal performance tends to decrease rapidly when stocking rate increases from 1 to 2 steers/hectare. Beef yields are generally lower from cattle grazing on this grass compared to grazing on Pangola, possibly reflecting the lower-quality forage. Grass tends to grow well in association with Stylo and Centro.

Advantages:
   a. Grows under both warm and cool, dry conditions but responds to irrigation.
   b. Seeds have little or no dormancy, therefore, can be seeded immediately after harvest.
   c. Responds to many major and secondary plant nutrients.

Disadvantages:
   a. Lower in quality than many other perennial grasses.
   b. Will not tolerate close grazing.
   c. Generally produces low-quality silage.

7. Molassesgrass (*Melinis minutiflora*)

Common synonyms: Mollassesgrass, Herbe du Brazil, Wynnegrass, Gordura, Melado, Calinguero, Yaragua (in Colombia, Venezuela, and Puerto Rico) (Parsons, 1972).

Common cultivars: Chania, Mbooni hill, and Kitale Commercial (all from Kenya). Roxo (red), Cabelo de negro, Francano, and Branco (all from Brazil).

Season of growth: Wide temperature tolerance from warm-season-tropical to cool-season-temperate areas ranging from 800 to 2500 m elevation.

Soil condition: Plants grow well on acid, low-fertility, and leached soils; at forest edges, open grasslands, and steep rocky slopes, especially on thin sandy soils. They do not tolerate waterlogged, heavy, clay soils.

Establishment: Plants establish best when sown on a clean-cultivated seedbed at a depth of 2–2.5 cm. Seed with 50% germination can be sown at 2–3 kg ha^{-1}. Vegetative planting has met with little success. The addition of P and K, but generally not N, may aid establishment.

Growth habit: A perennial, viscous bunch grass with many noded, erect, or gently ascending culms, about 1.5 m in height, with many branches. Leaves contain glandular hairs that exude a sticky, sweet-smelling substance. As plants grow, basal leaves die, with adult plants having only few leaves on top branches and stems.

Irrigation: Irrigation increases DM yields 26 and 49% on unfertilized and fertilized grass, respectively, during the dry season (Ladeira et al., 1966).

Management and utilization: Performs best when grazed. Rotational grazing is recommended at 40- to 60-day intervals, or each time plants attain a height of 35–45 cm (Whyte et al., 1959). Clipping studies indicate harvesting to a 12-cm stubble encourages lateral spread and increases ground cover. Plants express considerable competitive vigor; therefore, only a few legumes grow well in association, such as tropical kudzu and Stylo. Plants are susceptible to small-leaf or stunting virus. Dry-matter yields ranged from 6 to 8 Mg ha^{-1} in Colombia and may be doubled with 150 kg ha^{-1} N (Crowder et al., 1970). Crude protein ranged from 17 to 6% for 4- and 32-week-old herbage, respectively. Crude fiber is quite high, ranging from 30 to 40%, and ether extract is high, which is common for aromatic grasses. Regardless of high CF and low CP, ruminants consume the herbage relatively well. Copper in forage is high, ranging from 185 to 981 ppm for young and mature forage.

Advantages:
  a. Wide temperature tolerance from tropical (800 m) to temperate elevations (2500 m).
  b. Grows well on acid, low-fertility, and leached soils.
  c. Responds to irrigation and fertility.
  d. Suited for grazing rather than for hay.

Disadvantages:
  a. Will not tolerate waterlogged or heavy, clay soils.
  b. Can be destroyed by grass fires.
  c. Frequent and close clipping can destroy the plant stand.
  d. Little success with vegetative plantings.
  e. Susceptible to small-leaf or stunting virus.

8. Guineagrass (*Panicum maximum*)
   Greenpanicgrass (*P. maximum* var. *trichoglume*)

Common synonyms: Guineagrass, Capim coloniao, Sempreverde, Hierba de India, Privilegio, Zacaton, Fataque, Herbe de Guinee, Pasto Guinea.
Green panicgrass, slender guinea.
Common cultivars: Guineagrass, Coloniao (colonial), Boringuen, Broadleaf, Guinea, Hamil, Gatton, Semper verde, Sigor, Nchisi, King ranch.
Green panicgrass, Petrie, Sabi, Makueni, Embu.
Season of growth: Warm-season perennials; however, some green panic cultivars perform well during cool, dry seasons.
Soil condition: Prefers fertile, well-drained, light-textured soil and will not tolerate wet soil conditions. Green panic expresses medium to considerable drought tolerance and has been shown by Anderson (1970) to tolerate 5-10 days of flooding.
Establishment: Generally from seed, but it can be vegetatively established. This species is invaded easily by other species and weeds. With adequate moisture and fertility, it can be ready for grazing in about 4 months after planting.
Growth habit: Two types: (1) large or intermediate tufted and (2) small, low-growing bunch types. The green panic cultivar Embu has a long, creeping stem base containing many nodes.
Irrigation: Responds to irrigation, to a limited extent, during the cool, dry season.
Management and Utilization: Equally suitable for grazing or cutting and should be harvested at a plant height of 60-75 cm or about every 5 weeks. Plant height is the best criterion for harvesting or grazing a forage. Plants tolerate relatively close cutting or grazing and respond best at 7-10 cm stubble. It will not tolerate flooded conditions, but it is quite disease resistant. Plants tend to grow in light shade, under trees. Green panic cultivars Sabi and Makueni are excellent seed producers. Seed of excellent germination can be sown at 2 kg ha^{-1}. Fresh seed should not be sown because it requires a dormancy period of 6-18 months for postharvest maturation. Dry matter and CP yields respond well to N fertilizer at rates of 100-200 kg ha^{-1}. They also respond to other fertilizer elements if soil levels are deficient. Guineagrass tends to perform best with associated legumes, Centro and Stylo. Forage is palatable to all classes of livestock, and are high in CP and disgestibility when harvested as recommended. Grass can be utilized intensively with stocking rates of two steers per hectare. Realistic seed yields are about 25-30 kg ha^{-1} of pure live seed.

Advantages:
   a. Seeded or vegetatively planted.
   b. Grows in light shade under trees.
   c. Responds well to N fertilization.
   d. Palatable to all classes of livestock, with high CP and digestibility when harvested prior to infloresence.

Disadvantages:
   a. Flowering lasts over a long period of time. Bird damage and poor seed formation affect seed production and yield.
   b. Guineagrass will not tolerate flooding or saturated soil conditions.
   c. Easily invaded by other species or weeds.
   d. Freshly harvested seed has a dormancy period and may have to be stored for 6–18 months.

9. Bahiagrass (*Paspalum notatum*)
   Brown-seeded Paspalum (*P. plicatulum*)
   Common synonyms: Bahiagrass; Grama dulce, Forquinha; Gengibrillo; Pasto horqueta.
   Common cultivars: Bahiagrass—Pensacola, Tifhi-1, Common, Argentine, Batatai (Brazilian), Paraguay, Wilmington, Wallace, Tamba, Andre da Rocha, Capivari (Brazilian).
   Brown-seeded Paspalum—Rodd's Bay, Hartley.
   Season of growth: Bahiagrass is a warm-season perennial, producing much of its forage when the temperature is above 15°C. Brown-seeded Paspalum starts to flower at the end of the warm season, providing a longer period of vegetative growth. Plants of both species are frost sensitive; Pensacola tolerates temperatures to −15°C.
   Soil condition: Both species tolerate drought and wet soil conditions but will not tolerate long periods of flooding.
   Establishment: Mostly by seed but also by crowns or pieces of stolons. Seed germination is generally slow because water has difficulty penetrating the lemma and palea that clasp around the caryopsis. Following germination, development of young plants is generally slow.
   Growth habit: A creeping perennial with short stolons firmly pressed to the soil, forming a dense sod. The stolons contain numerous short internodes and develop roots at many nodes, which eventually produce shoots and leaves.
   Irrigation: Bahiagrass responds to irrigation if temperature is favorable for plant growth. Pensacola appears to be one of the most drought-tolerant varieties, whereas Argentine is more sensitive to drought.

Management and utilization: Little advantage is obtained in DM production when the rest period is extended beyond 2–3 weeks. Forage digestibility of Pensacola bahia drops rapidly following 3 weeks of regrowth. Crude protein content, determined after 2–7 weeks of regrowth in Florida, averaged 12%, which is similar to that for other grasses. Bahiagrass persists well under continuous close grazing and produces maximum yields when grazed to approximately 5 cm. It responds to N fertilizer; however, yield curves generally level off at N rates of 70–90 kg ha^{-1}. Since bahiagrass is quite aggressive, forming a dense mat, warm-season legumes have difficulty becoming established, resulting in only fair growth. Therefore, N fertilizer may be more suitable as a N source for bahiagrass. Research in Brazil indicates a close association between bahiagrass and a soil bacteria (*Azotobacter paspali*), which fixes atmospheric N in association with roots and rhizomes.

Advantages:
  a. Bahiagrass forms a dense sod and is valued for its productivity under low-fertility conditions, ease of establishment, and erosion control.
  b. Bahiagrass and brown-seeded Paspalum express considerable drought tolerance.
  c. Easily established from seed.
  d. Performs well with 50–60% of the fertilizer required by *Cynodon* spp.
  e. Argentine and Paraguay are desirable cultivars for landscaping.
  f. Contains no HCN-p.

Disadvantages:
  a. Produces less forage than most other tropical grasses.
  b. Reduced digestibility after 2–3 weeks' growth and digestibility remains low.
  c. Generally low forage production when temperatures drop below 25°C.
  d. Can be destroyed by mole crickets (*Scapteriscus vicinus* and *S. acletus*).

10. Elephantgrass (*Pennisetum purpureum*)
    Common synonyms: Elephantgrass, Napiergrass, Napier's Fodder, Herbe Elephant; Elefante, Pasto gigante.
    Common cultivars: Merker, Merkeron, Merkeron 534, French Cameroons, Napier, Capricorn, Mineiro, Uganda hairless, Uganda, Cubano, Domira, Panama, Ghana, Pungwe, and Urukwanu.

Season of growth: Warm-season perennial; however, it will produce considerable forage under cool conditions with temperatures above 10°C.

Soil condition: Established plants grow on soils from poorly drained to dry but will not tolerate continuous flooding or waterlogged soils. Good fertility is required for high production, but it grows on almost any soil with reduced growth and production.

Establishment: Stem cuttings; however, caution should be exercised in cutting or grazing during the first 6 months of development. Plants can be sprayed during plant emergence or later using 2.2 kg ha^{-1} atrazine and 2.2 kg ha^{-1} alachlor to control many broadleaf and annual grasses. Stem cuttings with three nodes, two placed in cultivated soil at a 45° angle, and one node placed above the soil give best regeneration. Cuttings can be stored up to 20 days under humid, tropical conditions and all winter under cool to cold subtropical conditions.

Growth habit: A robust, upright, bunch-type plant forming large, broad clumps, spreading by stem bases, rooting from nodes or by short rhizomes. Erect stems will branch at the upper parts and contain up to 30-noded, 2- to 6-m-tall culms.

Irrigation: Quite drought tolerant, but a 70% increase in DM yield is obtained with irrigation during the dry season.

Management and utilization: Best adapted for cutting (when 1.5–2.0 m tall) as a soilage crop; however, it provides excellent grazing when plants are 100–120 cm if not consumed closer to the soil than 30–45 cm. If cut frequently, plants may be killed, and if cut at maturity, nutritive value is drastically reduced. Severe grazing prior to a drought or other stress conditions reduces yields. Valued for high-herbage yields, competitive vigor, persistence, palatability (during the young stage), and good herbage quality. Has the ability to compete with and suppress *Imperata cylindrica*. Responds well to any level of fertilizer, N alone or in combination with P, K, and Mg. Satisfactory silage can be made from elephantgrass; however, DM losses are 9–12%, and TDN and DM consumption is lower than that of sorghum or maize silages. Leaf spot (*Helminthosporium ocillum*) is perhaps the most serious disease, along with red mites living on the underside of pubescent leaves. Although generally grown in pure stands, success has been obtained when grown in association with kudzu and *Desmodium intortum*. The percentage of DM (ranging from 12 to 18%) at early growth stages is generally much lower than that of most other tropical grasses. Young, immature plants are generally quite high in CP, averaging 23 and 12% for plants harvested 4 and 8 weeks, respectively. Seed setting is usually poor, with no dormancy.

Advantages:
  a. Grows on poorly drained ditch banks to droughty savanna soils.
  b. Best adapted as a soilage crop but can be made into silage.
  c. Grows from sea level to elevations of 2000 m or when temperature is above 10°C.
  d. Immature plants (4–8 weeks old) generally contain higher CP than other tropical grasses.

Disadvantages:
  a. Vegetatively planted, because seed production is usually poor.
  b. Palatability of stems is poor at an advanced stage of maturity.
  c. Not compatible with most tropical legumes.
  d. Low percentage of DM (12–18%) at young physiological stage.

11. Sugarcane (*Saccharum officinarum*)

   Common synonyms: Sugarcane, noblecane.

   Common cultivars: Many developed throughout the tropical and subtropical regions of the world, e.g., in Florida, Canal point (CP) 70-1133 and (CP) 78-1247.

   Season of growth: During the warm season for maximum growth when maximum–minimum temperature is 35–21°C.

   Soil condition: Moderately well drained, ranging from organic to mineral, with a texture from light sand to heavy clay.

   Establishment: Plants easily established by cutting stems so each vegetative portion contains 2–3 nodes, placing the entire stem into a furrow of cultivated soil, and covering with 10–15 cm of the moist soil. Generally 8 t are required to plant 1 ha. Regeneration of growth may require 1–3 months depending on moisture and temperatures.

   Growth habit: Sugarcane is a tall tufted plant, which does not contain stolons or rhizomes, and stores sugar in the stem. Well-fertilized plants may attain heights of 4–6 m, with a crown diameter of 1 m.

   Irrigation: Plants growing on mineral and organic soils generally respond to irrigation.

   Management and utilization: Sugarcane fodder can be used for cattle feed by utilizing immature plants or sugarcane tops that are a residue of plants harvested for sugar and bagasse (a plant residue following extraction of sugar). Immature plants may average 7,4, 35, and 60% CP, digestible protein, CF, and digestibility, respectively. Commercially cultivated sugarcane yields 50–70 Mg ha^{-1} DM with an average of about 10% tops. This fodder averages 5–6% CP and 32–35% CF.

Advantages:
    a. High-yielding crop that can be grown for a commodity like sugar, while utilizing all by-products for livestock feed.
    b. Limited land area required to obtain considerable fodder.

Disadvantages: Requires good fertility and adequate moisture for high yields.

12. Setaria (*Setaria anceps*)

Common synonyms: Setariagrass, napierzinho.

Common cultivars: Kazungula, Nandi, Nandi mark 2, Nandi mark 3, Narok, Bua river, and Toittskraal.

Season of growth: Moderately warm season; however, plants grow best at 20–25°C, which is lower than those temperatures in the hot tropics. Most cultivars exhibit some frost tolerance, and plants can survive −3 to 4°C.

Soil condition: Grows on a wide range of soils, except very acid or highly alkaline. It tolerates temporarily flooded or waterlogged soils.

Establishment: Established easily from vegetative crowns or from seed, when drilled into a clean seedbed. Vegetative crowns are split into rooted tillers, placed into moist soil, and packed. Seeding at 1.5 kg ha^{-1} with pure live seed to a depth of 2.5 cm results in good plant stands. Nitrogen should not be applied to young seedlings until good establishment is demonstrated to inhibit weed growth. Cool temperatures delay establishment.

Growth habit: Spreads by short rhizomes with relatively erect stems attaining a height of 1–2 m with glabrous leaves 40 cm long and 8–20 mm wide.

Irrigation: Responds to fertilizer and water during the dry season.

Management and utilization: Setaria grows well in association with greenleaf *Desmodium,* Siratro, and *Glycine*. It allows good development of legumes because of slow grass establishment. Several years after establishment the grass becomes more aggressive, competing vigorously with the legume. Grass–legume mixtures respond to P both at establishment and in older swards. Plants are sensitive to low-cutting stubble height, producing higher yields (28 Mg ha^{-1}) when cut at 15 cm than when cut at a stubble of 7.5 cm (23 Mg ha^{-1}) (Riveros and Wilson, 1970). Rotation grazing at 4- to 8-weekly intervals is generally recommended. Plants respond well to N fertilizer, producing about 30 kg ha^{-1} DM per kilogram of applied N. The Cu and zinc (Zn) requirement by the plant is generally low. Leaf diseases are not common, with the exception of *Pyricularia tirsa,* expressed as red spots, and a few infloresence diseases found in Zaire and Kenya. Attempts at making hay and silage generally have yielded poor results due to coarse stems for hay and low carbohydrates, resulting in poor lactic acid formation, for silage. Butterworth (1967) indicated CP ranged from 4.8 to 18.4% and CF ranged from 24 to 34%. Animal production is not outstanding and is similar to that of *Chloris guyana.* Setaria herbage could be quite acid, with pH dropping to 4.8 due to anhydrous oxalic acid. A high concentration of this acid can be toxic to cattle diagnosed as hypocalcemia.

Advantages:
  a. Responds rapidly when rainfall commences after a dry period.
  b. Resistant to grass fires.
  c. Grows under temperature conditions lower than most tropical grasses.
  d. Tolerates short periods of waterlogged soil.
  e. Responds to fertilizer and irrigation during drought.

Disadvantages:
  a. Not competitive at establishment when mixed with a vigorous grass like *Chloris gayana*.
  b. Unproductive during dry periods, producing only about 8% total yield.
  c. Sensitive to low-clipping or low-grazing stubble height.
  d. Generally poor-quality hay and silage.
  e. Generally low animal production compared with other tropical grasses.
  f. Contains a high concentration of anhydrous oxalic acid.

III. SELECTED WARM-CLIMATE PERENNIAL LEGUMES

1. Centro (*Centrosema pubescens*)
   Common synonyms: Centro, Jitirana.
   Common cultivars: Deodoro (Brazil), Belalto (Australia).
   Season of growth: Warm-season perennial well adapted to the humid tropics with an annual rainfall of about 1500 mm. Plants are frost sensitive, producing no growth under low temperatures.

Soil condition: Plants prefer well-drained soils and will not tolerate flooded conditions; however, they live under drought stressed conditions by shedding leaves and reducing transpiration. Plants grow on low soil fertility and tolerate considerable manganese (Mn) concentration.

Establishment: Rapidly from seed; however, it requires 8–12 months postharvest for seed to lose dormancy. The proportion of hard seed is quite high, averaging 60%. One method, in addition to scarification, to improve germination is freezing seed for a few weeks.

Growth habit: Develops a climbing growth habit with slender pubescent stolons. Some roots develop at each node, producing a dense sward.

Irrigation: Responds well to irrigation during drought.

Management and utilization: Quite resistant to disease; however, plants could be damaged by red spider (*Tetranychus* sp.) and Cercospora leaf spot. Popular because of vigorous and productive growth and good-quality forage. Forage ranges from 18 to 24% CP; however, grazing ruminants must become accustomed to the forage before it is well accepted. Forage also contains some oxalates, which could reduce palatability and availability of calcium and other minerals. Forage quality remains high over a long period of time, possibly because of little lignification in the stems. This legume grows well in associations with Pangola, paragrass, and guineagrass; however, due to the climbing habit, the grass component generally decreases over time. Centro has a specific *Rhizobium* requirement. Like many other tropical legumes, it has the ability to extract nutrients from soils with low mineral content. Dry-matter yields increase as stubble height increases and clipping frequency is prolonged.

Advantages:
  a. Fixes atmospheric N, requiring little or no N fertilizer for companion grass.
  b. Extracts nutrients from low-mineral-content soils and tolerates considerable Mn.
  c. Stolons will root at nodes.
  d. Forage quality remains high over a long period of time (6–12 weeks) as plants continue to mature.

Disadvantages:
  a. Seed harvesting is difficult because of shattering and because the period of flowering is long.
  b. Not tolerant of frequent close grazing.

c. Frost sensitive, producing little or no growth under low temperatures.
2. Silverleaf desmodium (*Desmodium uncinatum*)
 Greenleaf desmodium (*Desmodium intortum*)
 Common synonyms: Silverleaf desmodium, Tick clover, Spanish clover
 Greenleaf desmodium, Kuru vine.
 Common cultivars: Silverleaf (selected in Australia)
 Greenleaf syn. Beerwah (developed in Australia).
 Season of growth: Warm-season perennials growing up to elevations of 1000–1500 m. Optimum day/night temperatures are about 30/25°C, respectively. (Whiteman, 1968). Both species are short-day plants.
 Soil condition: Both species prefer moist, loamy or sandy soils and perform well on soils with a pH of 5.0. Neither species tolerates soil salinity. Greenleaf tends to be more drought sensitive than silverleaf.
 Establishment: Both can be seeded into a cultivated soil at 2 kg ha^{-1} followed by a firm rolling for good seed–soil contact. Greenleaf requires highly specific *Rhizobium*, but silverleaf performs well with *Rhizobium* from other *Desmodium* or *Glycine* genera.
 Growth habit: Silverleaf desmodium has a trailing growth habit with stems up to 5 m long containing many short, brownish, hooked hairs and a well-defined silvery midrib on the upper leaf surface with the remainder of upper leaf dark green. Greenleaf is an herbaceous, upright, ascending plant with many-branched often reddish-brown stems.
 Management and utilization: Resistant to short-term waterlogged soil. However, neither species tolerates intense close cutting or grazing. Silverleaf tolerates cool weather and frost conditions better than greenleaf. Both species are valued for their high herbage and good seed yields. Silverleaf is quite sensitive to low P and K. When P and K concentrations in forage drops below 0.23 and 0.80%, respectively, these elements must be applied to the soil. Silverleaf tolerates low levels of Cu and medium-high levels of Mn and Al. Both greenleaf and silverleaf grow well in association with Pangola, Setaria, and bahiagrass. Greenleaf fixes from 200 to 300 kg ha^{-1} N, which is about double that reported for silverleaf. Forage quality factors like CP and CF range from 18 to 24% and 28 to 31%, respectively, when plants are in flower. Silage and hay production has been attempted. Greenleaf produces adequate-quality hay and low-quality silage due to poor

packing and the low carbohydrates in forage. Greenleaf is quite tolerant to rootknot nematode, whereas silverleaf is quite susceptible. Flowers of both species can be consumed by Meloid beetles, thus reducing seed yields.

Advantages:
  a. Fixes atmospheric N, requiring little or no N fertilizer for companion grass.
  b. Tolerates short-term waterlogged soil.
  c. Tolerates acid soils to pH 5.0.
  d. Produces high-quality herbage and good seed yields.

Disadvantages:
  a. Greenleaf is not drought tolerant, often shedding its leaves during dry conditions.
  b. Both are not tolerant of soil salinity.
  c. Both contain 6–8% tannins in the leaves, which reduces palatability and rumen cellulose digestion.
  d. Neither tolerates continuous close cutting or grazing.

3. Leucaena (*Leucaena leucocephala*)

Common synonyms: Leucaena, Ipil-ipil, Koa haole, White popinac.

Common cultivars: Hawaii, Peru, El Salvador, Guatemala, and Australia.

Season of growth: Warm-season, perennial shrub or small tree. Growth occurs when the temperature is above 10°C, with little growth at elevations above 700 m.

Soil condition: Adapted to areas with an annual rainfall of 760–5000 mm. Will grow well under drought conditions; however, will not tolerate waterlogged soils. Plants grow well from pH 5.0 to 7.5 and are quite tolerant of low P. On soils of low mineral content, plants respond to calcium (Ca) and P.

Establishment: From seed relatively easily; however, seed must first be acid treated or other scarification methods used to obtain satisfactory germination. Seeding rates of 2–5 kg ha^{-1} are adequate provided seed is sown on a prepared seedbed. Companion grass should be seeded after Leucaena has been established.

Growth habit: Deep-rooted, upright shrub, forming an open type of sward that allows sunlight for grasses. This drought-tolerant shrub maintains green leaves during moisture-stressed conditions.

Irrigation: Plants respond well to irrigation, yielding as much as 34.6 Mg ha^{-1} (Anslow, 1957).

Management and utilization: This shrubby legume is very resistant

5. Forages for Grazing Systems in Warm Climates

to disease. Recently, however, a jumping plant louse (*Heteropsylla* spp.) was found to severely damage leucaena leaves. Leucaena has a vigorous plant growth, high yield, high CP (20–25%), and high seed production. Cattle will consume leaves and small twigs 6 mm in diameter. Lenient grazing should be practiced during the establishment year. In later years, as plants grow beyond the reach of ruminants, lopping may be necessary. The plant should be grazed by cattle only, since it contains an alkaloid (mimosine) that causes shedding of hair in horses and hogs, and it also is not recommended for grazing by sheep. Cattle should not have a complete diet of Leucaena since it could affect reproduction. Leucaena requires a specific *Rhizobium*. Plants can be cut for silage each time they attain a height of 90–150 cm and average 16–22% CP. Like most forage species, herbage quality is proportional to harvest frequency. Deficiencies in tryptophan and S-containing amino acids have been found. Dry-matter yields range from 3 to 20 Mg ha^{-1}.

Advantages:
   a. Fixes atmospheric N, requiring no N fertilizer for companion grass.
   b. Leaves remain green as a good protein source under drought stress.
   c. Leaves and stems have a high protein content.

Disadvantages:
   a. Will not tolerate fire.
   b. Establishment and growth of young plants is rather slow.
   c. Can be grazed only by cattle because of mimosine.
   d. Will not tolerate waterlogged soils.
   e. Little growth takes place below 10°C.
   f. Jumping plant louse found to severely damage leaves.

4. Siratro (*Macroptilium atropurpureum*)
   Common synonyms: Siratro.
   Common cultivars: Siratro.
   Season of growth: Warm season, produces much of its forage during long days, only flowers during short days. Maximum forage production obtained when day/night temperature ranges between 27/22 to 30/25°C, respectively (Hutton, 1970). Photosynthetic area damaged by frost, but plants tolerate temperatures to −5°C.
   Soil condition: Plants perform well on moderately dry to moist soils where there are short periods with standing water. It also performs well on infertile to highly fertile soils.
   Establishment: Propagated by seed; seedling plants establish easily

and rapidly when seeded at 3 kg ha^{-1} on clean land. Siratro generally contains considerable hard seed ranging from 20 to 30% germination. May be sown at same time as companion grass.

Growth habit: Perennial plants produce stems with a climbing or twining habit, rapidly forming a dense sward.

Irrigation: Valuable in producing seed flushes for commercial production (Hopkinson and Lach, 1973).

Management and utilization: Plants should not be closely grazed, but a lenient grazing scheme should be utilized with about 1 animal unit per hectare. Plants desire cowpea-type *Rhizobium* with nodulation increased by Ca and molybdenum (Mo), resulting in higher N content in plants. Desirable grasses to grow in association with Siratro are Pangolagrass, bahiagrass, and guineagrass. Many studies indicate Siratro produces the equivalent of 100–125 kg ha^{-1} N through increased forage when growing with a grass, compared with N-fertilized grass. Application of 30–40 kg ha^{-1} N, however, applied to grass–legume mixtures increases forage yields. Siratro harvested at a 60-day interval averages 58 and 10% digestibility and CP, respectively (Mislevy et al., 1981). These values were reduced due to leaf losses from the fungus *Rhizoctonia solani*, which is prevelant in Florida; however, plants appear to tolerate the disease. Palatability of Siratro is excellent by ruminants.

Advantages:
   a. Fixes atmospheric N, requiring little or no N fertilizer for it or its companion grass.
   b. Seedlings establish rapidly and easily even under adverse conditions.
   c. Excellent palatability.

Disadvantages:
   a. Will not tolerate saline soils.
   b. Severely defoliated by the fungus *Rhizoctonia solani*, resulting in reduced forage quality.

5. Perennial soybean (*Neonotonia wightii* syn. *Glycine wightii*)

   Common synonyms: Glycine, Soja perene.

   Common cultivars: Tinaroo (medium-late maturing), Cooper (early maturing), and Clarence.

   Season growth: Warm season, with plants showing good longevity. Optimum day/night temperatures are about 30/26°C. Plants express some frost tolerance.

   Soil condition: Plants will not tolerate wet soil conditions but have a fair drought tolerance over a short period of time. Optimum rainfall is about 1100 mm/year. Plants generally prefer heavy clay soils, and tolerance to soil salinity is rather low.

Establishment: From seed (7 kg ha^{-1}) with relative ease. The main problem is slow initial growth and slow nodulation (25–30 days). Generally, a good seed producer averaging 250 kg ha^{-1} seed. Generally, a high percentage of seed (80 to 90%) are hard, requiring some type of scarification.

Growth habit: A perennial having a climbing or twining habit and forming an intermediate type of sward. It grows from sea level to over 2000 m elevation. Plants flower at day lengths of 8–11 hr.

Management and utilization: It tolerates close (5 cm) cutting or grazing and responds to low levels of N at establishment time. It is desirable to establish this species with a perennial grass that has slow establishment vigor. It can be grown in association with guinea, Pangola, and setaria grasses, but with limited applied N to reduce grass competition. Like most tropical legumes, DM yields increase as grazing or clipping frequency decreases; however, a 5- to 6-week grazing frequency is recommended. Plants may respond to low applications of P, K, and Ca. Plants average about 15% CP and 65% digestibility at the flowering stage and are used for both hay and grazing. Plants harvested with guineagrass produce good-quality silage with a pH of 4–4.5 (Baker and Kyneur, 1962). Herbage yields are similar to those of most other tropical legumes after establishment. Animal performance on Glycine and grass association increases cattle gains about one-third over gains on unfertilized grass or is equal to gains in cattle fed on grass plus 100 kg ha^{-1} N.

Advantages:
   a. Fixes atmospheric N, requiring little or no N for companion grass.
   b. Good forage quality when harvested at flowering stage.
   c. Tolerates frequent and close grazing when well established.
   d. Expresses some frost tolerance, especially cv. Clarence.

Disadvantages:
   a. Exhibits poor seedling vigor, which can be improved with low amounts of N.
   b. Low tolerance for Mn and aluminum (Al).
   c. Will not tolerate wet soils.

6. Tropical kudzu (*Pueraria phaseoloides* var. *javanica*).
   Common synonyms: Tropical kudzu (Figs. 5.1 and 5.2), Puero.
   Common cultivars: None.
   Season of growth: Warm-season perennial that requires high tem-

Fig. 5.1. Kudzu (*Pueraria phaseoloides*) growing along the roadside in the Dominican Republic. (Courtesy of G. O. Mott, University of Florida, Gainesville, Florida.)

peratures but tolerates light frosts. It does not normally perform well at elevations above 1000 m. It grows well where annual rainfall ranges between 1200 and 1500 mm and tolerates long periods of drought (2–3 months).

Soil condition: It tolerates arid soils and grows at a pH between 4.0 and 5.5. Plants do well on soils low in P and Ca, but establish-

Fig. 5.2. Kudzu (*Pueraria phaseoloides*) at Belem Experiment Station, Brazil. (Courtesy of G. O. Mott, University of Florida, Gainesville, Florida.)

ment could be slow. It responds well to fertilizer, especially on heavy soil. Plants tolerate wet soils.

Establishment: Rather slowly from seed, requiring 6–8 months for a good stand (Rivera-Brenes et al., 1947). Seeded at 3 kg ha^{-1} is sufficient when sown on 1 × 1-m centers. It can also be established vegetatively from crowns and rooted stems, resulting in rapid ground cover in less than 4 months.

Growth habit: Plants have a very aggressive, climbing or twining growth habit and can result in invasion of nearby plants.

Management and utilization: During long periods of drought, plants will retain green leaves. It is used for hay, grazing, or silage, yielding 5–10 Mg ha^{-1}. It should be leniently grazed, allowing a considerable photosynthetic area, and clipped for hay or silage at a high stubble height. Close grazing or clipping eliminates plants. Once cattle are accustomed, palatability is good; however, CP is lower (11–19%) than in many other tropical legumes. The Ca and P content in herbage is generally good, but P in unfertilized forage reportedly is 0.16–0.17%. (Dirven and Ehrencron, 1963). Once established, a grass association is difficult unless grazing management to suppress the legume is utilized. Seed production in subtropical areas is generally poor or nonexistent. Plants produce more seed when grown on support structures. When seed is produced, there is usually a high percentage (80–90%) of hard seed, requiring some type of scarification. Generally, it can be inoculated successfully with the cowpea-type *Rhizobium*.

Advantages:
   a. Fixes atmospheric N, requiring little or no N for companion grass.
   b. Tolerates saturated soils with a high water table.
   c. Tolerates disease and long periods of drought.
   d. Grows well on acid soils (pH 4.0–5.5).
   e. Established from vegetative crowns or rooted stems.

Disadvantages:
   a. Establishes slowly from seed.
   b. Has poor ability to sustain close grazing, has low yields, and will not tolerate trampling.
   c. Seeds 80–90% hard, resulting in low germination.
   d. Forage quality generally lower than that of most tropical legumes.

7. Stylo (*Stylosanthes guyanensis*) (Fig. 5.3)
    Common synonyms: Stylo, Finestem stylo, Meladinho, Alfafa do Nordeste.

Fig. 5.3. Stylo (*Stylosanthes guyanensis*) in Australia. (Courtesy of G. O. Mott, University of Florida, Gainesville, Florida.)

Common cultivars: Schofield (late flowering), Oxley (early flowering), Cook (early flowering), Endeavour (Australia), Deodoro and Deodoro 2 (Brazil), and IRI-1022.

Season of growth: Warm-season perennial that tolerates a wide range of temperatures, including cooler temperatures than most tropically cultivated legumes will tolerate and survives temperatures of 0 to $-5°C$.

Soil condition: Grows on poor, dry soil, withstanding long periods of drought; however, it will not tolerate waterlogged soils.

Establishment: Propagation from seed is relatively easy; however, there is a high percentage of hard seed leading to poor germination unless scarified. Establishment has been successful by seeding directly into closely grazed native pasture at 4.5 kg ha^{-1} followed by disking. Broadleaf weed control using 2,4-D Amine at establishment does not harm seedlings (Bogdan, 1977). The addition of P to the soil improves establishment.

Growth habit: Upright-type plant forming a dense sward. Flowers are yellow to orange–yellow. Legume pod is a single ovoid joint with a minutely inflexed beak.

Management and utilization: Stylo has intermediate to good resistance to close clipping or grazing, and some varieties have high disease resistance. Plants appear to have low tolerance for shade. Produces more growth during drought stress than many other legumes. Generally cowpea *Rhizobium* can be used for cv. Schofield but cv. Oxley requires a selective strain of *Rhizobium*. Most stylo cultivars provide best N fixation at pH 6.0. It grows well in associations with guinea, Rhodes, Pangola, and jaragua grasses. Under good growing conditions, stylo should be ready for grazing at 6 months after seeding. Rotational grazing at about 5-weekly intervals is desirable. It is resistant to most serious plant pests and diseases. It increases grass–legume yields about 30–50% in addition to providing additional CP yield over grass alone. It contains oxalates at about 1.7% of DM (Ndyanabo, 1974). Some palatability problems occur when grazing stylo, indicating ruminants must be accustomed to the plant. Studies show young plants are less palatable than mature plants. Ruminant gains are quite good for stylo mixtures compared with gains on grass alone when fertilized with N.

Advantages:
 a. Fixes atmospheric N.
 b. Grows well on low-fertility soils; however, plants respond to additional P.
 c. Some varieties are resistant to most common diseases and insects.
 d. Produces more growth than most legumes under drought stress.
 e. Tolerates a wide range of temperatures, surviving 0 to $-5°C$.

Disadvantages:
 a. Contains nearly 2% oxalates on a DM basis, which can tie up calcium and other minerals.
 b. Will not tolerate waterlogged soils.
 c. Low shade tolerance.

REFERENCES

Anderson, E. R. (1970). *Proc. Int. Grassl. Congr., 11th*, pp. 591–594.
Anslow, R. C. (1957). *Rev. Agric. Sucr. Ile Maurice.* **36**, 39–49.
Baker, S. J., and Kyneur, G. W. (1962). *Proc. N. Queensl. Agrost. Conf.* **12**(4), 6.
Bogdan, A. V. (1977). "Tropical Pasture and Fodder Plants." Longman, New York.
Boote, K. J. (1976). *Proc.—Soil Crop Sci. Soc. Fla.* **36**, 15–23.
Bowden, B. N. (1963). *Emp. J. Exp. Agric.* **31**(123), 267–273.
Bowden, B. N. (1964). *J. Ecol.* **52**, 255–271.
Butterworth, M. H. (1967). *Nutr. Abstr. Rev.* **37**, 349–368.
Crowder, L. V., Chaverra, H., and Lotero, J. (1970). *Proc. Int. Grassl. Congr., 11th*, pp. 147–149.

Dirven, J. G. P. (1977). *Stikstof (Eng. Ed.)* **20**, 2–15.
Dirven, J. P. G., and Ehrencron, V. K. R. (1963). *Surinaomse Landbouw* **11**, 39–45.
Downes, R. W. (1969). *Planta* **88**, 261–273.
Food and Agriculture Organization of the United Nations (FAO) (1980). "Production Yearbook." FAO, Rome.
Hopkinson, J. M., and Loch, D. S. (1973). *Trop. Grassl.* **7**, 255–268.
Hunkar, A. E. S. (1969). *Grassl. Onderzoek Surin. Landb.* **17**(1), 37–40.
Hutton, E. M. (1970). *Adv. Agron.* **22**, 1–73.
Jones, R. M. (1969). *Trop. Grassl.* **3**, 57–63.
Ladeira, N. P., Sykes, D. J., Daker, A., and Gomide, J. A. (1966). *Rev. Ceres.* **13**(74), 105–116.
Laksesvela, B., and Said, A. N. (1978). *World Rev. Anim. Prod.* **14**, 49–57.
Loomis, R. S., Williams, W. A., and Hall, A. E. (1971). *Annu. Rev. Plant Physiol.* **22**, 412–468.
Ludlow, M. M. (1976). *In* "Water and Plant Life—Problems and Modern Approaches" (O. L. Lange, L. Kappen, and E. D. Schultz, eds.), Ecological Studies, Vol. 19, Springer-Verlag, Berlin and New York.
McDowell, R. E. (1972). "Improvement of Livestock Production in Warm Climates." Freeman, San Francisco, California.
McLeod, M. N., and Minson, D. J. (1974). *J. Agric. Sci.* **82**, 449–454.
Milford, R., and Minson, D. J. (1968). *Aust. J. Exp. Agric. Anim. Husb.* **8**, 413–418.
Minson, D. J. (1967). *Br. J. Nutr.* **21**, 587–597.
Minson, D. J., and Wilson, J. R. (1980). *J. Aust. Inst. Agric. Sci.* **46**, 247–249.
Mislevy, P., Blue, W. G., and Brolmann, J. B. (1981). *J. Environ, Qual.* **10**, 453–456.
Ndyanabo, W. K. (1974). *East Afr. Agric. For. J.* **39**, 210–214.
Parsons, J. J. (1972). *J. Range Manage.* **25**, 12–17.
Rivera-Brenes, L. (1947). *J. Agric. Univ. P. R.* **31**, 180–189.
Riveros, F., and Wilson, G. L. (1970). *Proc. Int. Grassl. Congr., 11th*, pp. 666–668.
Whiteman, P. C. (1968). *Aust. J. Exp. Agric. Anim. Husb.* **8**, 528–532.
Whyte, R. O., Moir, T. R. G., and Cooper, J. P. (1959). "Grasses in Agriculture." Agricultural Studies No. 42. FAO, Rome.
Wilson, J. R. (1981). *Proc. Int. Grassl. Congr., 14th*, pp. 470–472.
Wilson, J. R. (1982). *In* "Limitations to Animal Production from Pastures" (J. B. Hacker, ed.), pp. 111–131. Farmham Royal England: Commonwealth Agricultural Bureaux, England.
Wilson, J. R., and Ford, C. W. (1971). *Aust. J. Agric. Res.* **22**, 563–571.

6

Pasture Management for Optimum Ruminant Production

L. V. CROWDER
*The Rockefeller Foundation
and Department of Agronomy
University of Florida
Gainesville, Florida*

I.	Pasture Maintenance and Renovation.		104
	A.	Fertilization of Improved Pastures.	105
	B.	Factors Influencing Use of Fertilizers.	106
	C.	Recycling of Nutrients.	107
	D.	Nutrients Other Than NPK.	108
	E.	Weed Control.	108
	F.	Renovation Practices.	108
II.	Pasture Type and Grazing Management.		109
	A.	Native and Naturalized Grazing Lands.	109
	B.	Sown and Improved Pastures.	110
	C.	Pasture Supplementation.	111
III.	Grass–Legume Mixtures.		112
	A.	Role of Grasses.	113
	B.	Role of Legumes.	113
	C.	Animal Output from Grass–Legume Pastures.	114
IV.	Grazing Management Systems.		115
	A.	Comparison of Grazing Systems.	115
	B.	Grazing Pressure and Stocking Rate.	117
	C.	Grazing Behavior.	117
	D.	Patterns of Grazing.	118
	E.	Animal Selectivity of Herbage.	118
	F.	Grazing Action: Number and Size of Bites.	118
V.	Herbage Quality.		119
	A.	Factors Affecting Chemical Composition.	119
	B.	Factors Affecting Digestibility.	120
	C.	Forage Intake.	121
	D.	Palatability.	122

VI.	Pasture Use for Animal Production.	122
	A. Cow–Calf Operations.	122
	B. Fattening and Finishing.	123
	C. Milk Production.	124
	D. Calf Rearing and Herd Replacement.	124
	E. Dry Cows and Reserve Stock.	125
	References.	125

Animal productivity (meat, milk, wool, and reproduction) and herbage yields (dry matter and feed units) depend on skillful management or (1) the pasture sward and (2) the grazing animal. Output per animal, regardless of genetic potential, is directly related to the managerial skill of the animal husbandman. The factors under his control include (1) choice of grasses and legumes in pastures, (2) pasture maintenance, (fertilization, weed control, herbage accumulation), (3) type of grazing animal, (4) choice of grazing system, (5) intensity of grazing, (6) control of herbage quality and nutritive value, and (7) use of supplementary feeding. The successful husbandman judiciously manipulates these factors to optimize pasture and animal production (Crowder and Chheda, 1982).

I. PASTURE MAINTENANCE AND RENOVATION

The types of warm-season pastures include native or naturalized and improved (sown). Most native pastures (sometimes called grasslands) have been modified, so that the term *naturalized* is perhaps more appropriate. Native pastures are sometimes referred to as unimproved and comprise the major portion of pastures in the tropics and subtropics. They are usually dominated by grasses that are adapted to low-fertility soils, are susceptible to high grazing pressures, and respond poorly to applied fertilizers. They can be improved by bush and weed control, periodic resting to allow seed formation and regeneration of desirable species, and overseeding with so-called improved pasture species. Improved species are those that display superiority of some traits as compared to the commonly occurring species. Several legumes, particularly *Stylosanthes* spp., have been used for overseeding, but under many conditions, poor establishment occurs (Cook, 1980; Crowder and Chheda, 1982; Gillard, 1982; Whiteman, 1980). Thus, contribution of the improved species may be low (Fig. 6.1). Where moisture is adequate and economic returns favorable, consideration should be given to destroying the old sod and establishing new pastures with sown species.

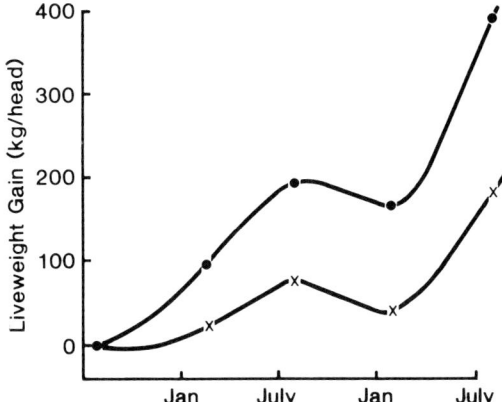

Fig. 6.1. Native grass pastures in Queensland, Australia, oversown with *Stylosanthes hamata* (Verano stylo): (●) adequate stand producing more than 600 kg/ha legume dry matter per year; (X) poor stand producing less than 600 kg/ha legume dry matter per year; this represents accumulative liveweight gains over a 2-year period. (Modified from Gillard et al., 1980.)

A. Fertilization of Improved Pastures

In determining fertilizer maintenance requirements of improved tropical pastures, there is a need for crop logging or monitoring of the soil–plant–animal relationships. The full effects of applied nutrients can only be achieved over time to obtain the full expression of dry matter production, balance of legume to grass, and overall animal production.

The effect of phosphate, whether inherent or applied, declines over time unless supplied periodically. Studies in Australia showed that annual topdressings of 10–25 kg/ha of phosphorus (P) kept the soil nutrient supply at a level for optimum plant growth, and boosted animal production several fold (Andrew and Bruce, 1975). The common forms of phosphate fertilizer are ordinary (single) superphosphate and triple superphosphate. They are particularly effective in soils having low to moderate P-fixation capacity. The single form is preferred in areas where sulfur (S) deficiency is likely to occur. The less soluble forms, such as rock phosphate, are more reactive in acid soils and cost one-third less in terms of elemental P (Hodges *et al.*, 1966).

The influence of potassium (K) on productivity of tropical pastures varies with plant species and critical plant level and ability of the plant to extract K from the soil. Most soils supply sufficient amounts of K during the first 2–3 years after pasture establishment, but after 3–4 years, the soil K begins to decline. Legumes appear to be at a disadvantage for utilization of K when grown in association with grasses. The latter have a greater absorptive capacity for this element. In fact, grasses absorb more K than required for maximum forage production, when the soil supply is plentiful. This luxury consumption is common to other crops but is important in growth of pasture grasses because of constant removal of immature herbage. Since the critical percentage of K in plants needed for continuous growth is higher than for P (0.6–1.2 and 0.16–0.20%, respectively), a greater soil reserve is needed (Andrew and Robins, 1971). Much of the K in forage consumed by grazing cattle is returned to the soil via excreta, thus an annual application of 10–20 kg/ha of K should be sufficient for maintenance of grasses and legumes.

In well-managed grass–legume pastures, application of N is not needed and, in fact, will reduce the legume component. With grass pastures, however, animal productivity is directly related to the amount of applied N.

B. Factors Influencing Use of Fertilizers

In most of the tropics and subtropics, the use of fertilizers on pastures is too small to be recorded on a country basis. By the mid-1960s in Australia, less than 1% of the 105 million hectares considered to be suitable for pasture improvement in the northern regions had been treated with fertilizers (Davies and Eyles, 1965). In Queensland, improved pastures received about 7% of the total superphosphate used in the state and the figure had climbed to 40% by the 1970s after fairly widespread use of pasture legumes (Williams and Andrew, 1970).

Factors to be considered in the use of pasture fertilizers include the following (Crowder and Chheda, 1982):

1. Economic return. With an intensified management system, applied fertilizer will be needed for optimal pasture and animal output. Points to consider include type and quality of cattle, price of animal product, cost of fertilizer, credit and return on investment, level of technology, and managerial skill of the operator.

2. Pasture species and botanical composition. Species differ in their fertilizer requirements, and some are more efficient than others in their extraction and utilization of nutrients. Grasses generally respond to applied N and legumes to P. If a high percentage of legume is to be maintained, phosphate must be applied at regular intervals.

3. Soil nutrients and their availability. The quantity and availability of soil nutrients vary with soil type, moisture, temperature, microbial activity, and previous land use. Some forms of nutrients are readily available, others are released by mineralization, and some are fixed within the soil fraction and may be released over time.

4. Quantity of nutrients removed. The extraction of soil nutrients is related to yield and chemical composition of the herbage. A grass yield of 10 t/ha of dry matter containing 1.5% of N would extract 150 kg of this element plus that found in the roots and stubble. A yield of 10 t/ha of most legumes will extract about 200 kg of N (part of which is fixed by the associated *Rhizobium*), 25 kg of P, and 200 kg of K, plus minor elements. The nutrients contained in the herbage will be lost under cutting conditions, unless animal manure is returned. The amount removed by grazing is relatively small, with about 80% of the nutrients being excreted in the feces and urine.

5. Loss of nutrients from the soil. Rapid mineralization from soil organic matter occurs when rains begin after the dry season, resulting in a flush of available nutrients, especially N. The subsequent heavy rainfall results in leaching of 50–60% of N and 20–40% of K. Part of these nutrients remain in the lower soil profiles and can be recycled by deep-rooted pasture species.

C. Recycling of Nutrients

Plant nutrients pass through a cycle from the soil or air to the plant and are returned to the soil in plant residue or through the grazing animal. In general, 60% or more of grass and legume roots occur in the upper 30 cm of the soil. Roots of some grasses penetrate to 2.0 m, and roots of legumes penetrate even deeper into the soil. The mobilization of plant nutrients from lower soil levels and their accumulation in the topsoil has been observed in several pasture species (Henzell and Ross, 1973; Simpson, 1961; Vallis *et al.*, 1982; Vine, 1968). For example, in Nigeria it was established that a 10 t/ha yield of *Cynodon nlemfuensis* dry matter "pumped-up" 24 kg/ha of N, 5 of P, 35 of K, 7 of calcium (Ca), and 5.5 of magnesium (Mg). This upward translocation counterbalanced losses from the topsoil that had been leached into the subsoil (Mackenzie and Chheda, 1970).

The dung and urine spots in a pasture represent an area of localized nutrient concentration with little lateral diffusion into the soil. Removal of nutrients from the excreta occurs from the time of deposition through volatilization, leaching, surface runoff of water, and absorption by plants and continues for about 100 days (Lotero *et al.*, 1966). Within the limited excreta zone, herbage production will be doubled or tripled, but this leaves much of the pasture (80–85%) unaffected. Persistence of droppings leads to fouling, since animals avoid grazing in the vicinity of the feces. Mechanical scattering distributes the droppings over a wider soil area but must be performed at rather frequent intervals to be effective.

D. Nutrients Other Than NPK

Application of limestone to pastures should be considered a source of nutrient Ca rather than an amendment, since most soils contain ample Ca for tropical and subtropical legumes. If lime is needed, it should be applied in modest amounts, as demonstrated by Spain et al. (1975), who found that 150 kg/ha of lime as $CaCO_3$ was sufficient to satisfy the needs of various legumes in Colombia. Liming also increased the availability of P and molybdenum (Mo). Sufficient Ca as a plant nutrient can also be supplied in the form of superphosphate and basic slag.

On some soil types and with intensive pasture production, other nutrients, such as S, Mg, boron (B), and cobalt (Co), may become deficient. The use of single superphosphate will take care of S needs, since it contains about 10% of this element. Many commercial fertilizers contain minor elements.

E. Weed Control

Invasion of weeds into a pasture indicates poor pasture management and usually results from inadequate fertilization and overgrazing. If unheeded, the undesirable plants frequently predominate, and the herbage available for grazing continues to decline. Many weedy species are highly adaptable to low soil fertility, readily invade pastures, and rapidly flourish under conditions of low fertility. A regular program of fertilization and judicious grazing management are the most effective, most efficient, and cheapest means of keeping weeds out of pastures. These promote growth and development of desirable vegetation in its competition with undesirable types. Mowing removes weedy vegetation, but most perennial plants generate new growth. Application of selective herbicides is usually not economical unless the pasture is intensively utilized. Spot application of irregular, infested areas might be considered, however.

F. Renovation Practices

Deterioration of a pasture occurs by reduction of the legume and desirable grass components, invasion of weedy species, prolonged drought, and soil erosion. A decision must then be made to renovate or destroy the old sod and reseed the entire pasture mixture. In the process of renovation any accumulated herbage may be heavily grazed, burned, or mowed, with plant residue removed. A simple approach would be to apply fertilizer and overseed, not disturbing the existing sod. This process is likely to be disappointing in terms of seedling establishment and improved sward density. A greater degree of success will be obtained by scarifying or lightly tilling the sod with a disk harrow set to cut into the soil surface. With bunch grasses, care must be taken not to uproot the entire crown. With stoloniferous and rhizomatous grasses, the disks must cut away enough of the grass to leave some soil exposed as a seedbed. Fertilizer may be applied before or after sod scarification. Seeds may be sown by broadcasting, using a grain drill with a grass seeding attachment or a regular grass seeder. The use of a sod or chisel seeder has an advantage of least sod disturbance, but the excess pasture residue must be removed or suppressed with a contact herbicide (Cook, 1980; Northwood and McCartney, 1969).

After reseeding, the old grasses regenerate growth and develop more rapidly than the sown grasses and legumes do. Light grazing will remove herbage of the old grasses and reduce competition of the newly sown pasture species.

The entire renovation procedure must be viewed in terms of the risks of seedling establishment, sward improvement, and economic returns as compared to complete destruction of the sod, land fitting, fertilizing, and reseeding. If mixed agriculture is practiced, the most desirable approach will probably be to grow a cash or food crop for several years and then establish a new pasture.

II. PASTURE TYPE AND GRAZING MANAGEMENT

A. Native and Naturalized Grazing Lands

A large majority of ruminants in the tropics and subtropics graze grass-based native and naturalized pastures, and relatively few have access to improved pastures. Under many conditions, an excess of herbage prevails during the rainy season, but an extreme shortage occurs during the dry season. Grazing animals find ample fair- to good-quality herbage in the early wet season and gain weight. When plants approach maturity, however, quality declines and animals lose weight (Blair Rains, 1982; Gillard, 1982; O'Donovan *et al.*, 1983; Norman and Stewart, 1964). This results in a "grow–stop" cycle of growth. Dirven (1970) estimated annual gains of about 20 and 90 kg/ha from cattle grazing in monsoonal and humid tropical climates, respectively. It is not uncommon to find losses of 15% of body weight during the dry season, as noted by Leeuw and Brinckman (1974) in Nigeria.

Carrying capacity of native and naturalized pastures is often based on the flush growth of the wet season, resulting in overstocking and overgrazing in the dry season. Generally, continuous grazing is practiced on such pastures, but other systems, such as deferred grazing, a type of rotational grazing, may be needed to balance the productive capacity of the land. Attention must be given to the control of animal numbers and to their distribution over the grazing land to assure conservation and proper utilization of herbage, or supplemental feed must be provided to reduce excessive loss of body weight in the dry season.

B. Sown and Improved Pastures

Improved grasses and legumes in warm climates have the potential to increase herbage yields and quality severalfold over that of native and naturalized pastures. Improved pastures can be intensely utilized under continuous or rotational grazing, but considerable managerial skill is needed to assure adequate supply of high-quality herbage and its utilization for optimum animal production.

Grasses that respond to applied N have the highest pasture potential for livestock production in warm regions with 1200–1500 mm annual rainfall, or where irrigation is available. A comparative estimate of beef production under different climatic conditions and management systems is given in Table 6.1. The data assume a six-month dry season for the monsoonal climate and sufficient rainfall for year-round plant growth in the humid tropics. Pasture production and animal output vary widely from region to

TABLE 6.1

Estimates of Liveweight Gains from Beef Cattle in Monsoon and Humid Tropical Climates[a,b]

Type of pasture and treatment	Climate	
	Monsoon (kg/ha)	Humid tropics (kg/ha)
Natural grazinglands		
Improved grazing management	20	90
Legumes oversown, fertilized	150	400
Cultivated pastures		
Grass–legume mixtures, fertilized	250	600
Nitrogen-fertilized grass	550	1650

[a] Modified from Dirven (1970).
[b] Calculations based on data from 21 grazing experiments on various continents.

region because of rainfall amount and distribution, botanical composition of the pasture, soil fertility and applied fertilizers, and type of cattle and stocking rate.

C. Pasture Supplementation

Herbage produced during the wet season often exceeds animal requirements and utilization. This forage could be used for supplementary dry-season feeding but loses most of its nutritive value with maturity. For instance, the crude protein content of mature, dried grass often drops below 1.5% (Fig. 6.2). Excess herbage can be (1) allowed to accumulate as standing hay, (2) cut, dried, and stored as hay, or (3) cut and ensiled.

In northern Australia, Townsville stylo (*Stylosanthes humilis*) and Verano stylo (*S. hamata*) yield up to 5.0 t/ha of dry material, have 12% crude protein, and lead to a five- to tenfold increase in carrying capacity compared to native pastures (Davies and Hutton, 1970; Gillard *et al.*, 1980). Herbage of grasses allowed to accumulate declines rapidly in quality and seldom provides maintenance requirements without supplementation.

Haymaking under tropical conditions is difficult and highly unsuccessful because of unreliable weather and poor herbage quality. Thus, hay is not a common feed source.

Silage from warm-season species does not compare favorably with that from temperate species because of coarse stems, high crude fiber, and low soluble carbohydrates. Fairly good silage can be made, however, using appropriate species (maize, sorghum, Rhodesgrass, elephantgrass), care in harvest (wilting of succulent grasses), and proper ensiling (well-con-

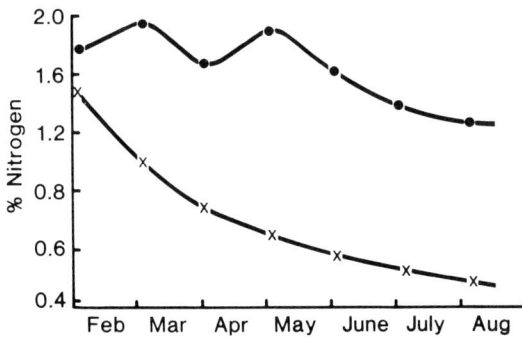

Fig. 6.2. Percentage of nitrogen of Townsville stylo (*Stylosanthes humilis*) (●) and spear grass (*Heteropogon contortus*) (X) during the cool dry season in Queensland, Australia. (Modified from Playne and Haydock, 1972.)

structed silo, fine chopping of forage, solid compaction, and exclusion of air) (Catchpoole and Henzell, 1971; Wilkinson, 1983, 1984).

The utilization of reserve, supplementary pastures usually augments animal performance. Separate areas can be sown with improved legumes (various *Stylosanthes* and *Desmodium* species and *Leucaena leucocephala*) or grasses (*Pennisetum americanum*) and grazed rotationally with regular pastures or utilized during distinct and strategic periods of the year.

Crop residues left after harvest, volunteer weeds and grasses, and other forage materials growing in fields and along boundaries, irrigation channels, and roadsides provide significant components of supplementary feed in warm climates. In addition, various industrial by-products, such as cereal brans, cottonseed, oil cakes, molasses, and citrus pulp, offer other sources of supplementary feed.

III. GRASS–LEGUME MIXTURES

The number of species comprising a pasture mixture varies from a single grass and legume up to five or more of each. Under most conditions, however, the more complex mixtures revert to simpler combinations. The percentage of legume considered to comprise an adequate proportion of the sward has not been established in the tropics and subtropics. In general, animal performance increases with the legume fraction up to 40% (Fig. 6.3), which is about the same as for the temperate zone.

Maintenance of legumes is extremely difficult and requires a regular fertilizer program, as well as judicious grazing management. A pronounced decline in most legumes occurs with increased grazing pressure. This affects the overall efficiency of herbage utilization. For example, in Australia, the digestible dry matter intake of Pangola grass–legume mixtures increased linearly with the legume fraction about 10% (Minson and Milford, 1966).

The benefits of a grass–legume combination in the pasture can be expressed as follows (Crowder and Chheda, 1982):

1. Forage production varies in stability, however, neither yields as much in mixture as when grown alone, assuming that the grass when grown alone is fertilized with N. The combined yields, however, generally exceed that of the legume grown alone.
2. Grasses utilize the nitrogenous compounds fixed by legumes.

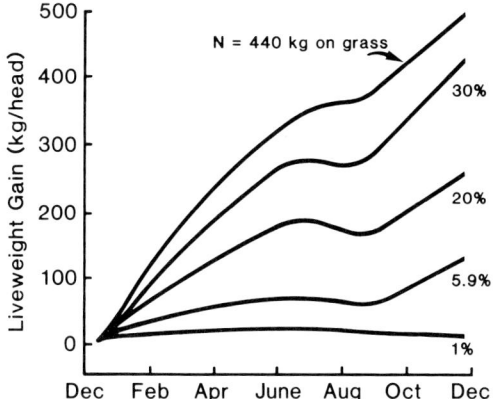

Fig. 6.3. Animal performance as influenced by percentage legume in grass–legume pastures and N-fertilized grass. (Modified from Evans, 1968.)

3. Crude protein of the grass herbage is higher than in unfertilized pure stands but that of the legume is likely to be lower than in pure stands.
4. Well-nodulated legumes make little or no demand on soil nitrogen.
5. Animal performance generally exceeds that on unfertilized grass.

A. Role of Grasses

Grasses generally occur as the predominate pasture species because of their greater aggression, competition for nutrients, resistance to trampling and grazing intensity, and tolerance of burning as compared to legumes. Perennial grasses in the pasture sward are desirable (Crowder and Chheda, 1982), because they (1) increase total herbage yield, (2) ensure stability of production, (3) increase the value of the pasture, and (4) suppress invasion of weeds.

Grasses regenerate growth more quickly with rains which follow the dry season and thus provide earlier grazing than legumes. Their other growth patterns are also different, resulting in more difficult pasture management of a grass–legume association than grass alone.

B. Role of Legumes

Legumes are desirable components of pastures for the following reasons (Crowder and Chheda, 1982):

1. They increase the crude protein content of herbage available for grazing. This is particularly important when grasses mature and become dry and lignified.

2. They extend the grazing period into the dry season. Many legumes remain green throughout much of the dry season, and their nutritive value is maintained when plants become brown or even die (Fig. 6.2).

3. They provide nitrogen for the companion grasses. Forage legumes have a symbiotic relationship with bacteria that transform atmospheric nitrogen into a form utilized directly by the legume and made available to associated plants when the nodules begin to deteriorate. Most legumes require a specific *Rhizobium* for effective nodulation. The bacteria are tolerant of acid soils and possess a high efficiency for extraction of calcium from soils low in this element. Application of lime is seldom needed, especially with superphosphate dressings, except on very infertile and acidic soils.

4. They increase voluntary intake of grass–legume mixtures as compared to pure grass herbage. This is due to reduced rumen retention.

5. They provide stability of the pasture system. Most legumes have a tap root that penetrates into the soil more deeply than roots of grasses do. In addition, an enlarged root section just below the crown of many legumes provides a nutrient and water reserve during stress periods.

C. Animal Output from Grass–Legume Pastures

Well-managed warm-season grass–legume pastures can provide the total digestible nutrients (TDN) and digestible protein required for maintenance and liveweight gain of grazing cattle. A 400-kg animal requires about 5.3 kg of TDN and 0.4 kg of total protein daily for maintenance, assuming 60% digestibility. In addition, about 3.5 kg of TDN and 0.5 kg of protein are needed per kilogram of liveweight gain. Such cattle require about 2.5 kg of dry matter daily (approximately 13.5 kg fresh weight) per 100 kg of liveweight gain. The 10% protein content in feed requirement of most cattle can be provided by herbage of grass–legume pastures during favorable growing seasons.

Legumes cannot provide the amount of nitrogen needed for maximum growth of the associated grass so that dry matter yields are lower than nitrogen-fertilized, high-yielding grasses. Nonetheless, the data in Table 6.1 show that addition of legumes to grasses notably increases liveweight gain as compared to grass alone. Still, the total production expected from grass–legume pastures does not approach the potential of nitrogen-fertilized grass. A resume of experiments carried out in various countries revealed that liveweight gain varied from 250 to 550 kg/ha (with an average of 410 kg/ha) with various grass-legume mixtures and from 260 to 1110 kg/ha (with an average of 710 kg/ha) with fertilization of different grass species (Aronovich *et al.*, 1970; Bryan and Evans, 1971; Grof and Harding, 1970; Jones, 1974; Magadan *et al.*, 1974; Mellor *et al.*, 1973; Stobbs,

1969; Vicente-C. *et al.*, 1964). On a ranch basis in Colombia and under more practical conditions, a comparative study of *Andropogon gayanus* alone and in mixture with *Centrosema pubescens* (both fertilized with phosphate at the initiation of the trial) showed that grass alone produced 435 kg and grass–legume mixture produced 530 kg/ha liveweight gain over a 3-year period (Ramirez, 1983). Under continuous grazing, the Zebu cattle accumulated 550 and 595 g/day/animal on the two types of pasture, respectively. A significant feature of the legume component was higher sustained animal weight during the 3.5- to 4.0-month dry season. Animals on grass alone lost an average of 150 g/day, while those on the grass–legume mixture lost 95 g/day.

An analysis of beef animal experiments in the tropics and subtropics revealed that (1) relatively few direct comparisons of grass and grass–legume pastures have been made in the same experiment, (2) a wide range in liveweight gains occurred, reflecting divergent environmental effects, (3) liveweight gains from grass–legume pastures exceeded grass alone at low levels of applied N, (4) the legume fraction in some instances was equivalent to about 100 kg/ha of nitrogen fertilizer, (5) liveweight gains increased proportionately with the percentage of legume in the herbage available for grazing, and (6) conception rate of cows increased and calf mortality decreased on grass–legume pastures as compared to grass alone (Koger *et al.*, 1970, in Florida; Edye *et al.*, 1972, and Guyton *et al.*, 1980, in Australia).

There are fewer data to substantiate the beneficial effects of legumes in mixtures or alone on milk production as compared to grass alone. The amount of feed to produce 1 kg of liveweight gain per day is equal to that needed to produce 8–9 kg of milk per day. In addition to maintenance requirements, lactating cows require 0.3 kg of TDN and 0.06 kg of protein per kilogram of 4.0% milk. A 550-kg cow producing 10 liters of milk daily needs about 14 kg of dry matter containing 7 kg of TDN and 1.1 kg of protein. A compilation of milk production records from various experiments in the tropics and subtropics showed daily milk yields of 6.5–15.0 kg/cow (Stobbs, 1971, 1975). The main factor limiting milk production on well-fertilized tropical pastures appears to be low intake of digestible energy (Table 6.2).

IV. GRAZING MANAGEMENT SYSTEMS

A. Comparison of Grazing Systems

Animal production per land area depends on two independent factors, namely animal performance and pasture carrying capacity. The manage-

TABLE 6.2

A Summary of Research Findings of Milk Production per Lactation from Tropical and Subtropical Australia and from Temperate Zone Pastures[a,b]

Diet	Dry matter digestibility (%)	Milk production (kg/head)
Tropical and subtropical pastures[c]		
Immature herbage	60–65	1800–2000
Seminative herbage	50–55	1000–1400
Temperate zone pastures	70–80	3300–3800
Concentrate rations[c]	80–85	4400–4900

[a] Modified from Stobbs and Thompson (1975).
[b] Data from Jersey cows only.
[c] Experiments in Australia.

ment systems commonly encountered are of three types: (1) continuous grazing, (2) rotational grazing, and (3) deferred grazing. Controversy exists as to whether continuous or rotational grazing is superior in terms of pasture production and animal performance despite many studies comparing the two systems. Generally, continuous grazing is advisable with native, naturalized, and sown pastures of low productivity. Under this system, the number of animals usually remains constant, resulting in under- or overstocking since plant growth and seasonal conditions are not predictable. It commonly favors production per head during the season of plant growth because of selective grazing but may result in spot grazing.

Rotational grazing is suited to intensive utilization of improved pastures and is designed to obtain more uniform grazing than the continuous system does. Forage yields are potentially higher, and this system eliminates, to a great extent, spot grazing. On highly productive pastures, e.g., where beef or milk cattle are grown on N-fertilized grass, on well-managed grass–legume pastures, and where irrigation is accessible, the rotational system might be more appropriate. It will, however, require closer supervision and greater managerial skill than continuous grazing (Blaser *et al.*, 1969; Grof and Harding, 1970; Wheeler, 1962). A modification of rotational grazing is the leader–follower system of dividing animals into two groups, i.e., high producers (milking cows and fattening cattle) and low producers (dry cows and reserves). The higher producers are allowed into the paddock for 3–7 days and are followed by the low producers, to consume the residue. An extreme form of the rotational system is strip or ration grazing, whereby a small section is separated, sometimes by an electric fence, and animals are permitted to intensely graze before moving into another area. Deferred grazing, i.e., setting aside a certain pasture area or paddock for accumulation of hay *in situ* and later utilization, is another variation of the rotational system.

B. Grazing Pressure and Stocking Rate

In an attempt to optimize pasture productivity, the animal husbandman selects a stocking rate (in terms of animals per hectare) based on knowledge or predictable herbage so as to effectively utilize a major portion of forage available for grazing. Several models have been devised to illustrate the relationship between stocking rates and production per animal as well as per unit area of land (Hart, 1972; Heath *et al.*, 1973; Jones and Sanderland, 1974). In general, increasing stocking rate beyond the optimum grazing pressure (or grazing intensity, i.e., animals per unit of available herbage) results in a decrease in output per animal but an increase in production per hectare, up to a critical point.

The stocking rate can be fixed or variable. Under a fixed system, the number of animals per hectare is chosen to bracket the optimum grazing pressure. The operator must then recognize that the quantity of forage will vary, selective grazing will result in localized differences of grazing pressures, and daily gains will fluctuate with the quantity and quality of herbage. Under a variable system, sometimes called "put-and-take" (Matches and Mott, 1975), the stocking rate is intentionally varied according to available herbage so as to achieve a constant stocking pressure. This requires extra animals during the flush season and a ready market, or reserve pasture, during the dry season.

The optimum carrying capacity (which is the stocking rate to achieve optimal daily gain per animal and liveweight per hectare) varies, so that the operator must give attention to the system of pasture management and the most appropriate stocking rate for the local situation (Humphreys, 1972). Factors to consider in making these decisions include (1) rate of herbage growth, (2) seasonal variation in total feed supply, (3) fertilization scheme, (4) nutritive value of herbage, (5) botanical composition and ground cover of the pasture, (6) accessibility of forage, (7) class of animals, and (8) nature of animal product.

C. Grazing Behavior

Tropical and subtropical pasture swards differ in dry matter yield, vertical development and structure, density or compactness of herbage, leaf–stem ratio, arrangement and accessibility of leaves, and ease of removal of herbage from the plant. Growth habits of grasses and legumes range from tall, upright and stemmy types to more prostrate open canopies or dense, small-leaved types. Fertilization increases plant growth rates and alters intermode length and leafiness. Grazing intensity modifies the amount of herbage on offer and canopy structure; e.g., a higher proportion of stems will be present under light grazing. Factors such as these affect grazing behavior (Crowder and Chheda, 1982).

D. Patterns of Grazing

Cattle on warm-season pastures spend from 7 to 12 hr per day grazing, 5–8.5 ruminating, and the remainder idling (lying, standing, drinking, or walking). The shorter times correspond to periods of abundant herbage and the longer ones to the dry season, when there is less herbage on offer or mature and stemmy forage prevails. A marked reduction in grazing occurs during midday, when temperatures are higher (Ebershon et al., 1983; Harker et al., 1954; Payne et al., 1951; Shultz et al., 1977; Stobbs and Cowper, 1972; Walker et al., 1981). Under environmental stress, the reaction of *Bos taurus* (e.g., Shorthorn, Hereford, Friesian) is more pronounced than that of *Bos indicus* (Zebu). The former suffer to a greater extent from hyperthermal stress, which reduces grazing activity under high ambient temperatures.

Cattle walk 2–3 km/day when grazing pastures having ample quantity and quality of forage. During periods when both decline, cattle spend more time grazing and walking in search of feed. Walking time is also influenced by pasture size, availability of forage or other feed, and density of the sward (Larkin, 1954; Lampkin and Quarterman, 1962).

E. Animal Selectivity of Herbage

In the process of grazing, cattle first eat from the sward in a horizontal plane and then in a verticle plane. They are selective in both directions but first consume stem tips and immature leaves, which are the most nutritious. Afterward, they search for younger leaves, followed by more mature ones (Gardener, 1980). In this way, the highest-quality diet is chosen. In fact, esophageal fistula studies show that ingested material contains more crude protein and less crude fiber than indicated by the chemical analyses of the cut samples (Brendon et al., 1967).

F. Grazing Action: Number and Size of Bites

Cattle draw herbage into the mouth by a swipe of the tongue and movements of the lower jaw and head. The number of bites varies from 40 to 65 bites per minute and from 30,000 to 60,000 bites per day but may be more variable on poor-quality pastures. The size of bite ranges from 0.05 to 0.80 g, depending on herbage availability and accessibility (Stobbs, 1974). An intake of 0.3 g per bite of organic matter is needed for a 400-kg animal to obtain adequate dry matter when grazing at 36,000 bites/day (Stobbs, 1973). Under conditions of low herbage quality and sparse pasture, animals graze for a longer period in an attempt to compensate for small bite size.

The number and size of bites are influenced by the following factors:

1. Plant species. These determine sward structure.
2. Leaf size. Large leaves interfere with biting behavior.
3. Leaf yield. A high leaf–stem ratio increases both biting rate and bite size.
4. Accessibility of leaves and ease of removal from plants. Leaves arranged on or near the outer canopy, and not mixed within, are more easily grasped and drawn into the mouth. Older leaves are more difficult to remove because of higher lignin content and increased cell wall structure. Fibrosity of legume petioles also affects ease of harvest (Stobbs and Hutton, 1974).
5. Sward and leaf bulk density. The density of the leaf–stem mixture and the leaf fraction vary widely with species and pasture management. For example, in Australia, leaf bulk density of *Desmodium intortum* averaged 32 kg/ha/cm compared to white clover, which exceeded 47 kg/ha/cm (Stobbs and Imrie, 1976). Less herbage, therefore, would be ingested per bite of *Desmodium*.

V. HERBAGE QUALITY

Nutritive value generally refers to chemical composition, digestibility, and the nature of digested products. The amount of forage consumed affects total nutrient intake and, therefore, animal response. Forage availability and acceptability, presence of undesirable substances, and rate of passage through the digestive tract influence intake by the animal. Herbage quality, therefore, includes nutritive value, level of animal consumption, and digestibility of available herbage (Crowder and Chheda, 1982).

A. Factors Affecting Chemical Composition

1. Soil and climatic conditions. The physical, chemical and biological properties of soil, rates at which nutrients are supplied and renewed, and

applied fertilizers affect plant chemical composition. Any soil nutrient below the critical level needed by pasture plants limits growth and chemical composition and must be corrected before production can be increased. Wide seasonal fluctuations in chemical composition commonly occur within the herbage of warm-season pastures. Absorption of nutrients is restricted by high moisture stresses and high temperatures, which induce physiological maturity and a decline in certain nutrients. The percentages of crude protein and various other plant constituents are directly related to the amount and distribution of rainfall.

2. Stage of plant growth. Chemical composition varies vertically within the plant and the pasture sward due to differences in maturity and cell wall structures along this axis. The top of plants and sward remain young and contain low amounts of cell wall components. Percentages of dry matter, lignin, and fiber increase downward along the axis, while cell contents and crude protein decrease. Leaves and stems are equal in nutritional value in the juvenile stage. Their dietary components, however, decline at different rates with age. The cell wall contents increase more rapidly in stems than in leaves.

3. Genotype. Genetic diversity of pasture plants causes considerable differences in the percentage of crude protein and various chemical components, crude fiber, and digestibility (Crowder and Chheda, 1977; Dougall and Bogdan, 1958). In addition, maturity date, leaf–stem ratio, growth habit, stress tolerance, responsiveness to fertilizer and management practices, and forage acceptability are some of the important genotypic factors that indirectly influence the chemical composition of herbage.

4. Grazing management. Grazing intensity determines in part the regeneration of juvenile herbage on offer. It might thus appear that frequent and close grazing would provide animals with higher levels of digestible nutrients. This practice, however, results in a cumulative reduction of herbage yields because of rapid exhaustion of carbohydrate reserves in the roots and stubble. In addition to grazing pressure, the animal husbandman must consider the differences in chemical composition of different pasture species in determining stocking rate so as to derive optimal benefit of nutritive value without adversely affecting herbage productivity.

B. Factors Affecting Digestibility

1. Climate. Climatic differences, particularly temperature changes, cause differences in digestibility of forage. Reduced digestibility of warm-season forages can be attributed to high temperatures that stimulate rapid plant growth and accelerate maturity. The average decline in digestibility

of many forages ranges from 0.80 to over 1.0% with each 1% increase in temperature (Denium and Dirven, 1975; Minson and McLeod, 1970).

2. Stage of plant growth and genotype. Digestibility of young and immature forages ranges from 75 to 85% but rapidly declines with advancing maturity. The rate of decline with age varies considerably between genera, species, and varieties as well as among genotypes within varieties (Reid et al., 1973). Grasses and legumes have comparable digestibilities at the immature stage of growth, but the rate of decline is lower in most legumes, perhaps due to the relatively lower contents of acid detergent fiber and hemicelluloses in legumes (Garcia and Ferrer, 1974).

3. Plant fraction. Leaves, leaf sheaths, and stems are equally and highly digestible during the early stage of plant growth. Stems mature more rapidly, however, while the cell content of leaves remains at a higher level and lignification proceeds at a slower rate (Raymond, 1969). Early flowering can also accelerate the decline in digestibility.

4. Protein and mineral content. If the ingested herbage contains less than about 7% crude protein, microbial activity in the rumen is depressed by lack of nitrogen. This causes an incomplete utilization of carbohydrates and a slow rate of passage of the digesta (Milford and Minson, 1966). A mineral content of herbage below that required by the animal adversely affects both intake and digestibility. For example, Thornton and Minson (1973) in Australia noted that dry matter digestibility of Pangola grass–legume pastures increased from 41.6 to 44.9% when the P concentration increased from 0.11 to 0.15%.

C. Forage Intake

Voluntary intake of herbage by the animal, or the total quantity of digestible nutrients consumed, is another measure of forage quality. Intake of warm-season forages, particularly grasses, is generally lower than temperate species and is a major factor contributing to lower animal productivity under tropical environments. In ruminants, herbage intake depends on the capacity of the digestive tract, particularly the rumen. The animal eats until a certain degree of gastro–intestinal "fill" is reached and then rests and/or ruminates until the "fill" is reduced as a result of digestion and movement of the ingesta through the digestive tract.

The most important factors influencing voluntary intake appear to be stage of plant growth, digestibility, and genotype (Jones et al., 1979). Under grazing conditions, availability of forage, low protein content, and mineral deficiencies, as well as environmental conditions, significantly alter intake. Animals differ in their selection of a species in the diet, and bite size can place limitations on intake. Supplements also alter voluntary intake of herbage.

D. Palatability

This is a rather confusing and ambiguous term used in reference to the appeal, attractiveness, or acceptability of forage. In effect, it alludes to a quality or state that is agreeable to the palate and connotes a gustatory or savory aspect. In a pasture mixture, the term also indicates animal preference among species or genotypes. In classifying range conditions, a palatability rating has been devised as follows: (1) Excellent-to-good conditions indicate a large proportion of highly preferred plants, a relatively dense plant cover, a thin mulch on the soil surface, and no active soil erosion. (2) Ranges in fair-to-poor condition are dominated by less palatable species and a number of undesirable plants. (3) Those in poor condition consist of sparse plant cover, poor plant growth, many weedy species, and heavy soil erosion (Stoddart *et al.*, 1975).

VI. PASTURE USE FOR ANIMAL PRODUCTION

A majority of pastures in the tropics and subtropics are utilized under an extensive beef cattle management system in which the entire herd, often including bulls, is kept together on a continuous basis. Under such a system, cattle may not be marketed until they are 6–8 years old. In recent years, however, attention has been given to more intensive pasture and herd management systems to fatten stock for market at an earlier age. In addition, a milking operation has developed around larger cities, prompting an effort to use improved pastures as a feed source, but often they deteriorate into an exercise lot with cows being fed concentrates and feed supplements. Under such a range of conditions, pastures for cattle production can be broadly categorized as (1) cow–calf, (2) fattening, (3) milking, (4) stock replacement, and (5) dry cow and reserve stock operations.

A. Cow–Calf Operations

A common practice under an extensive pasture management system is lack of, or little, attention to time of conception and calving. Furthermore, adult and young animals experience a "grow–stop" liveweight cycle, i.e., losing weight in the dry season followed by compensatory growth and weight accumulation in the rainy season. To improve this system, and particularly to offset the dramatic body weight change, the cow–calf operator should consider one or more of the following procedures:

1. Adopting a seasonal breeding program designed to correlate with the period of ample feed supply. This requires that calves be dropped just before or after the wet season, when the forage supply and quality are sufficient to meet production needs of the dam.
2. Reducing the number of grazing animals before onset of the dry season, thus maintaining a larger feed supply.
3. Improving the pasture to provide a more nutritious supply of herbage into the dry season, i.e., including a legume in the pasture sward, or allowing cows access to a legume reserve (protein bank).
4. Providing a feed supplement during the dry season.

A number of studies have shown that such practices increase conception and calving rates, increase milk production and prolong lactation, decrease calf mortality, and increase weaning weights (Grant, 1980; Holroyd *et al.*, 1979; Tierney and Taylor, 1983). In Australia, Evans and Biggs (1979) compared the breeding performance of Hereford cows grazing fertilized Pangola grass (250 kg of N per hectare per year) and Pangola–legume pastures over a 3-year period. Both pastures received 250 kg/ha of 9.6% superphosphate and 63 kg/ha of KCl. No significant differences were noted between pastures with 94 and 95% calving, 89 and 91% weaning at 200 days, and 0.85 and 0.79 kg/head mean daily gains from birth to weaning, respectively, for the two pastures.

B. Fattening and Finishing

Under an intensive management system, the forage production and nutritive value of herbage from improved pastures are satisfactory for growing young stock from about 200 kg liveweight to a slaughter weight of 400–450 kg/head (Burton and Bryan, 1983; Evans, 1970; Evans and Bryan, 1973; Harding and Grof, 1978; Hodges *et al.*, 1979; Mott *et al.*, 1970; Winks *et al.*, 1970). With proper pasture and herd management, liveweight gains of 500 g/head and more can be maintained. Pasture-fed cattle, however, are not as well finished as when provided a supplement. Winks *et al.* (1983), in Australia, demonstrated the value of molasses and mono-ammonium phosphate on the performance of steers grazing a grass pasture (*Panicum maximum* var. *trichoglume*) and grass-legume (same grass mixed with *Neonotonia wightii* cv. Tinaro). Over a 3-year period, the supplement increased liveweight gain from 595 kg/ha to 840 kg/ha, average of unsupplemented and supplemented on both pastures. Furthermore, the supplement increased dressing percentage, carcass fat in general, and depth of fat over the eye muscle, making animals suitable for the local market.

In Florida, Hargrove *et al.* (1982) showed that steers slaughtered from pastures had a lower dressing percentage, less fat, and less marbling as compared to animals provided 60 days of additional drylot feeding of a high energy ration. Intensive feeding offers a flexible means of profitably fattening by increasing edible meat up to 50% and improved finishing of range-reared animals. This can be accomplished by using by-products, such as bran, brewery residue, molasses, oil cake, cannery wastes, as demonstrated in East Africa by drylot feeding for about 80 days (Oram, 1975).

The economics of short-term intensive feeding depend on the type of cattle, level of management, expertise in pasture production, availability and cost of supplementary feed, price of beef, and local or export demand for quality meat.

C. Milk Production

Much of the milk produced in the tropics comes from crossbred and unimproved cows on native and naturalized pastures often producing less than 1 liter/day for no more than 100 days. Yet, intensively managed pastures offer a potential feed resource for dairy production. Grasses alone, even though succulent and containing relatively high percentages of crude protein, are insufficient to allow full expression of the milk potential of high-producing cows (Dale and Holden, 1968; Cowan *et al.*, 1975; Flores *et al.*, 1979; Stobbs and Thompson, 1975). Supplements should be based on animal performance, with a ratio of 1.0 kg of concentrate (which is about 0.45 kg of protein) to 1.2 or 3 kg of milk, to significantly increase yields. Concentrate feeding, however, may not be economical except at moderate levels (Antonio-P. *et al.*, 1983; Caro-C. and Vicente-C., 1980; Hamilton *et al.*, 1970; McDowell *et al.*, 1977; Yazman *et al.*, 1979).

D. Calf Rearing and Herd Replacement

In general, herd replacement of beef in the tropics is a consequence of the operator keeping young stock which have "eye appeal." With more intensive herd management, more attention is given to breeding and selection. Relatively few studies exist for beef calf rearing, but a number have been aimed at replacements for the dairy herd (Carlo and Velez, 1978; Moss *et al.*, 1983; Randel and Mendoza, 1983; Yazman *et al.*, 1983). In the tropics temperatures are sufficient for year-round plant growth, but outdoor rearing of calves is not highly feasible because (1) grass-dominant pastures mature rapidly and become undesirable for animal growth and development, (2) internal parasites accentuate the difficulty of maintaining calves under grazing conditions, and (3) calves are exposed to considerable climatic stress at a stage of growth when they are least resistant. In most of the tropics, heifers grow slowly, but sexual and body-conformation maturity occur at approximately the same physiological age. Furthermore, they are subjected to the same "grow and stop" pattern of development as their dams. It has been demonstrated in the Philippines, however, that calves can be reared by alternating limited grazing and confinement on a slatted floor or in a graveled pen (Payne et al., 1967). Also, in Puerto Rico, they are reared under confinement to about 150–200 kg of body weight and then fed almost entirely on pasture. Under this system, it is possible to obtain Holstein heifers weighing 480 kg or more at 38–40 months of age (Carlo and Velez, 1978).

E. Dry Cows and Reserve Stock

Pastures for dry cows and reserve stock need not be the same quality as for highly productive and growing types. Certainly, they are not intensively managed but should be well maintained to keep out weeds. Herd management is usually such that animal weight is sustained or loss is kept to a minimum.

References

Andrew, C. B., and Bruce, R. C. (1975). *Trop. Grassl.* **9**, 133–139.
Andrew, C. B., and Robins, M. F. (1971). *Aust. J. Agric. Res.* **22**, 693–706.
Antonio-P., M., Fernandez-Van Cleve, J., Arroyo-A., J. A., and Quinones-T., R. (1983). *J. Agric. Univ. P. R.* **67**, 476–485.
Aronovich, S., Serpa, A., and Riberio, H. (1970). *Proc. Int. Grassl. Congr., 11th*, pp. 789–793.
Blair Rains, A. (1982). *Outlook Agric.* **11**, 96–103.
Blaser, R. E., Bryant, H. T., Hammes, Jr., R. C., and Boman, R. L. (1969). *Va. Polytech. Inst. Res. Div. Bull.* **45**.
Brendon, R. M., Torell, D. T., and Marshall, B. (1967). *J. Range Manage.* **20**, 317–320.
Bryan, W. W., and Evans, T. R. (1971). *Trop. Grassl.* **5**, 89–98.
Burton, R. O. J., and Bryan, W. B. (1983). *Proc. Int. Grassl. Congr., 14th*, pp. 741–783.
Carlo, D., and Velez, J. (1978). *J. Agric. Univ. P. R.* **62**, 311–320.
Caro-C., R., and Vicente-C., J. (1980). *J. Agric. Univ. P. R.* **64**, 47–53.
Catchpoole, V. R., and Henzell, E. F. (1971). *Herb. Abstr.* **41**, 213–221.
Cook, S. J. (1980). *Trop. Grassl.* **14**, 181–187.
Cowan, R. T., Byford, I. F. R., and Stobbs, T. H. (1975). *Aust. J. Exp. Agric. Anim. Husb.* **15**, 740–746.
Crowder, L. V., and Chheda, H. R. (1977). *In* "Food Crops of the Lowland Tropics" (C. L. A. Leakey and J. B. Wills, eds.), pp. 127–129. Oxford Univ. Press, London and New York.

Crowder, L. V., and Chheda, H. R. (1982). "Tropical Grassland Husbandry." Longman, London.
Dale, A. B., and Holden, J. M. (1968). *Proc. Aust. Soc. Anim. Prod.* **7,** 86–91.
Davies, J. G., and Eyles, A. G. (1965). *J. Aust. Inst. Agric. Sci.* **31,** 77–93.
Davies, J. G., and Hutton, E. M. (1970). *In* "Australian Grasslands" (R. E. Moore, ed.), pp. 273–302. Australian National Univ. Press, Canberra.
Denium, B., and Dirven, J. P. G. (1975). *Neth. J. Agric. Sci.* **23,** 26–82.
Dirven, J. P. G. (1970). *Proc. Int. Potash Inst. Congr., Antebes, 9th,* pp. 403–409.
Dougall, H. W., and Bogdan, A. V. (1958). *East Afr. Agric. J.* **24,** 17–23.
Ebershon, J. P., Evans, J., and Limpus, J. F. (1983). *Trop. Grassl.* **17,** 76–81.
Edye, L. A., Ritson, J. B., and Haydock, K. P. (1972). *Aust. J. Exp. Agric. Anim. Husb.* **12,** 7–11.
Evans, T. R. (1968). *Trop. Grassl.* **2,** 192–195.
Evans, T. R. (1970). *Proc. Int. Grassl. Congr., 11th,* pp. 803–807.
Evans, T. R., and Biggs, J. (1979). *Trop. Grassl.* **13,** 129–134.
Evans, T. R., and Bryan, W. W. (1973). *Aust. J. Exp. Agric. Anim. Husb.* **13,** 530–536.
Flores, J. F., Stobbs, T. H., and Minson, D. J. (1979). *J. Agric. Sci.* **92,** 351–357.
Garcia, R., and Ferrer, F. (1974). *Proc. Int. Grassl. Congr., 12th,* pp. 187–193.
Gardener, C. J. (1980). *Aust. J. Agric. Res.* **31,** 379–392.
Gillard, P. (1982). *World Anim. Rev.* **44,** 2–8.
Gillard, P., Edye, L. A., and Hall, R. L. (1980). *Aust. J. Agric. Res.* **31,** 205–220.
Grant, J. (1980). *Modern Farm.* **8,** 31–36.
Grof, B., and Harding, W. A. T. (1970). *Trop. Grassl.* **4,** 85–95.
Guyton, R. F., Cathupoulis, T. E., and Baylor, J. E. (1980). *Turrialba* **30,** 189–195.
Hamilton, R. D., Lambourne, L. F., Roe, R., and Minson, D. J. (1970). *Proc. Int. Grassl. Congr., 11th,* pp. 860–864.
Harding, W. A. T., and Grof, B. (1978). *Queensl. J. Agric. Anim. Sci.* **35,** 11–21.
Hargrove, D. D., West, R. L., and Prichard, D. L. (1982). *J. Anim. Sci.* **55** (Suppl. 1) (abstr.), 9.
Harker, K. W., Taylor, J. D., and Rollinson, D. H. L. (1954). *J. Agric. Sci.* **44,** 193–198.
Hart, R. H. (1972). *Herb. Abstr.* **42,** 345–353.
Heath, M. E., Metcalfe, D. S., and Barnes, R. E. (1973). "Forages: The Science of Grassland Agriculture." Iowa State Univ. Press, Ames.
Henzell, E. F., and Ross, P. J. (1973). *In* "Chemistry and Biochemistry of Herbage" (G. W. Butler and R. W. Bailey, eds.), Vol. 2, pp. 227–245. Academic Press, London.
Hodges, E. M., Kirk, W. G., and Peacock, F. M. (1966). *Proc. Int. Grassl. Congr., 9th,* pp. 915–918.
Hodges, E. M., Peacock, F. M., and Chapman, Jr., H. L. (1979). *Proc.—Soil Crop Sci. Soc. Fla.* **38,** 51–54.
Holroyd, R. G., O'Rourke, P. K., and Alan, J. (1979). *Aust. J. Exp. Agric. Anim. Husb.* **19,** 389–394.
Humphreys, L. R. (1972). *Ext. Bull., ASPAC Food Fert. Technol. Cent.* **7.**
Jones, R. J. (1974). *Proc. Aust. Soc. Anim. Prod.* **10,** 340–343.
Jones, R. J., and Sanderland, R. L. (1974). *J. Agric. Sci.* **83,** 335–342.
Jones, R. J., Ludlow, M. M., Troughton, J. H., and Blount, C. G. (1979). *J. Agric. Sci.* **92,** 97–100.
Koger, M., Blue, W. G., Killinger, G. B., Green, R. E. L., Myers, J. M., Gammon, N., Warnick, A. C., and Crockett, J. R. (1970). *Bull.—Fla., Agric. Exp. Stn.* **740.**
Lampkin, G. H., and Quarterman, J. (1962). *J. Agric. Sci.* **8,** 119–123.
Larkin, R. M. (1954). *Queensl. J. Agric. Sci.* **11,** 115–141.

Leeuw, P. N. de, and Brinckman, W. L. (1974). *In* "Animal Production in the Tropics" (J. K. Loosli, V. A. Ayenuga, and G. M. Babatunde, eds.), pp. 124–133. Heineman Educational Books, Ibadan, Nigeria.
Lotero, J., Woodhouse, Jr., W. W., and Peterson, R. G. (1966). *Proc. Int. Grassl. Congr., 9th,* pp. 168–190.
McDowell, R. E., Cesteno, H., Rivero-A., J. D., Soldevila, M., Roman-G.,F., and Arroyo-A., J. A. (1977). *J. Agric. Univ. P. R.* **61**, 204–216.
MacKenzie, J. A., and Chheda, H. R. (1970). *Niger. Agric. J.* **7**, 91–97.
Magadan, P. B., Javier, E. Q., and Madamba, J. C. (1974). *Proc. Int. Grassl. Congr. 12th,* pp. 293–299.
Matches, A. G., and Mott, G. O. (1975). *In* "Proceedings of the World Conference of Animal Production" (P. L. Reid, ed.), pp. 203–208. Sydney Univ. Press, Sydney.
Mellor, W., Hibberd, M. J., and Grof, B. (1973). *Queensl. J. Agric. Anim. Sci.* **30**, 259–266.
Milford, R., and Minson, D. J. (1966). *In* "Tropical Pastures" (W. Davies and C. L. Skidmore, eds.), pp. 106–114. Faber and Faber, London.
Minson, D. J., and McLeod, M. N. (1970). *Proc. Int. Grassl. Congr., 11th,* pp. 719–732.
Minson, D. J., and Milford, R. (1966). *Aust. J. Agric. Res.* **17**, 411–423.
Moss, R. J., and Buchanan, I. K. (1983). *Queensl. Agric. J.* **109**, 153–156.
Mott, G. O., Quinn, L. R., Bisschoff, W. V. A., and da Rocha, G. O. (1970). *IRI Res. Inst. [Bull.]* **36**.
Norman, M. J. T., and Stewart, G. A. (1964). *J. Aust. Inst. Agric. Sci.* **30**, 39–46.
Northwood, P. J., and McCartney, J. C. (1969). *East Afr. Agric. For. J.* **35**, 185–189.
O'Donovan, P. B., Marquis da Silva, J., and Euclides, V. P. B. (1983). *World Anim. Rev.* **47**, 30–37.
Oram, P. A. (1975). *In* "Proceedings of the Third World Conference of Animal Production" (R. L. Reid, ed.), pp. 309–330. Univ. of Melbourne, Melbourne.
Payne, W. J. A., Laing, W. I., and Raivoka, E. N. (1951). *Nature (London)* **167**, 610–611.
Payne, W. J. A., Van der Does, C., Kroenberg, J. B. M., Aquino, A. R., Salvatoria, S. A., and Dimayuga, E. C. (1967). *Proc. Annu. Conf. Phil. Anim. Sci., 4th,*
Playne, M. F., and Haydock, K. P. (1972). *Aust. J. Exp. Agric. Anim. Husb.* **12**, 365–370.
Ramirez P., A. (1983). *Pastos Trop., Bol. Inf.* **5**(3), 5–7.
Randel, P. F., and Mendoza, N. (1983). *J. Agric. Univ. P. R.* **67**, 476–485.
Raymond, W. F. (1969). *Adv. Agron.* **21**, 1–108.
Reid, R. L., Post, A. J., Olsen, F. J., and Mugerwa, J. S. (1973). *Trop. Agric. (Trinidad)* **50**, 1–15.
Shultz, E., Shultz, T. A., Garnendias, J. C., and Chicco, C. F. (1977). *Agron. Trop. (Maracay, Venez.)* **27**, 319–333.
Simpson, J. R. (1961). *East Afr. Agric. For. J.* **36**, 158–163.
Spain, J. M., Francis, C. A., Howler, R. H., and Calvo, F. (1975). *In* "Soil Management in Tropical America" (E. Bornemissza and A. Alvaro, eds.), pp. 308–329. N. C. State Univ., Raleigh.
Stobbs, T. H. (1969). *Trop. Agric. (Trinidad)* **46**, 187–194.
Stobbs, T. H. (1971). *Trop. Grassl.* **5**, 159–170.
Stobbs, T. H. (1973). *Aust. J. Agric. Res.* **24**, 809–813.
Stobbs, T. H. (1974). *Proc. Aust. Soc. Anim. Prod.* **10**, 299–302.
Stobbs, T. H. (1975). *Trop. Grassl.* **9**, 141–150.
Stobbs, T. H., and Cowper, L. J. (1972). *Trop. Grassl.* **6**, 107–112.
Stobbs, T. H., and Hutton, E. M. (1974). *Proc. Int. Grassl. Congr., 12th,* pp. 510–517.
Stobbs, T. H., and Imrie, E. C. (1976). *Trop. Grassl.* **10**, 99–106.
Stobbs, T. H., and Thompson, P. A. C. (1975). *World Anim. Rev.* **13**, 27–31.

Stoddart, L. A., Smith, A. D., and Box, T. W. (1975). "Range Management." McGraw-Hill, New York.
Thornton, R. F., and Minson, D. F. (1973). *Aust. J. Exp. Agric. Anim. Husb.* **13,** 537–543.
Tierney, T. J., and Taylor, W. J. (1983). *Trop. Grassl.* **17,** 97–105.
Vallis, D., Haryor, L. A., Catchpoole, V. R., and Weier, K. L. (1982). *Aust. J. Agric. Res.* **33,** 97–107.
Vicente-C., J., Caro-C., R., Pearson, R. W., Abruna, F., Figarella, J., and Silva, S. (1964). *Bull.—Agric. Exp. Stn., Rio Pedras, P. R.,* **187.**
Vine, H. (1968). *In* "The Soil Resources of Tropical Africa" (R. P. Moss, ed.), pp. 89–119. Cambridge Univ. Press, London and New York.
Walker, B., Rutherford, M. T., and Whiteman, P. C. (1981). *Proc. Int. Grassl. Congr., 14th,* pp. 681–684.
Wheeler, J. L. (1962). *Herb. Abstr.* **32,** 1–7.
Whiteman, P. C. (1980). "Tropical Pasture Science." Oxford Univ. Press, London and New York.
Wilkinson, J. M. (1983). *World Anim. Rev.* **45,** 36–42.
Wilkinson, J. M. (1984). *World Anim. Rev.* **46,** 35–40.
Williams, C. H., and Andrew, C. S. (1970). *In* "Australian Grasslands" (R. M. Moore, ed.), pp. 321–338. Aust. Natl. Univ. Press, Canberra.
Winks, L., O'Grady, P., Edgely, W., and Stokol, J. (1970). *Proc. Aust. Soc. Anim. Prod.* **8,** 450–454.
Winks, L., Walker, R. W., O'Rourke, P. K., Lorton, I. D., Holmes, A. E., and Shaw, K. A. (1983). *Trop. Grassl.* **17,** 64–76.
Yazman, J. A., McDowell, R. E., Cesteno, H., Roman-G., F., and Arroyo-A., J. A. (1979). *J. Agric. Univ. P. R.* **63,** 281–293.
Yazman, J. A., Velez-S., J., Arroyo-A., J. A., and McDowell, R. E. (1983). *J. Agric. Univ. P. R.* **67,** 79–94.

7

Providing Energy–Protein Supplementation during the Dry Season

E. J. GOLDING
Department of Animal Science
University of Florida
North Florida Research and Education Center
Quincy, Florida

I. Development of the Production System. 130
   A. Optimal versus Maximal Production. 130
   B. Decisions Concerning Supplemental Feeding. 131
   C. Factors That Influence the Decisions. 134
II. Improving Ruminant Production Potential of Dry-Season Forage. . . 135
   A. Forage Quantity. 136
   B. Forage Quality. 137
III. Sources of Supplemental Crude Protein or Energy. 139
   A. Crude Protein Sources. 139
   B. Energy Sources. 146
IV. Management for Forage Conservation and Efficient Dry-Season Feeding. 154
   A. Forage Conservation. 154
   B. Breeding-Season Length. 158
   C. Frequency of Supplementation. 158
      References. 158

In warm climates, annual dry seasons vary in length and intensity, but in many instances both quantity and quality of forage available in dry-season pastures are diminished to levels that require supplemental feeding of ruminants with energy and/or protein (Fig. 7.1). There are many factors that influence both the decision of whether or not to supplement ruminants during the dry season, and that of what types and amounts of supplements to feed. Therefore, the objective of this presentation is not to

Fig. 7.1. Unsupplemented cattle from Colombia awaiting slaughter during a prolonged, tropical dry season. (Courtesy of L. R. McDowell, University of Florida, Gainesville, Florida.)

outline a series of dogmatic, set plans for dry-season supplementation of ruminants. Instead, an attempt will be made to set forth (1) factors that should be considered when deciding whether and how to carry out such feeding, (2) potential sources of supplemental protein and energy, and (3) ideas related to management for forage conservation and efficient supplementation of ruminants during the dry season.

I. DEVELOPMENT OF THE PRODUCTION SYSTEM

A. Optimal versus Maximal Production

Among producers, objectives of ruminant production systems may be many and varied. However, in the rational economic sense, the objective of such a system should be to obtain an optimal amount of animal products per hectare. The optimal amount of products is the amount that will yield the highest net monetary return per hectare of land owned. Production level should be optimal and not maximal, since maximal production per hectare will yield maximal net revenue per hectare only if each unit of product can be sold at a price greater than its cost. Since this is seldom the case, an optimal point of production generally must be located at some point below that of maximal production.

B. Decisions Concerning Supplemental Feeding

Successful feeding of ruminants is conducted by providing correct amounts of economical, useful nutrients to the right kind of animals (calves, steers, dry cows, etc.) at the right time. Deciding how to supplement growing, male ruminants and nonreplacement females during the dry season appears more difficult than deciding how to supplement replacement and mature females. Decisions regarding the former two groups of animals must be made based on local conditions, experience, and production and financial records, but decisions related to replacement and mature females can be based on meeting required weights or nutrient requirements at specific times.

1. Growing Males and Nonreplacement Females

Allden (1970) concluded that there was little likelihood that grazing ruminants that experienced low planes of nutrition in early life would have reduced production capabilities in later life. Thus, decisions related to dry-season supplementation of growing, male ruminants and nonreplacement females, except where required to ensure survival, should be based purely on economics. With these types of animals, the overriding principle is that, over the animal's lifetime, total value of additional production due to supplementation should be greater than total cost of supplementation. The optimal point of supplementation would be where this difference was maximal. Zero supplementation would produce zero additional cost initially, and no supplemental feeding may be required under conditions where animals do not become too distressed during the dry season. Due to compensatory gain, such animals have reached the same weight as have better-fed animals, while consuming the same amount of feed (Winchester and Howe, 1955; Winchester et al., 1957; Meyer and Clawson, 1964). However, during long and/or intense dry seasons, a point probably will be reached where growing ruminants on low nutritional planes have higher total feed requirements (Meyer and Clawson, 1964; Allden, 1970). Thus, they no longer have potential to exhibit feed-to-gain ratios equal to those of better-fed animals, of equal final weight, over combined stress and recovery periods. Since higher feed-to-gain ratios imply higher feed costs anyway, supplementation probably should commence upon reaching this point, so that animals do not take an overly long period of time to reach market weight. Due to fluctuating interest rates and varying profit margins caused by seasonal markets, time may be an important factor in determining how these types of animals are supplemented during the dry season. Several feeding methods used to maintain steers during the dry season on forage-based diets are shown in Table 7.1. However, feeding for maintenance might not always be the most economical dry-season feeding method for steers.

2. Replacement Females

Over the animal's lifetime in the herd, it was found economically advantageous in Florida for replacement beef heifers to calve first at 2 years, rather than at 3 years of age (Lee *et al.*, 1982). If replacement females are to be bred after passing through only one dry season, they must be supplemented during that time to make rather substantial gains (perhaps more than 0.45 kg/day, for heifers), in order to achieve target weights (Allden, 1970; Foot *et al.*, 1983; Wiltbank, 1983) required for successful breeding. Determination of actual required dry-season daily gain can be made by subtracting liveweight at the start of the dry season, plus expected gain on wet-season pasture prior to breeding, from the target weight, and dividing by the average number of days in the dry season. Breeding of young, replacement females should be done about 1 month prior to that of mature females, so that young animals have ample chance of breeding back on time. Replacement heifers that are managed to give birth after passing through two annual dry seasons probably could be supplemented only at a survival level during the first dry season (Allden, 1970). Heifers should then be supplemented during the second dry season to achieve target weight at about 27 months of age.

Underfeeding of pregnant heifers prior to calving has (1) lowered weights of calves at birth and of heifers at rebreeding, (2) delayed return to estrus, and (3) reduced pregnancy percentages at 60 days following the breeding season (Bellows *et al.*, 1972). Dry-season supplementation of pregnant heifers almost certainly will be required. However, care must be taken not to allow these animals to become overly fat prior to calving. Young, growing females also should be fed a higher plane of nutrition after parturition than that fed to mature females, and such feeding may be required toward the end of an abnormally long dry season.

3. Mature Females

Figure 7.2 shows that mature beef cows have well-defined but variable metabolizable-energy requirements throughout the year, depending upon the cow's physiological status. Some dry-season supplementation of ma-

TABLE 7.1
Dry-Season Feeding Methods for Steers

Feeding method	Forage	Supplement	Amount of supplement (kg/day)	Initial weight (kg)	Days	Weight change (kg/day)	Reference
1	Standing hay	Molasses–urea (10:1)	0.9	279	141	0.00	von la Chevallerie, 1965
2	Standing hay	Fish meal (40%) + salt (60%)	0.65	291	150	−0.02	Bishop and Grobler, 1971
3	Standing hay	Urea (15%) + corn meal (30%) + bone meal (25%) + salt (30%)	0.36	291	150	−0.03	Bishop and Grobler, 1971
4	Standing hay	Hay (*Medicago sativa*, 33%; native grass, 67%)	5.6	226	112	0.15	Smith, 1981
5	Standing hay	Soybean meal	0.45	230	126	0.17	Smith, 1981
6	Chopped hay[a]	Meat and bone meal	0.45	260	56	0.03	Moran, 1975
7	Chopped hay[a]	Poultry litter (60%) + sorghum meal (40%)	1.22	260	56	0.12	Moran, 1975
8	Silage[a,b]	None	0.00	231	90	0.01	M. Ventura, unpublished
9	Silage[a,b] + 4% molasses[c]	None	0.00	226	90	0.06	M. Ventura, unpublished

[a] Forage fed ad libitum, in pens.
[b] *Panicum maximum*, direct-cut.
[c] Molasses added during ensiling.

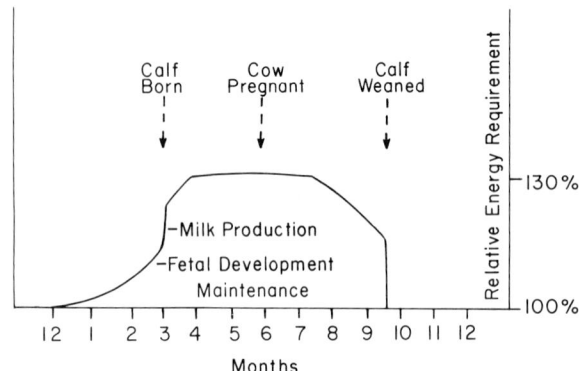

Fig. 7.2. Yearly variation in the metabolizable-energy requirement of the mature beef cow. (Modified from Shirley, 1982.)

ture females undoubtedly will be required. However, planning the calving season to coincide approximately with the beginning of the rainy season will decrease the amount of required energy supplementation. Also, such planning will result in minimal amounts of required protein supplementation during the dry season. This is due to the fact that the last two-thirds of pregnancy, when cows' crude protein (CP) requirements vary between only 7 and 8% of dry matter (DM) [National Research Council (NRC), 1984], will occur during this time. However, care must be taken to meet the mature female's nutritional requirements during the dry season, because allowance of large weight losses during this time has resulted in reduced calving percentages and weaning weights, and increased death losses (Bembridge, 1963). Failure to meet cows' energy requirements during the last third of pregnancy has produced delays in onset of estrus after calving (Wiltbank et al., 1962). This would tend to lengthen the calving interval, and eventually might result in a cow's not breeding back during a controlled breeding season.

C. Factors That Influence the Decisions

The following are factors that should be considered when developing a ruminant production system and, therefore, when deciding whether and how to provide supplemental feed during the dry season: (1) input costs, (2) selling prices of products, (3) financial situation of the operational owner, (4) kinds of outputs produced, (5) types of animals present, (6) quantity of available forage, (7) forage quality, (8) possibility of conserving excess forage during the wet season, (9) availability of supplemental feeds, and (10) availability of irrigation water. As can be seen from the number of factors involved, development of a ruminant production system is a complicated process, and the system must be open constantly to evaluation and change.

II. IMPROVING RUMINANT PRODUCTION POTENTIAL OF DRY-SEASON FORAGE

Daily weight gains of from about 0.25 to 0.4 kg have been reported for unsupplemented zebu steers grazing *Panicum maximum* during the dry season (Mott *et al.*, 1967). However, the more frequent result during the dry season is for deficiencies in forage quantity and/or quality to produce reductions in ruminant liveweight (Fig. 7.3). On native pastures in Australia, dry-season liveweight losses of from 20 (Norman and Arndt, 1959) to 30% (Cohen and O'Brien, 1974) are common. Daily losses ranging as high as from 0.46 (Norman, 1963b) to 0.79 kg (Smith, 1961) have been reported, and such losses have represented about 50–60% of wet-season weight gains (Norman, 1965). In this latter study, mean annual liveweight gain by steers was about 50 kg.

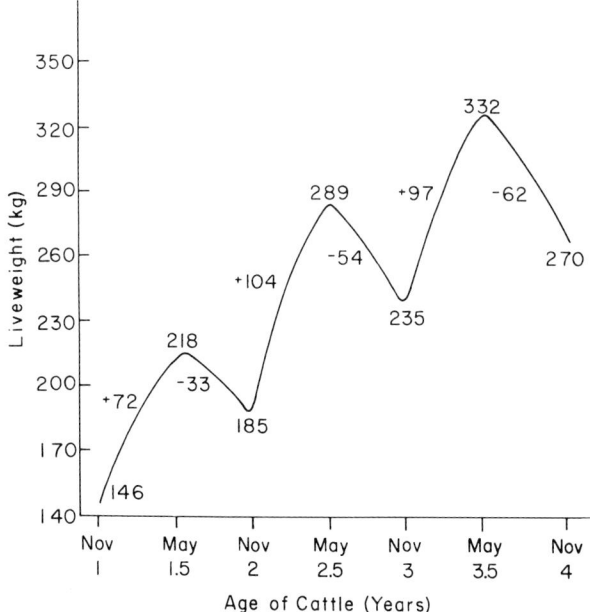

Fig. 7.3. Seasonal liveweight changes in Australian cattle 1–4 years old supplemented only with minerals during the dry season. (Modified from Norman, 1965.)

At times, producers must take steps to lessen effects of reduced quantity and/or quality of dry-season forage on ruminant performance. Apart from supplemental feeding, some agronomic practices may aid the situation.

A. Forage Quantity

On ungrazed plots, Norman (1963a) showed that accumulated DM on offer declined throughout the dry season, due to plant respiration and nutrient translocation, except in response to dry-season precipitation. Decreases commenced within 4 weeks of cessation of the rainy season. Haggar (1970) found virtually no growth of pasture grass after the first month of the dry season. Over 2 years, data of Quinn et al. (1961) showed that dry-season stocking rates on continuously grazed *P. maximum* pastures declined to about 40 and 31% of their wet-season levels, for pastures that did not receive and that received nitrogen (N) fertilizer, respectively. However, attention to fertilization practices and to botanical composition of pastures may help increase dry-season stocking rates.

1. EFFECTS OF NITROGEN FERTILIZATION

Pastures fertilized with 100 or 200 kg of N per hectare per year had dry-season stocking rates that averaged 58 and 85% higher, respectively, than the stocking rate on a pasture that received no N (Quinn et al., 1961). Mott et al. (1965, 1967) obtained similar results while using only the lower N level. However, during a dry season that was very dry, no response in stocking rate was found to N fertilization (Quinn et al., 1963). In the experiment of Quinn et al. (1961), percentage increases in dry-season stocking rates due to N fertilization were greater in a wetter-than-average dry season. Also, applying N at the start of the dry rather than during the wet season produced higher dry-season stocking rates, as well as higher mean annual rates. Again, differences were larger during a wetter-than-average dry season.

2. EFFECTS OF BOTANICAL COMPOSITION

Quinn et al. (1963) showed differences in dry-season stocking rates in Brazil among six grasses, with the Colonião variety of *P. maximum* supporting almost twice the rate of the Tanganyika variety. In comparison with native pasture, introduction of *Brachiaria decumbens* to savannas of Colombia has increased dry-season carrying capacity by more than tenfold (J. E. Velásquez, personal communication). In Australia, Roe et al. (1970) reported that an introduction of *Setaria sphacelata* produced 7.1 tons of DM per hectare during the dry season and early spring in small plots, while several temperate grasses produced from 8.5 to 10.1 tons. Colman (1971) suggested that temperate grasses for dry-season feeding might be maintained in pure stands in the subtropics, thus avoiding decreases in persistence due to competition from warm-season grasses.

With respect to legumes, the Clare covariety of *Trifolium subterraneum* yielded 4 tons of DM per hectare during the dry season, while the Hannaford covariety of *Medicago truncatula* produced 4.1 tons (Roe et al., 1970). Also in Australia, the Verano covariety of *Stylosanthes hamata* outyielded *S. humilis* during and following the second year after seeding at two sites in the dry, northern tropics. Yields at two other sites were equal, or in favor of Verano (Gillard et al., 1980).

Compared to native pastures alone, association of native grasses with *S. humilis* supported lactating cows during the dry season at twice the stocking rate (Holroyd et al., 1977, 1979). Animal performance on the association of grasses with the legume was higher in several instances. Winks et al. (1983), with an association of the Trichoglume variety of *P. maximum* and the Tinaroo covariety of *Neonotonia wightii*, showed that more than 2 tons of DM per hectare could be available at a given time during the dry season, when the legume was maintained in the stand. When the legume yielded well and comprised a minimum of about one-half of associated DM on offer, forage availability was increased by supplementing grazing steers with molasses.

B. Forage Quality

Forage nutritive value (Henderson and Robinson, 1982) and quality (Norman, 1963b; Deinum, 1984) may not decrease much during the parts of the dry season that are totally devoid of rain. However, CP content of native pasture (Norman, 1963b) or standing hay (Norman, 1963a; Haggar, 1970) already may be well below 7% at the start of the dry season. When this is not the case, such values generally are reached as the season progresses (Karue, 1974; O'Donovan et al., 1983). Such low CP levels have limited ruminant intake of warm-season forages (Milford and Minson, 1966). Smith (1961), in separate experiments where forage DM availability was not limiting, reported that heifers grazing standing hay with 2.6% CP lost 0.23 kg of body weight per day during 109 days of the dry season, while mature oxen on similar forage with 1.3% CP lost 0.79 kg of body weight per day during 73 days. On abundant *Cenchrus ciliaris* of less than 3.8% CP content, cattle lost 45 kg of body solids between mid-June and late October (McCown and McLean, 1983). Measured liveweight loss was 18 kg, with the remainder of body solids loss being masked by increases in tissue water and gut contents. Grazing ruminants at times may be able to improve the CP contents of their diets early in the dry season by selecting leaves (Smith, 1961; Haggar, 1970). This apparently becomes less feasible as the dry season progresses (Arndt and Norman, 1959; Cohen, 1978). As is the case with forage quantity, certain agronomic practices may aid in increasing forage quality during the dry season.

1. Effects of Fertilization

In a series of grazing experiments that employed the put-and-take method of maintaining grazing pressure at near-optimal levels, Quinn *et al.* (1961) showed that steers grazing *P. maximum* that received no N, or that received N only during the wet season, gained less weight during the dry season than did animals on pastures fertilized at the termination of rains. Quinn *et al.* (1963) reported that when fertilized and unfertilized pastures supported equal stocking rates, dry-season daily weight gains were higher for five of six grasses fertilized with N and phosphorus (P) at the start of the dry season. Increased performance on fertilized pastures may have been due to higher forage CP contents. To terminate the series, Mott *et al.* (1965, 1967) found that, during the dry season, steers grazing *P. maximum* pastures fertilized with 100 kg of N per hectare per year at the start of the dry season outgained steers that grazed unfertilized pastures. Mott *et al.* (1967) reported that applying N at the start of the dry season increased CP content of dry-season forage by about 3–5 percentage units.

Edye *et al.* (1971, 1972), working over 4 years in an area of Australia characterized by a long dry season, found that fertilization of *S. humilis–Heteropogon contortus* pastures with superphosphate significantly increased conception and calving rates of cows, calf weight weaned per cow, and calf-weaning rate. Effects of stocking rate were not significant. Also, fertilization significantly increased calf weaning weight during 2 years when the effect of stocking rate was not significant.

2. Effects of Botanical Composition

Differences were found by Quinn *et al.* (1963) among six warm-season grasses, whether fertilized or unfertilized, with respect to the amount of dry-season weight gain supported at nearly optimal stocking rates. O'Donovan *et al.* (1982) reported that *B. decumbens* had 6.9% CP about halfway through the dry season. This forage supported significantly higher intake than did native pasture, which contained about one-half as much CP. Compared to that of native pasture, quality of *B. decumbens* diminished at a slower rate during the dry season.

Holroyd *et al.* (1977, 1979) found that when cows lactated during a prolonged dry season, conception rate was higher and calving interval was shorter on a native grass–legume association, at twice the stocking rate, than on native pasture alone. Over 3 years, calf daily gain from birth to weaning generally was higher on the association. Winks *et al.* (1983) reported that, even during the dry season, CP content of the grass component of a grass–legume association never dropped below 6.9%, while that of the legume always was above 10.6%. *Stylosanthes humilis* maintained an N content of better than 1% (6.25% CP) until well into the dry season (Playne and Haydock, 1972).

III. SOURCES OF SUPPLEMENTAL CRUDE PROTEIN OR ENERGY

A. Crude Protein Sources

Potential sources of supplemental CP for forage-fed ruminants are many and varied, but generally can be classified into three groups: (1) nonprotein, (2) plant, and (3) by-product, from processing industries. Sources used by producers will depend largely upon local availability and price, but also should be determined as much as possible by the production system, which dictates amount of production required from supplemented animals.

Since most true proteins generally are considered to contain about 16% N, animal scientists habitually calculate CP percentages of feedstuffs by multiplying N percentages by 6.25. Thus, CP includes all of a feedstuff's nitrogenous components, regardless of their true chemical descriptions.

1. Sources of Nonprotein Nitrogen

There are many potential nonprotein sources of supplemental N (Briggs, 1983). Such N sources per se provide no true protein to ruminants. However, upon hydrolysis in the rumen, nonprotein nitrogen (NPN) provides ammonia, which can be converted to microbial protein by rumen microorganisms. Preston (1982) implied that, when the production system requires only maintenance or a low level of animal productivity, NPN can be the sole supplier of supplemental CP for ruminants consuming tropical forages. Compounds that supply NPN may be cheaper, and their use at appropriate times will have a sparing effect on other protein-containing substances that can be better utilized in other areas of animal production or for human consumption.

a. Urea. Urea, obtained by combining ammonia and carbon dioxide in the presence of heat and pressure, is by far the most common NPN source utilized to supply supplemental CP to forage-fed ruminants (Briggs, 1983). Since feed-grade urea generally contains about 46% N, its CP equivalent is approximately 287%.

A large volume of research has been conducted to elucidate the effects of urea supplementation of forage of very low CP content on ruminant productivity. Reports by Smith (1961), Campling *et al.* (1962), Coombe and Tribe (1962), Chalupa (1968), Rodrigues *et al.* (1978), Stephenson *et al.* (1981), and Umunna (1982) represent results generally obtained from urea supplementation. Burroughs *et al.* (1971) developed the concept of urea fermentation potential (UFP), which aids in calculation of the amount of urea efficiently utilized in a given diet. They postulated that ruminants consuming low-quality forages containing more than 6–7% CP would not be benefited by urea supplementation, unless the supplement also contained a feed with a positive UFP. Mixing molasses with urea in a lick tank, a common practice in tropical areas where sugarcane is grown, provides positive UFP (Burroughs *et al.*, 1971). However, starch is reported to be a better source of readily available carbohydrates for urea utilization than is molasses (Bell *et al.*, 1953; Bloomfield *et al.*, 1958). Nevertheless, conflicting reports suggest that such may not always be the case (Coombe and Tribe, 1962; Elias *et al.*, 1968). Feeding a molasses–urea mixture from a lick tank also takes advantage of the fact that urea utilization has been enhanced by frequent feeding (Bloomfield *et al.*, 1961; Campbell *et al.*, 1963). Addition of sulfur to the diet also has increased urea utilization by ruminants fed low-quality forages (Siebert and Kennedy, 1972).

Initial ruminant response to supplementation with urea probably will be low, since rumen microorganisms apparently require a period of time before reaching maximum efficiency in converting ammonia to microbial protein. This time period has ranged from 7 (Oltjen *et al.*, 1969) to 50 days (Smith *et al.*, 1960), but usually lasts about 2–3 weeks.

b. Biuret. Biuret, produced by controlled heating of urea, is the second most commonly used NPN source for CP supplementation of ruminants (Briggs, 1983). Feed-grade biuret varies in N content according to purity, but generally is in the range of 37–40% N. This gives biuret a CP equivalent of from 231 to 250%. Biuret is considered by many researchers to be generally nontoxic, and therefore it offers possibilities for use in mineral mixtures fed to ruminants consuming low-protein forages. It is felt that the few reports of slight toxicity of biuret resulted from contamination with urea, since frequently in the past it was difficult to obtain pure biuret.

With forage-fed ruminants, biuret has performed as well as has urea

(Oltjen et al., 1969, 1974; Kondos and Mutch, 1975). However, biuret is hydrolyzed more slowly in the rumen than is urea, and this may produce better utilization of biuret N when energy is adequate (Kondos and Mutch, 1975). Oltjen et al. (1969) found that biuret outperformed urea when both were fed twice daily, but Coleman and Barth (1977) reported equal performance under such conditions. Biuret appeared more palatable to ruminants than did urea (Oltjen et al., 1969).

As with urea, rumen microorganisms apparently must adapt to biuret before utilizing it efficiently. This has taken from 21 (Oltjen et al., 1969) to about 40 days (Coleman and Barth, 1977).

c. Poultry Waste. Wastes from caged layers and from broiler houses are included here because they contain about 50–60% of their CP as NPN (Bhattacharya and Fontenot, 1966; Bhattacharya and Taylor, 1975). These authors and Swingle et al. (1977) found that uric acid comprised slightly more than one-half of this NPN. This may favor more efficient use by ruminants of NPN from poultry waste than from urea, since Oltjen et al. (1968) reported that uric acid was degraded more slowly in the rumen than was urea. Oltjen and Dinius (1976) found that steers fed high-forage diets gained weight more rapidly and efficiently when supplemented with caged-layer manure than when given urea or biuret. Positive results from supplementing ruminants consuming low-quality roughages with poultry waste were reported by Moran (1975), Gihad (1976), and Swingle et al. (1977).

Bhattacharya and Taylor (1975) reported that CP content averaged about 30% of DM for both types of poultry waste. However, this is quite variable (Tinnimit et al., 1972; Oliphant, 1974; Swingle et al., 1977) and decreases with increased storage time of layer manure (Flegal et al., 1972). The CP content of broiler litter will fluctuate with the type of litter base (peanut hulls, rice hulls, wood shavings, etc.) and with the number of batches of broilers raised per base (Martin et al., 1983). Also, CP content of poultry waste may be lower in warm climates [21.3% of DM in studies done in Venezuela (E. J. Golding, unpublished results); 22.1% of DM in Zambia (Gihad, 1976); 24.2% of DM in Malaysia (Devendra, 1976)], and will be decreased if waste is contaminated with soil during recovery.

Palatability of poultry waste may be a problem, at least initially. This may be overcome by coarse grinding, especially of layer manure, and mixing with molasses. Palatability of wet broiler litter apparently can be increased by deep stacking (storing in piles about 1.5–2.5 m deep, in a dry place, away from combustible materials) for about 6–8 weeks (Kunkle, 1982). Cattle in Brazil ate about 8 kg/day of litter that had been stored in pits for 18 months (T. J. Cunha, personal communication). It was observed that the longer the litter was stored, the better cattle consumed it. Heat generated during storage will kill many potentially dangerous bacteria in wet broiler litter. Deep stacking probably will not improve palatability of layer manure, and bacteria normally present in this material should be reduced to safe levels simply by drying (Kunkle, 1982).

As poultry waste may contain residues of antibiotics and/or other feed additives often fed to poultry, waste should not be fed to lactating dairy cows, and should be removed from diets of other ruminants at least 14 days prior to slaughter.

d. Anhydrous Ammonia. In recent years, much interest has been focused on use of anhydrous ammonia as an additive for increasing N contents of low-quality roughages. To date, most such work has been conducted with cereal straws. Additions of from 2.5 to 4%, by weight, of anhydrous ammonia have increased considerably the original straw–N contents of less than 1% (Moller and Hvelplund, 1982; Orskov *et al.*, 1983; Alibes *et al.*, 1984). Such increases could be of great utility in raising the quality of low-quality roughages for ruminants (Milford and Minson, 1966), especially when the production system requires only maintenance or near-maintenance animal productivity. Efficiency of ammonia utilization for increasing N contents of low-quality roughages may be enhanced in warm climates (Alibes *et al.*, 1984). Methods and effects of adding anhydrous ammonia to low-quality forages deserve investigation in subtropical- and tropical-forage research programs.

2. Sources of Plant Protein

a. Oilseed Meals. Biologically, good-quality oilseed meals are excellent sources of supplemental CP for dry-season feeding of ruminants. However, such CP sources generally are expensive, and their use often is limited by low, and/or sporadic, local availability. Also, because quality-control standards are lacking in many developing countries, excess hulls often are added to oilseed meals. Meals that contain an excessive amount of hulls should be avoided, since they would have lowered CP and increased crude fiber contents.

Depending upon level of feed intake, diet type, and/or oilseed-processing method, employing oilseed-meal supplements results in variable amounts of true protein reaching the small intestine for digestion and absorption (Owens and Bergen, 1983). Termed bypass protein, such true protein is neither dissolved by rumen fluid nor fermented by rumen microorganisms. Heating or treatment with formaldehyde increases the bypass nature of true protein (Owens and Bergen, 1983). Amino acids absorbed from bypass protein may complement those provided by microbial protein

(Nipper et al., 1984). Use of supplemental feeds containing high levels of bypass protein appears to increase not only the quantity but also the quality of protein available for ruminant production. Therefore, use of oilseed meals during the dry season probably should be restricted to situations where the production system calls for significant amounts of ruminant production. If oilseed meals cannot be mixed with other dietary components prior to feeding, efficiency of meal utilization probably will be enhanced by (1) providing sufficient space per animal at the feeder and (2) dividing supplemented animals into age groups. Such steps aid in assuring that younger, more timid animals receive their allotted share of supplement. Also, consumption of oilseed meals can be regulated by mixing with salt at a level that gives desired meal intake. For this method, plenty of water should be available to animals, or adverse effects may result.

On an as-fed basis, soybean meal usually contains about 44–50% CP, depending upon methodology used during bean processing. The higher levels are attained by dehulling beans prior to oil extraction (Wolf, 1983), but variability in dehulling efficiency often yields a product of variable CP content (R. O. Myer, personal communication). Diets containing soybean meal usually have been utilized more efficiently by ruminants than have diets containing cottonseed meal (Buysse et al., 1966) or peanut meal (Buysse et al., 1966; Kay et al., 1966).

Cottonseed meal generally contains 41% CP, regardless of method of oil extraction (Wolf, 1983). However, this may be lowered by addition of finely ground hulls. Beneficial effects of supplementing steers consuming low-quality, tropical forage with cottonseed meal have been reported by Hennessy and Murison (1982). Work by Silvestre et al. (1977a) suggested that, in the absence of supplemental energy, CP from cottonseed meal might promote more efficient weight gain on sugarcane diets than would that from fish meal or meat meal. Whole cottonseed contains about 21.5% CP in DM (Wolf, 1983) and can be fed to ruminants. However, due to whole cottonseed's high oil content, care must be taken to avoid consumption of such large amounts that animals scour. Gossypol contained in cottonseed should not present problems to animals with a functional rumen (Wolf, 1983).

Peanut meal generally contains ground hulls and averages about 46% CP and 12% crude fiber (Wolf, 1983). Peanut meal should not be stored for very long periods of time in warm climates, since it tends to become rancid. Advantages of including peanut meal in dry-season feeding systems for ruminants were reported by Smith (1961). Efficiency of CP utilization by ruminants may be slightly lower for peanut meal than for cottonseed meal (Bowers et al., 1965).

b. Legumes. To provide supplemental, dry-season CP, forage legumes can be made into hay or silage, grazed in association with grasses, or grazed in pure stands for short periods as "protein banks." Alfalfa, or lucerne (*M. sativa*), hay has been used at dietary levels of from 10 to 20%, to increase intake of low-quality hay (Siebert and Kennedy, 1972). Finely chopped alfalfa increased heifer performance when used to supplement low-quality, native-grass hay (Morris, 1958). Effects of grass–legume associations on quantity and CP content of dry-season forages were discussed previously. Periodic grazing of legume "protein banks" during the dry season has shown promise on the savannas of Colombia (J. E. Velásquez, personal communication), and continued investigation into this aspect of legume utilization appears warranted in warm climates.

c. Nitrogen-Fertilized Forage. As discussed previously, N fertilization of tropical-grass pastures at the start of the dry season has increased both quantity and CP content of dry-season forage.

d. Others. Other plant sources of supplemental CP that may be available locally for dry-season use include copra meal (20.7–21.3% CP, depending upon the oil-extraction process; NRC, 1982), dehydrated tomato pomace (21.6%), flax or linseed meal (34.3–34.6%), sunflower seed meal with (23.3%) or without hulls (41.4–46.3%), and sesame seed meal (45.5%). Ravindran *et al.* (1983) reported that cassava-leaf meal contained 20.2% CP in DM, though this apparently varies with plant variety (Barrios and Bressani, 1967). In areas where cassava is grown commercially, research is needed on methods of harvesting cassava leaves for CP supplementation of ruminants, and on the effects of such harvesting on tuber growth. Meyreles *et al.* (1977) increased daily weight gain of cattle by replacing up to 45% of a sugarcane–urea diet with chopped forage cassava (Zenon variety; 15.8% CP in DM).

3. BY-PRODUCT PROTEIN SOURCES

By-product protein sources presented here generally contain high levels of bypass protein, due to heat application during processing. Therefore, like oilseed meals, these sources probably should be used to supplement ruminants during the dry season only when the production system requires relatively high levels of animal output.

When least-cost diets for meat production were formulated by considering abilities of potential ingredients to balance amino acid supply at the duodenum, by-product protein sources were preferred over plant protein sources (Dennison and Phillips, 1983). However, when utilizing feeds containing protein that is slowly degraded in the rumen, dietary inclusion of a rapidly fermented N source, such as urea, may be required to meet ammonia requirements of rumen microorganisms (Waller, 1978; Krause and Klopfenstein, 1978). Preston (1982) discussed advantages often obtained from feeding CP sources with high levels of bypass protein to ruminants consuming tropical forages.

a. Animal Protein. The actual CP content of animal protein sources varies depending upon composition of raw materials used in production (Young, 1983). The CP contents of three supplemental sources derived from animal parts generally not included by meat packers in dressed carcasses are as follows: meat meal, 51.4%; meat and bone meal, 50.4%; and blood meal, 79.8% (NRC, 1982). It is important to be aware of high calcium (Ca), P, and other minerals in animal protein sources, and what their effect may be in the total diet. When fed in conjunction with urea to steer calves, both meat meal and blood meal performed at least as well as did soybean meal plus urea, and better than did urea fed alone (Stock *et al.*, 1981). Positive results from supplementing ruminants consuming low-protein forages with meat meal, or meat and bone meal, were reported by Morris (1958) and Moran (1975), respectively.

Depending upon fish species used in production, fish meal varies considerably in CP content (36.4–72%; NRC, 1982) and in protein solubility in the rumen (Barlow and Windsor, 1983). However, since about 90% of all fish meal is produced from oily fish (Barlow and Windsor, 1983), these variations may be reduced in practice. These authors also stated that fish meal of very dark brown or blackish brown color may have been overheated during production or storage, which would reduce protein quality. The N in fish meal was utilized more efficiently by ruminants than was that in soybean meal or peanut meal (Whitelaw *et al.*, 1963; Bowers *et al.*, 1965; Kay *et al.*, 1966). Good-quality fish meal was reported to be an excellent source of supplemental CP for ruminants consuming bagasse- and/or molasses-based diets by Randel (1970), Preston and Willis (1974), and Preston (1982).

b. Brewer's Grains. Dried brewer's grains contain 27.1% CP (NRC, 1982). However, CP content of brewer's grains varies among breweries, depending upon unique composition of the original grain mix (Asplund, 1983). Still, brewer's grains contain more CP in DM than did the original grain mix. Due to high moisture content, use of wet brewer's grains for supplementing ruminants probably is economical only in areas situated close to breweries. Purchase price of this by-product should reflect its moisture content.

When storing wet brewer's grains in a pile, problems of surface spoilage and with flies can be overcome by spraying the pile's surface with propionic acid and covering with black plastic (Tindall, 1983). If DM content cannot be determined frequently, wet brewer's grains should be fed by volume, rather than by weight. This is done because a given volume will contain about the same amount of DM, even though moisture content varies somewhat (Holter, 1983). Advantages of supplementing roughage-fed, growing calves with CP from wet or dried brewer's grains were reported by Crickenberger and Johnson (1982) and Klopfenstein and Abrams (1981), respectively.

B. Energy Sources

Sources of energy for supplementing ruminants during the dry season can be classified generally as either plant sources or by-products.

1. PLANT SOURCES OF ENERGY

a. Cereal Grains. Wheat (McInnes *et al.*, 1968) and oats (Bogdanovic, 1983; Hodge *et al.*, 1983) are used in Australia for drought feeding of sheep. However, dry-season feeding of cereal grains to ruminants that return to full-time grazing at the start of the wet season will not be considered here as generally practical. In warm climates, cereal grains usually are expensive, due to low levels of production (Preston and Willis, 1974) and high levels of importation. Also, at the national level, humans and nonruminant animals should have higher priorities for consumption of cereal grains than do ruminants. Exceptions would occur when other energy sources were not available for (1) survival feeding, (2) growing out of replacement females, and (3) taking economic advantage of seasonal markets. However, in most instances, prior economical provisions probably could be taken to eliminate such necessities.

b. Legumes. See previous sections.

c. Fertilized Forage. See previous sections.

d. Sugarcane. Among grasses, sugarcane (*Saccharum officinarum*) appears unique in maintaining a relatively constant nutritive value as it matures. This is due to increasing sugar content and decreasing digestibility of the fibrous component (Pate, 1979a). Thus, sugarcane need not be harvested and stored at a relatively early age in order to negate deleterious effects of advancing maturity on nutritive value. However, once cut, fresh sugarcane should be fed within 24 hr, to reduce negative effects of fermentation on ruminant performance (Preston, 1977).

Sugarcane will support different animal production levels (i.e., fit into different production systems), depending upon the manner in which it is supplemented. Reports by Preston and Willis (1974), Silvestre *et al.* (1977b), Preston and Leng (1978), Fernandez *et al.* (1979), Meyreles *et al.* (1979), and Preston (1982) give in-depth explanations of results of different methods of supplementing sugarcane for ruminant feeding. The bottom line is that for weight gain of about 1.0 kg/day, sugarcane diets must include adequate urea and minerals, medium-to good-quality long forage, and bypass protein and energy. Removal from the diet of either bypass nutrients or long forage reduced gains of young cattle by about one-half (Meyreles *et al.*, 1979), while removal of both reduced performance to about the maintenance level (Rodrigues *et al.*, 1978; Meyreles *et al.*, 1979). Energy may be the more limiting bypass nutrient (Silvestre *et al.*, 1977b). When fed without supplemental bypass nutrients and long forage, chopped-whole cane outperformed chopped or derinded stalks (Fernandez *et al.*, 1979). This probably was due to inclusion of leaves in chopped-whole cane.

e. Hay. Baled hay often is difficult to produce during the rainy season, when forage generally is in abundance. To avoid rain damage, hay has been baled at high moisture content, and successfully preserved with propionic acid (Knapp *et al.*, 1976; Davies and Warboys, 1982) or anhydrous ammonia (Knapp *et al.*, 1975; Weiss *et al.*, 1982). With *P. maximum* in Venezuela, E. J. Golding (unpublished results) found that hay baled at 27% moisture and preserved with 1% of the bale weight of propionic acid had digestible organic matter intake by sheep equal to that of untreated hay baled at 10% moisture. Untreated hay baled with 27% moisture was badly decomposed after 60 days of storage. Drying agents such as potassium carbonate and related chemicals may prove useful in warm climates for increasing drying rate of cut hay (Wieghart *et al.*, 1980; Johnson and Thomas, 1983; Johnson *et al.*, 1983).

Many ruminant producers disdain, as too low in quality to bother baling, mature forage left standing after rains have ceased. However, low-quality hay many times is of great value as a feed reserve late in the dry season (Smith, 1961), and also for production-system phases that require only low levels of ruminant performance. Anhydrous ammonia may prove useful for increasing the quality of such hays.

Dry, standing forage (standing hay) has been a valuable dry-season source of energy for ruminants (Smith, 1961; Norman, 1963b; Norman and Stewart, 1964). Norman (1963b) reported that nutritive value of standing hay apparently did not decline much during dry weather, but that a decrease was produced by periodic, early rains, before arrival of the true wet season. However, the decline in steer liveweight that led to the conclusion of reduced standing-hay nutritive value following early rain apparently was caused, at least in part, by decreased gut fill (McLean *et al.*, 1983). Reports of scientific attempts at efficient production and utilization of standing hay in warm climates are lacking in the literature. Investigation of how time of last cutting or grazing during the wet season, fertilization practices, and method of harvesting affect production and efficiency of utilization of standing hay could be of great value in increasing its utility.

f. Silage. Wilkinson (1983a) cited evidence that silage may be of lower quality than hay made from the same forage. However, silage can be made during the wet season, when attempts at haymaking could result in loss of the entire forage crop. Also, due to reduced nutrient losses and increased overall efficiency of utilization compared to hay, silage may support higher levels of ruminant production per hectare (Waldo, 1977). Ruminant production on silage-based diets has been improved markedly by supplementation with bypass protein (Alvarez *et al.*, 1977; Garstang *et al.*, 1979), or with good-quality hay or dried grass (Castle and Watson, 1975; Thomas *et al.*, 1983). As with wet brewer's grains, silage should be fed by volume rather than by weight, if silage DM content cannot be determined frequently (Holter, 1983).

Literature cited by Steen (1984) indicates that wilting of forage prior to ensiling generally increases ruminant DM intake, but does not always increase rate of weight gain. However, wilting will increase concentration of water-soluble carbohydrates, and has decreased silage acetic acid concentration (Davies, 1963). These considerations appear important in improving fermentation and utilization, respectively, of silages made from tropical forages (Wilkinson, 1983a,b; Fig. 7.4). Care must be taken not to overwilt forage, however, since this apparently increases heat damage of protein during fermentation (Waldo, 1977).

Addition of formic acid, formaldehyde, or a combination of these chemicals to direct-cut forage during ensiling increased mean rate and efficiency of ruminant production relative to untreated silages (Waldo, 1977). Use of formic acid also improved the animal production characteristics of wilted silage. In Venezuela, addition of up to 4% molasses to direct-cut *P. maximum* during ensiling increased steer daily gain only from 0 (no molasses) to 0.1 kg/day, and was deemed uneconomical (M. Ventura, unpublished results).

g. Crop Residues. In grain-growing areas, crop residues (straw, stalks, cobs, etc.) can provide large amounts of feed DM for the dry

Fig. 7.4. Dairy cattle consuming wilted, tropical grass silage in Barbados. (Courtesy of L. R. McDowell, University of Florida, Gainesville, Florida.)

season (Fig. 7.5). Crop residues, due to advanced plant maturity and corresponding high fiber content at the time of grain harvest, generally are quite low in quality. However, though ruminants consuming crop residues generally have been supplemented with energy and/or protein, there is some evidence that quality may be improved by chemical treatment of residues. In work cited by Klopfenstein (1978) and in that of Paterson *et al.* (1982), diets containing residues treated with sodium hydroxide supported better ruminant performance than did diets that contained untreated residues. Similar results were found with ammonia by O'Shea *et al.* (1981) and Alibes *et al.* (1984). Orskov *et al.* (1983) reported that ammonia-treated barley straw outperformed straw plus urea. Also, there is some evidence that ensiling with urea may improve quality of crop residues (Mbatya, 1983).

When supplemented with small amounts of energy, untreated crop residues plus urea (Otchere *et al.*, 1977; McLennan *et al.*, 1981), or ammonia-treated residues (Saenger *et al.*, 1982), have maintained ruminants. When further supplemented with energy and bypass protein, such diets have supported weight gains above 0.5 kg/day, as did residues treated with

Fig. 7.5. Ethiopian cattle consuming harvested cereal straw. (Courtesy of L. R. McDowell, University of Florida, Gainesville, Florida.)

sodium hydroxide and supplemented with urea and energy (Smith *et al.*, 1984).

Since ambient temperatures generally are high, and because it adds NPN without requiring mechanical processing of residues, ammonia may be more practical than is sodium hydroxide for treating crop residues in warm climates.

h. Cassava Tubers. Production of tubers of *Manihot esculenta* Crantz is quite high and widespread in many countries with warm climates (Oyenuga, 1955; Preston and Willis, 1974; Coursey and Halliday, 1974). When used as meal (Ahmed, 1977) or when chopped whole (Silvestre *et al.*, 1977b), cassava tubers contained about 3% CP and 68% starch, in DM. Fresh tuber peels were equivalent to these other feeds in CP (or slightly higher in CP; Oyenuga, 1955; Barrios and Bressani, 1967) and gross-energy contents (Larsen and Amaning-Kwarteng, 1976). However, fresh peels contained a potentially toxic level of hydrocyanic acid. Acid content of peels was reduced to a safe level for cattle by sun-drying or ensiling.

Literature cited by Ahmed (1977) suggests that cassava could replace all of the energy value of barley, oats, or corn in supplemental rations for ruminants. As concerns corn, this possibility is supported by work of De Alba *et al.* (1954). Larsen and Amaning-Kwarteng (1976) reported that yearling bullocks on dry-season pasture alone maintained weight during 7 weeks. However, animals that received supplemental rations of urea, molasses, and 85% sun-dried or ensiled cassava-tuber peels gained about 0.33 kg/day. Otchere *et al.* (1977) supplemented one group of young, grazing sheep with sun-dried peels (78%), molasses, and urea during late rains and the following dry season. Fourteen weeks into the next rainy season, these sheep still weighed about 11 and 20% more, respectively, than did unsupplemented sheep, or those that had been supplemented with rice straw (78%), molasses, and urea. Chopped-whole, sun-dried tubers were a good source of supplemental energy for sugarcane–urea diets, but not for molasses–urea diets (Silvestre *et al.*, 1977b). Calf daily gain was improved by using cassava meal to replace about 20% of molasses in a molasses–urea supplement (Labbe *et al.*, 1975). With young bulls and steers consuming cut, low-quality *P. maximum,* use of cassava meal to replace about 78% of molasses in a molasses–urea supplement produced equal weight gain and efficiency of N utilization (Shultz *et al.*, 1970).

i. Bananas. In Costa Rica, green bananas not suitable for export have been used successfully as an energy supplement for cattle grazing *P. maximum* pastures (Alpizar and Vohnout, 1974; Ruiz *et al.*, 1974). However, since bananas are relatively low in DM content, their economical dry-season use probably would be restricted to areas near export sites, or where banana plants were irrigated locally. Also, dry-season availability of bananas probably would be restricted to such areas.

Ffoulkes and Preston (1978) prepared isonitrogenous diets by adding molasses–urea to either banana forage (80% pseudostem : 20% leaves, fresh basis) or sugarcane (relatively mature, chopped whole). For both diets, digestible DM intake by young Zebu bulls was similar. Replacing 33% of sugarcane DM with banana forage produced digestible DM intake twice as high as that of either of the single-forage diets.

2. BY-PRODUCT SOURCES OF ENERGY

a. Sugarcane Molasses. In some areas, the price of molasses has risen drastically in recent years, due to increased use by the rum-producing industry. However, molasses generally is available for feeding to ruminants in warm climates. In many such areas, average annual yields of molasses per unit of land are higher than are those of cereal grains (Preston and Willis, 1969, 1974).

Molasses is an extremely versatile energy source, having been used (1) as the base for ruminant diets supporting weight gains of from slightly above maintenance to almost 1.0 kg/day (Preston and Willis, 1969, 1974; Silvestre *et al.*, 1977b; Gaya *et al.*, 1979); (2) for mixing with urea, to form protein–energy supplements for animals grazing low-quality, dry-season pastures (Holroyd *et al.*, 1977; Winks *et al.*, 1982); and (3) alone, as a supplement to dry-season, grass-legume pastures (Winks *et al.*, 1983) or *P. maximum* pastures grazed by Zebu steers (Mott *et al.*, 1965, 1967). To achieve high rates of gain, molasses was supplemented with urea and bypass protein, and fresh forage was provided at 1.5% of body weight. Forage quality was important in determining daily gain, and gain was reduced by ad libitum forage feeding, or by reducing bypass protein.

When feeding molasses ad libitum, it initially should be diluted with water to about 15° Brix, to avoid toxicity problems due to initial high consumption of molasses (Preston and Willis, 1969).

b. Sugarcane Bagasse. Fresh bagasse (which contains about 50% moisture) comprises approximately 25% of the sugarcane stalk. Even when dry, bagasse generally outyields cereal grains on a per unit of land basis, in cane-growing countries (Preston and Willis, 1974). Unfortunately, nutritive value of bagasse is very low, falling below even that of cereal straws (Molina *et al.*, 1983). However, though fed in mixed diets, there is evidence that nutritive value of bagasse is enhanced by treatment with sodium hydroxide (Molina *et al.*, 1983; Joshi *et al.*, 1984), or with steam (Pate, 1979b; Joshi *et al.*, 1984). The N content of bagasse was increased 512% by ammoniation with urea (Hamad and El-Saied, 1982). This increase was greater than that for corn stalks, rice straw, or rice hulls.

When included in a molasses–urea diet, ad libitum–fed bagasse successfully replaced two-thirds of the animal requirement for fresh forage (Sansoucy *et al.*, 1973). Using ad libitum–fed bagasse to completely replace fresh forage lowered rate of gain.

c. Broiler Litter. This poultry waste contains about 15–20% ash. Broiler litter is a good source of supplemental Ca and P, while having approximately the digestible-energy value of good-quality hay (Bhattacharya and Fontenot, 1966; Bhattacharya and Taylor, 1975).

Pregnant ewes performed normally when fed diets containing 50 (Webb *et al.*, 1973) or 63% (Galmez *et al.*, 1970) broiler litter. However, the former workers found that copper toxicity could be a problem if litter containing high levels of this element was fed. According to Kunkle (1982), pregnant beef cows grazing residual pasture can be fed supplemental rations containing 75–90% broiler litter and 10–25% ground grain, molasses, etc. Such a system has been used commercially in Venezuela to maintain steers grazing dry-season pastures of mature *P. maximum*. If dry-season grazing is not available, litter-supplemented cattle should be fed 1.0–1.5 kg of roughage per day. Moran (1975) maintained pen-fed steers consuming low-quality hay by supplementing with a pelleted mixture of 60% litter and 40% crushed sorghum grain.

d. Rice Bran, Polishings, and Hulls. Rice bran, which makes up about 10% of whole rice grain, varies in nutritive value depending upon the amounts of hulls and oil it contains (Moran, 1983). This study reported that feeding 1.2 kg of rice bran per head per day to cattle and buffalo bulls consuming good-quality forage reduced forage intake, while increasing rate of gain and efficiency of feed utilization.

Polishings constitute about 3% of whole rice grain (Moran, 1983) and contain 10% CP, 11% lipid, and up to 60% starch (Preston, 1982). Rice polishings were an excellent source of bypass nutrients when included in sugarcane-based diets (Preston *et al.*, 1976; Ferreiro *et al.*, 1977; Preston, 1982).

Young bulls gained about 0.2 kg/day on a diet that contained only small amounts of green grass and dry, protein-energy supplement, and where 85% of DM consumed was constituted by rice hulls mixed with molasses–urea (Joshi and Khan, 1984).

e. Coffee Pulp. Dehydrated coffee pulp (dried remains of coffee pods, after seed extraction) contained about 12% CP, and high levels of acid-detergent fiber, lignin, caffeine, and tannin (Vargas *et al.*, 1982). When dehydrated coffee pulp comprised 60% of the diet, steers lost weight for the first 4 weeks, but maintained weight over 84 days. At 20 and 40% inclusion levels, daily gains were 1.3 and 0.8 kg, respectively. Intake of DM and of digestible CP declined as the level of coffee pulp increased, and the results confirmed those of Cabezas *et al.* (1974). Other effects of feeding high levels of coffee pulp included reduced digestibility (Vargas *et al.*, 1977a) and utilization (Vargas *et al.*, 1977b) of N. Also, Cabezas *et al.* (1977) and Vargas *et al.* (1982) reported animal toxicity due to high intake of caffeine and tannin.

f. Cocoa Husks. Discarded husks of *Theobroma cacao* pods constitute an abundant source of useful, dry-season energy for ruminants, in areas where cocoa is produced (De Alba *et al.*, 1954; Adeyanju *et al.*,

1975; Otchere *et al.*, 1983). De Alba *et al.* (1954) found that when a dairy concentrate contained bypass nutrients and 50% meal from artificially dried cocoa husks, cows produced more milk than when the concentrate contained an equal percentage of 20%-moisture corn. Though efficiency of feed utilization was somewhat low, Bateman and Larragan (1966) reported that yearling steers gained 1.14 kg/day when fed limited amounts of sorghum silage and supplemented ad libitum with a mixture of bypass nutrients, molasses, and 60% sun-dried, cocoa-husk meal. Otchere *et al.* (1983) obtained a maintenance diet for wethers by feeding 60% sun-dried, coarsely ground cocoa husks, 35% wheat bran, and 3% peanut cake. Doamekpor (1977) observed that a diet containing 75% cocoa husks had nearly enough digestible energy to maintain wethers. Theobromine concentration in dried cocoa husks should not be high enough to affect ruminants adversely (Bateman and Larragan, 1966; Otchere *et al.*, 1983).

IV. MANAGEMENT FOR FORAGE CONSERVATION AND EFFICIENT DRY-SEASON FEEDING

A. Forage Conservation

In warm climates, pastures containing sparse, dry, mature grasses often provide the only source of forage for ruminants during the dry season. Also, abundant wet-season forage of relatively high quality many times is underutilized to the point that much simply is wasted. Reasons for such occurrences include the belief that forage conservation is too expensive. Also, during the wet season, stocking rates are habitually low. These low stocking rates result from faulty ideas concerning production systems, and from herd sizes dictated by dry-season pasture carrying capacities.

Whether forage is conserved as hay or as silage, conservation schedules during the wet season should be related to (1) forage quality and quantity, (2) necessity of reducing stocking rates as forage growth rate declines, and/or (3) utility of producing standing hay. To minimize machinery breakdowns and resultant time losses, areas to be conserved should be clear of tree trunks, rocks, and holes. Also, both land and forage should be dry enough to eliminate bogging down and clogging of equipment.

1. Costs

While forage conservation in developing areas generally has the potential to become a relatively high-cost operation, this need not be the usual result. In Venezuela, molasses at 3% of fresh-forage weight was used as an additive to produce direct-cut *P. maximum* silage. This silage cost about one-half as much per kilogram of DM as did dry-season supplemental feed, based on low-quality concentrates, distributed by local cattlemen's associations. Hay was produced at about one-fourth the supplemental-feed cost. These calculations did not include costs of transporting and storing supplemental concentrates. Since conservation operations always included high daily labor costs, cost per unit of conserved DM was reduced drastically by increasing the rate of conservation (tons of forage conserved per day). To this end, excellent care was taken of machinery, and provisions were made always to have replacements for high-risk parts (belts, chains, pins, blades, etc.) close at hand.

Producers often are influenced negatively by the relatively high initial cost of forage conservation machinery. Joint ownership of machinery through partnerships, cattlemen's associations, or co-ops can be of great help in reducing these costs per individual. Careful planning and much cooperation are needed in allocating use of jointly owned machinery. In too many cases, forage crops of joint owners of equipment are ready for harvesting at the same time. At the national level, tax breaks and reduced import charges for agricultural machinery, and for replacement parts, also could be of aid.

2. Wet-Season Stocking Rates

Opportunity to exercise good dry-season management of ruminant production systems that include use of conserved forage requires a high level of wet-season pasture management. Also, a correct understanding of the relative importance of production-system criteria is required. One reason for low wet-season stocking rates and resultant forage losses is that many ruminant producers in warm climates feel that gain per animal is the most important production criterion. Thus, they employ low stocking rates in order to obtain maximal gain per animal. Such producers forget that if net economic return per hectare is to be maximized, then animal production per hectare must be optimal. Since this latter point generally does not fall within the range of very low stocking rates (Mott, 1962), producers who stock low during the wet season in order to achieve maximal gain per animal are foregoing potential economic gain, as well as wasting valuable forage. Correct management strategy for the period of rapid forage growth during the rainy season would be to increase the stocking rate on a fraction of the available pasture land, to the point of optimal animal production per hectare of that land. The correct stocking rate will vary from ranch to ranch, but the resultant concentration of animals on fewer hectares will lead to more efficient utilization of forage on that land, as well as allow conservation of forage produced on land not required for grazing.

3. Hay Production

Hay often is difficult to produce during the rainy season. This is due to field conditions that retard drying, and to danger of losing the entire cut crop through repeated wetting. Wet forage that lies on a pasture for a relatively long time retards regrowth, delays fertilization, and becomes tangled in regrowth. This latter problem leads to fouling of subsequent hay crops. Perhaps, in the future, use of drying agents, and/or preservatives for high-moisture baling, will alleviate some of these problems.

Still, as practiced in Venezuela, making hay exhibited several advantages when compared to direct-cut silage production. During favorable weather, *P. maximum* hay was produced by cutting at about 10–12 cm above ground with a cutter-conditioner when the first seed heads appeared. Cut material was dried in the swath for 2–3 days and was baled from the swath without raking. When weather permitted, the same large area was conserved faster as hay than as silage, with cutter and baler working simultaneously in different fields. This helped hay exhibit the lesser cost per kilogram of DM. Also, hay required less manual labor during feeding and, once stored, was of more predictable quality at feeding time.

In the Dominican Republic, hay is made on a small scale by cutting forage by hand, drying, and then baling as shown in Fig. 7.6.

4. Silage Production

Some mud from tractor and wagon wheels at times may be introduced into horizontal silos. However, direct-cut silage can be made during the wet season at any time that allows machinery to work in the field. Since it requires considerably less drying time than does conventional hay, production of wilted silage also should be possible during the wet season. Harvest dates may require slight shifting to accommodate wilting. In Venezuela, it was found that, compared to making hay, initial cost of machinery for direct-cut-silage production was lower (though operational costs were higher), and silage machinery was simpler to operate and to repair.

Horizontal silos should have long and narrow dimensions (Wilkinson, 1983a), so that the area of the silo face is minimized, while rate of silage extraction along the long axis is maximized. Rapid silage removal from a small face tends to reduce spoilage of silage from air infiltration during feeding. A silage production system that gave good results in Venezuela included use of portable side panels to construct $8 \times 24 \times 1.5$-m bunker silos, each of which accommodated about 200 tons of direct-cut forage. Each side panel was 4 m long, and was constructed of heavy boards spaced 50

Fig. 7.6. Small-scale method of baling hay practiced in the Dominican Republic. (Courtesy of L. R. McDowell, University of Florida, Gainesville, Florida.)

cm apart and covered with chain-link fence. Panels were supported by heavy fence posts set vertically, with angled supports for posts. Once the silo was filled, sides were removed and repositioned for refilling. This allowed use of the same materials for making any number of silos. Also, silos could be placed at any desired location, though they always were located near feeding areas. A tractor was used to remove chopped forage from wagons by pulling a wire cable attached to a movable, false wagon front mounted on rails. Forage was chopped fine, and compacted continuously with tractor wheels. Initially, molasses was added at 3% of fresh-forage weight (molasses was weighed before it was mixed with water). However, this was found uneconomical when *P. maximum* was cut before seed head appearance, and when air was excluded efficiently from the silo by compaction and covering with thick, black plastic (M. Ventura, unpublished results). Clear plastic decomposed after a few weeks of exposure. Plastic was buried at the silo edges after side panel removal and was protected from wind damage by weighting on top with old, rubber tires. Temporary fences around silos excluded animals, and care was taken to repair quickly any rips in the plastic. When the system was perfected so that silos were filled in 5 or 6 days, fermentations always were acceptable, and the encompassing spoilage layer averaged about 10–15 cm in depth.

B. Breeding-Season Length

Dry-season supplementation of ruminants will be easier and more efficient if animals are grouped by type (steers, replacement heifers, mature cows, etc.). Also, homogeneity within type, with regard to physiological stage, will aid in determining type and level of supplementation. Use of a controlled breeding season of some 60–90 days will facilitate attainment of such homogeneous groups of animals. This also will facilitate matters during weaning, pregnancy checking, vaccinating, marketing, etc.

C. Frequency of Supplementation

Hennessy *et al.* (1981) obtained equal steer weight gains during 140 days of winter and spring grazing in Australia by offering 600 g/head/day of a pelleted mineral and protein meal mix, or by offering 2.1 kg of mix twice weekly. The protein meal used in the pellets contained cottonseed meal, meat meal, and fish meal (8:1:1, respectively). Moran (1975) and Graham *et al.* (1983) split weekly amounts of protein supplements or urea–molasses, respectively, into twice-weekly feedings. Daily weight gains of steers consuming low-quality forages were increased to, or slightly above, the maintenance level. In the latter work, water was added to urea–molasses to regulate intake. Contrary to findings of Oltjen *et al.* (1974) with steers, Meaker and Liebenberg (1984) reported that an interrupted feeding (4 days with and 3 days without) of a urea-containing supplement had no negative effects on cow performance when compared with continuous feeding. If effective, such labor- or feed-saving tactics might reduce the cost of dry-season supplementation of ruminants.

REFERENCES

Adeyanju, S. A., Ogutuga, D. B. A., Ilori, J. O., and Adegbola, A. A. (1975). *Nutr. Rep. Int.* **11,** 351–357.
Ahmed, F. A. (1977). *East Afr. Agric. For. J.* **42,** 368–372.
Alibes, X., Muñoz, F., and Faci, R. (1984). *Anim. Feed Sci. Technol.* **10,** 239–246.
Allden, W. G. (1970). *Nutr. Abstr. Rev.* **40,** 1167–1184.
Alpizar, J., and Vohnout, K. (1974). *Mem. Asoc. Latinoam. Prod. Anim.* **9,** 123 (abstr.).
Alvarez, F. J., Priego, A., and Preston, T. R. (1977). *Trop. Anim. Prod.* **2,** 27–33.

Arndt, W., and Norman, M. J. T. (1959). *In* "Division of Land Research and Regional Survey, Technical Paper No. 3," pp. 1–20. C.S.I.R.O., Australia.
Asplund, J. M. (1983). *In* "Handbook of Nutritional Supplements" (M. Rechcigl, Jr., ed.), Vol. II, pp. 149–161. CRC Press, Boca Raton, Florida.
Barlow, S. M., and Windsor, M. L. (1983). *In* "Handbook of Nutritional Supplements" (M. Rechcigl, Jr., ed.), Vol. II, pp. 253–272. CRC Press, Boca Raton, Florida.
Barrios, E. A., and Bressani, R. (1967). *Turrialba* **17**, 314–320.
Bateman, J. V., and Larragan, A. (1966). *Turrialba* **16**, 25–28.
Bell, M. C., Gallup, W. D., and Whitehair, C. K. (1953). *J. Anim. Sci.* **12**, 787–797.
Bellows, R. A., Varner, L. W., Short, R. E., and Pahnish, O. F. (1972). *J. Anim. Sci.* **35**, 185 (abstr.).
Bembridge, T. J. (1963). *Rhod. Agric. J.* **60**, 98–103.
Bhattacharya, A. N., and Fontenot, J. P. (1966). *J. Anim. Sci.* **25**, 367–371.
Bhattacharya, A. N., and Taylor, J. C. (1975). *J. Anim. Sci.* **41**, 1438–1457.
Bishop, E. J., and Grobler, J. (1971). *Farming S. Afr.* **47**(2), 21–23.
Bloomfield, R. A., Muhrer, M. E., and Pfander, W. H. (1958). *J. Anim. Sci.* **17**, 1189 (abstr.).
Bloomfield, R. A., Welsch, C., and Garner, G. B. (1961). *J. Anim. Sci.* **20**, 926 (abstr.).
Bogdanovic, B. (1983). *Anim. Prod.* **37**, 459–460.
Bowers, H. B., Preston, T. R., MacLeod, N. A., McDonald, I., and Philip, E. B. (1965). *Anim. Prod.* **7**, 303–308.
Briggs, M. H. (1983). *In* "Handbook of Nutritional Supplements" (M. Rechcigl, Jr., ed.), Vol. II, pp. 87–97. CRC Press, Boca Raton, Florida.
Burroughs, W., Trenkle, A. H., and Vetter, R. L. (1971). *Proc. Annu. Texas Nutr. Conf., 26th*, pp. 103–121.
Buysse, F., Boucque, Ch., and Eeckhout, W. (1966). "Ministry of Agriculture Research Centre Publication No. 1," Ghent.
Cabezas, M. T., Murillo, B., Jarquin, R., Gonzalez, J. M., Estrada, E., and Bressani, R. (1974). *Turrialba* **24**, 160–167.
Cabezas, M. T., Vargas, E., Murillo, B., and Bressani, R. (1977). *Proc. Int. Symp. Feed Comp., 1st*, pp. 112–117. Utah State Univ. Press, Logan.
Campbell, J. R., Howe, W. M., Martz, F. A., and Merilan, C. P. (1963). *J. Dairy Sci.* **46**, 131–134.
Campling, R. C., Freer, M., and Balch, C. C. (1962). *Br. J. Nutr.* **16**, 115–124.
Castle, M. E., and Watson, J. N. (1975). *J. Br. Grassl. Soc.* **30**, 217–222.
Chalupa, W. (1968). *J. Anim. Sci.* **27**, 207–219.
Cohen, R. D. H. (1978). Ph.D. thesis, Univ. of New England, Armidale, Australia.
Cohen, R. D. H., and O'Brien, A. D. (1974). *Trop. Grassl.* **8**, 71–79.
Coleman, S. W., and Barth, K. M. (1977). *J. Anim. Sci.* **45**, 1180–1187.
Colman, R. L. (1971). *Trop. Grassl.* **5**, 181–194.
Coombe, J. B., and Tribe, D. E. (1962). *J. Agric. Sci.* **55**, 125–141.
Coursey, D. G., and Halliday, D. (1974). *Outlook Agric.* **8**, 10–14.
Crickenberger, R. G., and Johnson, B. H. (1982). *J. Anim. Sci.* **54**, 18–22.
Davies, M. H., and Warboys, I. B. (1982). *Grass Forage Sci.* **37**, 165–167.
Davies, T. (1963). *J. Agric. Sci.* **61**, 309–328.
De Alba, J., Garcia, H., Perez Cano, F., and Ulloa, G. (1954). *Turrialba* **4**, 29–34.
Deinum, B. (1984). *Proc. Gen. Meet. Eur. Grassl. Fed. 10th*, pp. 338–350.
Dennison, C., and Phillips, A. M. (1983). *South Afr. J. Anim. Sci.* **13**, 229–235.
Devendra, C. (1976). *Malays. Agric. J.* **50**, 513–522.

Doamekpor, S. K. (1977). B.S. dissertation, Univ. of Ghana, Legon.
Edye, L. A., Ritson, J. B., Haydock, K. P., and Davies, J. G. (1971). *Aust. J. Agric. Res.* **22**, 963–977.
Edye, L. A., Ritson, J. B., and Haydock, K. P. (1972). *Aust. J. Exp. Agric. Anim. Husb.* **12**, 7–12.
Elias, A., Preston, T. R., Willis, M. B., and Sutherland, T. M. (1968). *Cuban J. Agric. Sci.* **2**, 55–63.
Fernandez, A., Ffoulkes, D., and Preston, T. R. (1979). *Trop. Anim. Prod.* **4**, 95 (abstr.).
Ferreiro, H. M., Preston, T. R., and Sutherland, T. M. (1977). *Trop. Anim. Prod.* **2**, 56–61.
Ffoulkes, D., and Preston, T. R. (1978). *Trop. Anim. Prod.* **3**, 125–129.
Flegal, C. J., Sheppard, C. C., and Dorn, D. A. (1972). *Proc. Cornell Agric. Waste Manage. Conf., 4th*, pp. 295–300.
Foot, J. Z., McIntyre, J. S., and Heazlewood, P. G. (1983). *Aust. J. Exp. Agric. Anim. Husb.* **23**, 374–382.
Galmez, J., Santisteban, E., Haardt, E., Crempien, C., Villalta, L., and Torell, D. (1970). *J. Anim. Sci.* **31**, 241 (abstr.).
Garstang, J. R., Thomas, C., and Gill, M. (1979). *Anim. Prod.* **28**, 423 (abstr.).
Gaya, H., Nasseeven, R., Hulman, B., and Preston, T. R. (1979). *Trop. Anim. Prod.* **4**, 148–153.
Gihad, E. A. (1976). *J. Anim. Sci.* **42**, 706–709.
Gillard, P., Edye, L. A., and Hall, R. L. (1980). Aust. J. Agric. Res. **31**, 205–220.
Graham, T. W. G., Wood, S. J., Knight, J. L., and Blight, G. W. (1983). Trop. Grassl. **17**, 11–20.
Haggar, R. J. (1970). *J. Agric. Sci.* **74**, 487–494.
Hamad, M. A., and El-Saied, H. (1982). *J. Sci. Food Agric.* **33**, 253–254.
Henderson, M. S., and Robinson, D. L. (1982). *Agron. J.* **74**, 943–946.
Hennessy, D. W., and Murison, R. D. (1982). *Aust. J. Exp. Agric. Anim. Husb.* **22**, 140–146.
Hennessy, D. W., Williamson, P. J., Lowe, R. F., and Baigent, D. R. (1981). *J. Agric. Sci.* **96**, 205–212.
Hodge, R. W., Bogdanovic, B., and Kat, C. (1983). *Aust. J. Exp. Agric. Anim. Husb.* **23**, 266–270.
Holroyd, R. G., Allan, P. J., and O'Rourke, P. K. (1977). *Aust. J. Exp. Agric. Anim. Husb.* **17**, 197–206.
Holroyd, R. G., O'Rourke, P. K., and Allan, P. J. (1979). *Aust. J. Exp. Agric. Anim. Husb.* **19**, 389–394.
Holter, J. B. (1983). *J. Dairy Sci.* **66**, 1403–1408.
Johnson, T. R., and Thomas, J. W. (1983). *J. Sci. Food Agric.* **34**, 534–540.
Johnson, T. R., Thomas, J. W., and Rotz, C. A. (1983). *J. Dairy Sci.* **66**, 1052–1056.
Joshi, A. L., Rangnekar, D. V., Badve, V. C., and Waghmare, B. S. (1984). *Indian J. Anim. Sci.* **54**, 149–152.
Joshi, D. C., and Khan, M. Y. (1984). *Indian Vet. J.* **61**, 158–162.
Karue, C. N. (1974). *East Afr. Agric. For. J.* **40**, 89–95.
Kay, M., Preston, T. R., MacLeod, N. A., and Philip, E. B. (1966). *Anim. Prod.* **8**, 43–45.
Klopfenstein, T. (1978). *J. Anim. Sci.* **46**, 841–848.
Klopfenstein, T., and Abrams, S. M. (1981). *Neb. Beef Cattle Rep.* EC 81-218, Univ. of Nebraska, Lincoln, pp. 2–6.
Knapp, W. R., Holt, D. A., and Lechtenberg, V. L. (1975). *Agron. J.* **67**, 766–769.
Knapp, W. R., Holt, D. A., and Lechtenberg, V. L. (1976). *Agron. J.* **68**, 120–123.
Kondos, A. C., and Mutch, B. (1975). *J. Agric. Sci.* **85**, 359–368.
Krause, V., and Klopfenstein, T. (1978). *J. Anim. Sci.* **46**, 499–504.

Kunkle, W. E. (1982). *Anim. Sci. Fact Sheet, AS 24.* Univ. of Florida, Gainesville.
Labbe, S., Urdaneta, R., Perozo, T., Olivares, R., and Avedaño, A. (1975). *Agron. Trop.* (*Maracay, Venez.*) **25**, 201–205.
Larsen, R. E., and Amaning-Kwarteng, K. (1976). *Ghana J. Agric. Sci.* **9**, 43–47.
Lee, R. W., Koger, M., Warnick, A. C., and Greene, R. E. L. (1982). *Proc. Annu. Beef Cattle Short Course, 14th*, pp. 34–38. Univ. of Florida, Gainesville.
McCown, R. L., and McLean, R. W. (1983). *J. Agric. Sci.* **101**, 25–31.
McInnes, P., Austin, P. J., and Jenkins, D. L. (1968). *Aust. J. Exp. Agric. Anim. Husb.* **8**, 401–404.
McLean, R. W., McCown, R. L., Little, D. A., Winter, W. H., and Dance, R. A. (1983). *J. Agric. Sci.* **101**, 17–24.
McLennan, S. R., Wright, G. S., and Blight, G. W. (1981). *Aust. J. Exp. Agric. Anim. Husb.* **21**, 367–370.
Martin, J. H., Jr., Loehr, R. C., and Pilbeam, T. E. (1983). *Agric. Wastes* **7**, 13–38.
Mbatya, P. B. A. (1983). *Anim. Feed Sci. Technol.* **9**, 181–183.
Meaker, H. J., and Liebenberg, G. C. (1984). *South Afr. J. Anim. Sci.* **14**, 115–117.
Meyer, J. H., and Clawson, W. J. (1964). *J. Anim. Sci.* **23**, 214–224.
Meyreles, L., MacLeod, N. A., and Preston, T. R. (1977). *Trop. Anim. Prod.* **2**, 73–80.
Meyreles, L., Rowe, J. B., and Preston, T. R. (1979). *Trop. Anim. Prod.* **4**, 255–262.
Milford, R., and Minson, D. J. (1966). *Proc. Int. Grassl. Congr., 9th* **1**, pp. 815–822.
Molina, E., Boza, J., and Aguilera, J. F. (1983). *Anim. Feed Sci. Technol.* **9**, 1–17.
Moller, P. D., and Hvelplund, T. (1982). *Z. Tierphysiol. Tierernaehr. Futtermittelkd.* **48**, 46–57.
Moran, J. B. (1975). *J. Aust. Inst. Agric. Sci.* **41**, 63–65.
Moran, J. B. (1983). *J. Agric. Sci.* **100**, 709–716.
Morris, J. G. (1958). *Queensl. J. Agric. Sci.* **15**, 161–180.
Mott, G. O. (1962). *In* "Forages" (H. D. Hughes, M. E. Heath, and D. S. Metcalfe, eds.), 2nd ed., pp. 108–118. Iowa State Univ. Press, Ames.
Mott, G. O., Quinn, L. R. C., Bisschoff, W. V. A., and da Rocha, G. L. (1965). *Proc. Int. Grassl. Congr., 9th* **2**, pp. 981–988.
Mott, G. O., Quinn, L. R., Bisschoff, W. V. A., and da Rocha, G. L. (1967). *Pesqui. Agropecu. Bras.* **2**, 441–459.
National Research Council (NRC) (1982). "United States–Canadian Tables of Feed Composition," 3rd rev. ed. Natl. Acad. Sci., Washington, D.C.
National Research Council (NRC) (1984). *"Nutrient Requirements of Domestic Animals, No. 4.* Nutrient Requirements of Beef Cattle," 6th rev. ed. Natl. Acad. Sci., Washington, D.C.
Nipper, W. A., Sorbet, R. H., Jr., Adkinson, R. W., and Stutts, J. A. (1984). *Louisiana Agric.* **27**, 3.
Norman, M. J. T. (1963a). *Aust. J. Exp. Agric. Anim. Husb.* **3**, 119–124.
Norman, M. J. T. (1963b). *Aust. J. Exp. Agric. Anim. Husb.* **3**, 280–283.
Norman, M. J. T. (1965). *Aust. J. Exp. Agric. Anim. Husb.* **5**, 227–231.
Norman, M. J. T., and Arndt, W. (1959). *In* "Division of Land Research and Regional Survey, Technical Paper No. 4," pp. 1–12. C.S.I.R.O., Australia.
Norman, M. J. T., and Stewart, G. A. (1964). *J. Aust. Inst. Agric. Sci.* **30**, 39–46.
O'Donovan, P. B., Euclides, V. P. B., and Marques da Silva, J. (1982). *Pesqui. Agropecu. Bras.* **17**, 1655–1670.
O'Donovan, P. B., Marques da Silva, J., and Euclides, V. P. B. (1983). *World Anim. Rev.* **47**, 30–37.
Oliphant, J. M. (1974). *Anim. Prod.* **18**, 211–217.

Oltjen, R. R., and Dinius, D. A. (1976). *J. Anim. Sci.* **43,** 201–208.
Oltjen, R. R., Slyter, L. L., Kozak, A. S., and Williams, E. E. (1968). *J. Nutr.* **94,** 193–202.
Oltjen, R. R., Williams, E. E., Slyter, L. L., and Richardson, G. V. (1969). *J. Anim. Sci.* **29,** 816–822.
Oltjen, R. R., Burns, W. C., and Ammerman, C. B. (1974). *J. Anim. Sci.* **38,** 975–983.
Orskov, E. R., Reid, G. W., Holland, S. M., Tait, C. A. G., and Lee, N. H. (1983). *Anim. Feed Sci. Technol.* **8,** 247–257.
O'Shea, J., Lawlor, M. J., and Hopkins, J. P. (1981). *Irish J. Agric. Res.* **20,** 101–103.
Otchere, E. O., Dadzie, C. B. M., Erbynn, K. G., and Ayebo, D. A. (1977). *Ghana J. Agric. Sci.* **10,** 61–66.
Otchere, E. O., Musah, I. A., and Bafi-Yeboa, M. (1983). *Trop. Anim. Prod.* **8,** 33–38.
Owens, F. N., and Bergen, W. G. (1983). *J. Anim. Sci.* **57,** (Suppl. 2), 498–518.
Oyenuga, V. A. (1955). *Emp. J. Exp. Agric.* **23,** 81–95.
Pate, F. M. (1979a). *Fla. Beef Cattle Res. Rep.* pp. 42–45. Univ. of Florida, Gainesville.
Pate, F. M. (1979b). *Fla. Beef Cattle Res. Rep.* pp. 46–50. Univ. of Florida, Gainesville.
Paterson, J. A., Klopfenstein, T. J., and Britton, R. A. (1982). *J. Anim. Sci.* **54,** 1056–1066.
Playne, M. J., and Haydock, K. P. (1972). *Aust. J. Exp. Agric. Anim. Husb.* **12,** 365–372.
Preston, T. R. (1977). *Trop. Anim. Prod.* **2,** 125–142.
Preston, T. R. (1982). *J. Anim. Sci.* **54,** 877–884.
Preston, T. R., and Leng, R. A. (1978). *World Anim. Rev.* **27,** 7–12.
Preston, T. R., and Willis, M. B. (1969). *Outlook Agric.* **6,** 29–35.
Preston, T. R., and Willis, M. B. (1974). "Intensive Beef Production," 2nd ed. Pergamon, Oxford.
Preston, T. R., Carcaño, C., Alvarez, F. J., and Gutierres, D. G. (1976). *Trop. Anim. Prod.* **1,** 150–161.
Quinn, L. R., Mott, G. O., and Bisschoff, W. V. A. (1961). *IBEC Res. Inst. (Bull.)* **24,** 1–35.
Quinn, L. R., Mott, G. O., Bisschoff, W. V. A., and da Rocha, G. L. (1963). *IBEC Res. Inst. (Bull.)* **28,** 1–36.
Randel, P. F. (1970). *J. Dairy Sci.* **53,** 1722–1726.
Ravindran, V., Kornegay, E. T., and Cherry, J. A. (1983). *Nutr. Rep. Int.* **28,** 189–196.
Rodrigues, V., Gonzalez, R., and Guzman, J. (1978). *Trop. Anim. Prod.* **3,** 277 (abstr.).
Roe, R., Jones, R. M., and Rees, M. C. (1970). *Div. Trop. Pastures Tech. Pap. (Aust. C.S.I.R.O.),* 45–54.
Ruiz, M. E., Vohnout, K., Isidor, M., and Jimenez, C. (1974). *Mem. Asoc. Latinoam. Prod. Anim.* **9,** 124 (abstr.).
Saenger, P. F., Lemenager, R. P., and Hendrix, K. S. (1982). *J. Anim. Sci.* **54,** 419–425.
Sansoucy, R., Nielsen, S. A., Delaitre, C., and Preston, T. R. (1973). Cited by Preston, T. R., and Willis, M. B. (1974). "Intensive Beef Production," 2nd ed. Pergamon, Oxford.
Shirley, B. (1982). *Beef Mag.* **19,** 14.
Shultz, T. A., Chicco, C. F., Shultz, E., and Carnevali, A. A. (1970). *Agron. Trop. (Maracay, Venez.)* **20,** 185–194.
Siebert, B. D., and Kennedy, P. M. (1972). *Aust. J. Agric. Res.* **23,** 35–44.
Silvestre, R., MacLeod, N. A., and Preston, T. R. (1977a). *Trop. Anim. Prod.* **2,** 81–89.
Silvestre, R., MacLeod, N. A., and Preston, T. R. (1977b). *Trop. Anim. Prod.* **2,** 151–157.
Smith, C. A. (1961). *J. Agric. Sci.* **57,** 311–317.
Smith, E. F. (1981). *Kans. Agric. Exp. Stn. Tech. Bull.* **638,** 1–16.
Smith, G. S., Dunbar, R. S., McLaren, G. A., Anderson, G. C., and Welch, J. A. (1960). *J. Nutr.* **71,** 20–26.
Smith, T., Grigera-Naon, J. J., Broster, W. H., and Siviter, J. W. (1984). *Anim. Feed Sci. Technol.* **10,** 189–197.

Steen, R. W. J. (1984). *Grass Forage Sci.* **39**, 35–41.
Stephenson, R. G. A., Edwards, J. C., and Hopkins, P. S. (1981). *Aust. J. Agric. Res.* **32**, 497–509.
Stock, R., Merchen, N., Klopfenstein, T., and Poos, M. (1981). *J. Anim. Sci.* **53**, 1109–1119.
Swingle, R. S., Araiza, A., and Urias, A. R. (1977). *J. Anim. Sci.* **45**, 1435–1441.
Thomas, C., Tetlow, R. M., Gibbs, B. G., and Gill, M. (1983). *Anim. Prod.* **37**, 195–202.
Tindall, B. (1983). *Feed Manage. Mag.* **34**, 38–43.
Tinnimit, P., Yu, Y., McGuffey, K., and Thomas, J. W. (1972). *J. Anim. Sci.* **35**, 431–435.
Umunna, N. N. (1982). *J. Agric. Sci.* **98**, 343–346.
Vargas, E., Cabezas, M. T., and Bressani, R. (1977a). *Agron. Costarric.* **1**, 51–56.
Vargas, E., Cabezas, M. T., and Bressani, R. (1977b). *Agron. Costarric.* **1**, 101–106.
Vargas, E., Cabezas, M. T., Murillo, B., Braham, J. E., and Bressani, R. (1982). *Arch. Latinoam. Nutr.* **32**, 973–989.
von la Chevallerie, M. K. S. L. (1965). *Proc. S. Afr. Soc. Anim. Prod.* **4**, 120.
Waldo, D. R. (1977). *J. Dairy Sci.* **60**, 306–326.
Waller, J. C. (1978). Ph.D. dissertation, Univ. of Nebraska, Lincoln.
Webb, K. E., Jr., Phillips, W. A., Libke, K. G., Harmon, B. W., and Fontenot, J. P. (1973). *J. Anim. Sci.* **36**, 218 (abstr.).
Weiss, W. P., Colenbrander, V. F., and Lechtenberg, V. L. (1982). *J. Dairy Sci.* **65**, 1212–1218.
Whitelaw, F. G., Preston, T. R., and MacLeod, N. A. (1963). *Anim. Prod.* **5**, 227–235.
Wieghart, M., Thomas, J. W., and Tesar, M. B. (1980). *J. Anim. Sci.* **51**, 1–9.
Wilkinson, J. M. (1983a). *World Anim. Rev.* **45**, 36–42.
Wilkinson, J. M. (1983b). *World Anim. Rev.* **46**, 35–40.
Wiltbank, J. N. (1983). *Vet. Clin. North Am. Large Anim. Pract.* **5**, 41–57.
Wiltbank, J. N., Rowden, W. W., Ingalls, J. E., Gregory, K. E., and Koch, R. M. (1962). *J. Anim. Sci.* **21**, 219–225.
Winchester, C. F., and Howe, P. E. (1955). *U.S. Dep. Agric. Tech. Bull.* **1108**, pp. 1–34.
Winchester, C. F., Hiner, R. L., and Scarborough, V. C. (1957). *J. Anim. Sci.* **16**, 426–436.
Winks, L., O'Rourke, P. K., and McLennan, S. R. (1982). *Aust. J. Exp. Agric. Anim. Husb.* **22**, 252–257.
Winks, L., Walker, R. W., O'Rourke, P. K., Loxton, I. D., Holmes, A. E., and Shaw, K. A. (1983). *Trop. Grassl.* **17**, 64–76.
Wolf, W. J. (1983). *In* "Handbook of Nutritional Supplements" (M. Rechcigl, Jr., ed.), Vol. II, pp. 163–175. CRC Press, Boca Raton, Florida.
Young, R. H. (1983). *In* "Handbook of Nutritional Supplements" (M. Rechcigl, Jr., ed.), Vol. II, pp. 217–251. CRC Press, Boca Raton, Florida.

8

Contribution of Tropical Forages and Soil toward Meeting Mineral Requirements of Grazing Ruminants

L. R. McDOWELL
Department of Animal Science
University of Florida
Gainesville, Florida

I.	Introduction.	165
II.	Tropical Forages as Sources of Minerals.	166
	A. Biological Availability of Forage Minerals.	166
	B. Grazing Selectivity and Forage Mineral Intake.	170
	C. Factors Affecting Forage Mineral Content.	170
III.	Soils as Sources of Minerals	176
	A. Reasons for Deliberate Soil Consumption.	177
	B. Beneficial Effects of Soil Consumption.	178
	C. Detrimental Effects of Soil Consumption.	182
	References.	185

I. INTRODUCTION

According to most nutritionists, the principal factors limiting performance of grazing animals are the low protein content of grasses (Milford and Haydock, 1965), low-energy intake due to the high fiber content of forages (Moore and Mott, 1973), and mineral deficiencies or imbalances (McDowell, 1976). For unsupplemented grazing livestock, protein and energy are supplied almost exclusively by pastures, while minerals are derived principally from forage, soil, and water. Although highly variable, all mineral elements essential as dietary nutrients occur to some extent in water. Water as a source of minerals for livestock is discussed in Chapter

3 (this volume), with the present chapter studying mineral supplies derived from forages and soil consumption.

II. TROPICAL FORAGES AS SOURCES OF MINERALS

Frequently, livestock producers in tropical countries do not supplement grazing livestock with minerals, with the possible exception of salt. Grazing livestock must, therefore, depend largely upon forages to supply their mineral requirements. However, only rarely can forages completely satisfy each of the mineral requirements for grazing animals. Table 8.1 summarizes the mineral concentrations of 2615 Latin American forages (McDowell et al., 1977). Borderline or deficient levels of certain elements were noted for many entries: cobalt (Co), 43%; copper (Cu), 47%; magnesium (Mg), 35%; phosphorus (P), 73%; sodium (Na), 60%; and zinc (Zn), 75%.

A number of publications from different tropical regions of the world have well documented the inadequacy of tropical forages in meeting the mineral requirements of grazing ruminants (French, 1957; McDowell, 1976; Conrad and McDowell, 1978; Underwood, 1981; Mtimuni, 1982). Table 8.2 illustrates forage mineral deficiencies for livestock in Malawi, while Table 8.3 evaluates forage mineral concentrations for the Dominican Republic, Bolivia, Colombia, and Guatemala. In these regions, the minerals most often deficient were P, Na, calcium (Ca), Cu, selenium (Se), and Zn. Potassium, iron (Fe), and manganese (Mn) were least likely to be deficient in these tropical forages. From the Malawi data (Mtimuni, 1982), forages were lower in potassium (K), P, Cu, and Se during the dry season when compared to the wet season.

A. Biological Availability of Forage Minerals

There is an extreme lack of data on the biological availability of minerals, particularly trace elements from forage sources. From temperate pastures (growing perennial ryegrass and white clover) fed fresh to sheep, apparent availability of Mg, Ca, P, Na, and K was found to be 31, 21, 17, 95, and 97%, respectively (Grace, 1972).

Perdomo et al. (1977) studied the apparent digestibility in sheep of eight minerals at three stages of regrowth for four tropical forages (Table 8.4). Digestibility and retention of Ca, P, Na, and Cu by sheep tended to decline with maturity, while availability of Mg and Fe tended to increase at 56 days of age for the four species. The percentage of digestibility

TABLE 8.1

Mineral Breakdown and Concentrations of 2615 Latin American Forages (dry basis)[a]

Element	Percentage of forages with entries (analyses of forages)[b]	No. of entries	Requirements of ruminents[c]	Percentage of total entries	Low level	Higher level
Calcium	42.9	1123	0.18–0.60%	Concentrations, %	0–0.30	Over 0.30
				Percentage of total	31.1	68.9
Cobalt	5.4	140	0.05–0.01 ppm	Concentrations, ppm	0–0.10	Over 0.10
				Percentage of total	43.1	56.9
Copper	9.0	236	4–10 ppm	Concentrations, ppm	0–10	Over 10
				Percentage of total	46.6	53.4
Iron	9.8	256	10–100 ppm	Concentrations, ppm	0–100	Over 100
				Percentage of total	24.1	75.9
Magnesium	11.1	290	0.04–0.18%	Concentrations, %	0–0.20	Over 0.20
				Percentage of total	35.2	64.8
Manganese	11.2	293	20–40 ppm	Concentrations, ppm	0–40	Over 40
				Percentage of total	21.0	79.0
Molybdenum	5.1	133	0.01 ppm or less	Concentrations, ppm	0–3	Over 3
				Percentage of total	86.4	13.6
Phosphorus	43.2	1129	0.18–0.43%	Concentrations, %	0–0.30	Over 0.30
				Percentage of total	72.8	27.2
Potassium	7.6	198	0.60–0.80%	Concentrations, %	0–0.80	Over 0.80
				Percentage of total	15.1	84.9
Sodium	5.6	146	0.10%	Concentrations, %	0–0.10	Over 0.10
				Percentage of total	59.5	40.5
Zinc	6.8	177	10–50 ppm	Concentrations, ppm	0–50	Over 50
				Percentage of total	74.6	22.4

[a] Latin American Tables of Feed Composition (McDowell et al., 1977).
[b] Less than 1% of the other minerals were included.
[c] McDowell et al., (1977).

TABLE 8.2

Forage Mineral Concentrations in Malawi for the Wet and Dry Seasons[a]

Mineral level	Critical level based on ruminant needs	Dry season forages[b]		Wet season forages[c]	
		Mean	% deficient for animal	Mean	% deficient for animal
Ca, %	0.3	0.63	12.5	0.25	81.0
K, %	0.8	0.98	56.7	1.98	3.2
Mg, %	0.2	0.27	31.3	0.17	76.7
Na, %	0.06	0.05	96.9	0.05	95.7
P, %	0.25	0.19	75.0	0.25	55.5
Fe, ppm	50	195	3.1	207	6.3
Co, ppm	0.1	0.41	12.5	0.29	13.6
Cu, ppm	8	3.1	90.6	11.8	46.8
Mn, ppm	20	245	3	98	6.3
Mo, ppm	—	0.08	—	0.61	—
Se, ppm	0.10	0.05	95.8	0.10	55.4
Zn, ppm	40	18.9	93.8	44.4	95.7

[a] Modified from Mtimuni (1982).
[b] Means based on 21 total forage samples from two regions.
[c] Means based on 48 total forage samples from three regions.

TABLE 8.3

Percentage of Forage Minerals below Critical Concentrations Needed by Ruminants in Samples Collected during the Dry Season in the Dominican Republic, Bolivia, Colombia, and Guatemala

Mineral level	Critical level based on ruminant needs[a]	Dominican Republic[b]	Bolivia[c]	Colombia[d]	Guatemala[e]
Ca, %	0.30	24	57	100	71
P, %	0.25	83	100	92	57
K, %	0.60–0.80	0	1	15	13
Na, %	0.06	78	100	100	88
Mg, %	0.20	33	64	56	76
Fe, ppm	30	0	0	0	0
Zn, ppm	30	86	81	74	49
Cu, ppm	10	64	100	100	92
Mn, ppm	30–40	10	0	0	24
Co, ppm	0.10	26	48	31	1
Mo, ppm	>6	0	0	0	0
Se, ppm	0.10	48	47	74	49

[a] Critical levels (McDowell and Conrad, 1977).
[b] Based on 69 samples (Jerez et al., 1984).
[c] Based on 84 samples (Peducassé et al., 1983).
[d] Based on 36 samples (Vargas et al., 1984).
[e] Based on 84 samples (Tejada, 1984).

TABLE 8.4

Percentage of Apparent Digestibility of Various Nutrient Elements in Four Forages Harvested at Three Ages of Regrowth and Fed Fresh to Sheep[a]

Forage	Element	Age of forage regrowth, days (% digestibility of forage mineral elements)			Mean ± standard error
		28	42	56	
Pangola	Ca	10.9[c]	−40.0[b]	23.3[c]	−1.92 ± 29.4[f]
	P	43.0[c]	11.6[b]	41.4[c]	32.00 ± 10.8[e,f]
	Mg	41.4[b]	32.3[b]	66.3[c]	46.70 ± 16.4[f]
	K	80.6[b]	80.9[b]	88.6[c]	83.40 ± 6.0[e]
	Na	75.0	68.4	79.8	74.10 ± 7.2[f]
	Fe	40.6[b]	47.0[b,c]	65.6[c]	50.90 ± 13.3[g]
	Cu	71.9[c]	40.9[b]	64.1[c]	58.90 ± 14.2[e]
	Zn	27.6[b]	37.3[b]	74.5[c]	46.50 ± 22.0[f]
Guineagrass	Ca	62.6[d]	8.87[b]	37.0[c]	36.20 ± 23.0[g]
	P	62.4[c]	20.9[b]	21.9[b]	35.00 ± 22.3[e,f]
	Mg	25.4[b]	44.5[c]	56.5[c]	42.20 ± 15.0[f]
	K	88.8	87.2	89.2	88.40 ± 2.5[f]
	Na	69.6	58.5	63.9	64.00 ± 10.1[e]
	Fe	−29.0[b]	−12.9[b]	44.0[c]	0.93 ± 39.9[e]
	Cu	69.2[c]	51.0[b]	51.1[b]	57.10 ± 10.4[e]
	Zn	36.1[b,c]	23.8[b]	50.9[c]	36.90 ± 18.6[e,f]
Stargrass	Ca	58.9[c]	20.0[b]	33.1[c]	37.30 ± 18.2[g]
	P	63.1[c]	20.8[b]	26.6[b]	36.90 ± 22.1[f]
	Mg	29.9[b]	54.3[c]	57.5[c]	47.20 ± 16.1[f]
	K	90.1[c]	88.6[c]	18.6[b]	87.40 ± 3.9[f]
	Na	62.6	70.0	63.1	65.20 ± 9.1[e]
	Fe	7.63[b]	30.3[c]	51.1[c]	29.70 ± 21.2[f]
	Cu	60.2[c]	60.7[c]	48.6[b]	59.80 ± 13.3[e,f]
	Zn	50.5	34.8	49.6	45.00 ± 14.3[e,f]
Sheepgrass	Ca	26.6[c]	−40.6[b]	12.7[c]	−0.44 ± 43.2[f]
	P	35.5	21.5	22.1	26.40 ± 15.5[e]
	Mg	62.2[c]	45.6[b]	64.5[c]	57.40 ± 13.4[g]
	K	85.6	82.9	88.0	87.90 ± 3.9[f]
	Na	66.7[c]	82.9[c]	24.8[b]	58.10 ± 32.0[e]
	Fe	34.3[c]	34.7[c]	25.3[b]	31.40 ± 19.0[f]
	Cu	73.7[c]	70.8[c]	52.7[b]	65.70 ± 16.6[f]
	Zn	55.1[c]	26.4[b]	18.2[b]	33.20 ± 30.1[e]

[a] Perdomo et al. (1977).
[b-g] Different superscript letters in a row indicate that variations in values are significant for effect of age of regrowth ($P < 0.05$); and different superscript letters in a column for mean ± standard error values for specific elements indicate that variations are significant ($P < 0.05$).

ranged for each mineral as follows: Ca (12.7–37.0), P (21.9–41.4), Mg (56.5–64.5), K (18.6–89.2), Na (24.8–79.8), Fe (25.3–65.6), Cu (48.6–64.0), and Zn (18.2–74.5). The data of Perdomo and co-workers (1977) clearly illustrate the wide variation in apparent digestibility among the various minerals, species, and stage of maturity.

B. Grazing Selectivity and Forage Mineral Intake

Both amount of mineral in forages and biological availability of minerals need to be considered in formulating rations. Likewise, selective grazing patterns of livestock in relation to choosing not only particular species of forage but parts of the plant need consideration. It is frequently suggested that selective grazing by the animal can provide a different pattern of mineral element intake than that indicated by analyses of herbage sampled by hand. Often this argument is presented with an inference that there is consequently an even lower likelihood of inadequate intakes of trace elements in cases of marginal herbage content (Egan, 1975). However, selective grazing does not necessarily result in ingestion by the animal of a selected herbage mixture providing higher content of an otherwise inadequately supplied mineral element.

The principles of dietary selection are not clearly understood. Nevertheless, there is recorded evidence that grazing animals tend to select higher protein and more highly digestible fractions from the pasture available (Arnold, 1966). In condition of reduced availability of mineral elements from the soil, the younger leaves of growing plants, which constitute a highly digestible fraction, are often the leaves of lowest content of most trace elements, which are poorly or only slowly mobilized within the plant (Lonergan, 1975). Kincaid and Conrath (1983) reported that significant portions of Ca, P, Fe, Zn, and Cu are associated with the fibrous versus the leaf portion of alfalfa hay and silage. Of eight minerals studied, P, Na, K, Mg, and Zn were all in higher concentration in stems than in leaves of dwarf *Pennisetum purpureum* (Montalvo et al., 1983)

C. Factors Affecting Forage Mineral Content

Wide ranges of mineral concentrations in tropical forages have been observed, including reports by Kayongo-Male et al. (1974), McDowell et al. (1977), and Perdomo et al. (1977). The concentration of mineral elements in plants from diverse world regions are dependent upon the interaction of a number of factors including soil, plant species, stages of maturity, yield, pasture management, and climate.

1. Soils and Fertilization

Soil is the source of all mineral elements found in plants. Most naturally occurring mineral deficiencies in livestock are associated with specific regions and are directly related to both soil mineral concentration and soil characteristics. Of the total mineral concentration in soils, only a fraction is taken up by plants. The "availability" of minerals in soils depends upon their effective concentration in soil solution (Reid and Horvath, 1980). The mineral content of soils depends not only on parent material but on a complex of pedogenic factors, which include podzolization, laterization, calcification, and salinization. Within the resulting soil profile, translocation further occurs by processes of surface erosion, leaching, evaporation with transport to the soil surface, extraction by plant roots and redeposition of minerals on the surface, and redistribution to the upper horizons of the soil (Beeson and Matrone, 1976).

Young and alkaline geological formations are more abundant in most trace elements than are the older, more acid, coarse, sandy formations (Hartmans, 1970). In tropical regions under conditions of heavy rainfall and high temperature, there has been marked leaching and weathering of soils, making them deficient in plant minerals (Pfander, 1971).

Soils derived from basic igneous rocks usually have higher readily soluble contents of the important trace elements than do corresponding soils from acid igneous or metamorphic parent materials. The amounts of readily soluble trace elements are generally smaller in coarse-textured than fine-textured soils. For instance, the amount of Co extracted by 2.5% acetic acid from a granite soil containing 8% clay was 0.14 ppm compared with 0.39 ppm in a similar granite soil containing 24% clay (Reith, 1973.) Aubert and Pinta (1977) concluded that the distribution of trace elements generally relates to the humus and clay content of soils, with an accumulation in upper horizons with high organic matter content.

The fertilization of grasses and legumes is a practice capable of increasing the mineral level of these plants. However, there are frequently cases in which the plants do not respond to fertilization; i.e., the mineral composition is not increased and, at times, the inverse reaction occurs. As an example, the results obtained for *Panicum maximum,* fertilized or nonfertilized, respectively, cut at 28 days of age were as follows: 2.40 and 2.38% nitrogen (N), 0.14 and 0.15% P, 2.33 and 2.41% K, 0.34 and 0.45% Ca, 0.23 and 0.29% Mg, and 34 and 32 ppm Zn (Gomide, 1978). One frequent response to N fertilization of pastures is a change in botanical composition, with a loss of the legume component and a consequent decline in total herbage concentration of elements such as Ca and Mg. In general, N fertilization appears to increase the concentration of P and K in the plant when these elements are in adequate supply in the soil and to decrease uptake when soil reserves are low (Whitehead, 1970). Potassium fertilization, with or without N, has been shown to bring about marked changes in the uptake of other ions by plants, with possible consequences to the health of animals. Numerous studies (Reid and Hovath, 1980) have demonstrated a depression of Ca, Mg, and Na concentration in herbage with the application of K fertilizers.

The effects of P fertilization on plant mineral uptake are dependent to a considerable degree on the properties of the soil. Many soils fix phosphate, so that only a small proportion of the fertilizer is available for crop production, this is particularly the case in tropical areas (Reid and Horvath, 1980). Whitehead (1966) concluded that fertilization often had little effect on the P concentration of herbage, unless the soil was severely deficient. With infertile soils, P is frequently the major factor limiting pasture growth, and under these conditions, the regular use of fertilizer may not only increase yield substantially but also double the concentration of P in herbage (Reith, 1973). Similar responses have been obtained with tropical forages in Australian studies (Andrew and Robins, 1969). From pastures grown on soils from Florida, Kirk et al. (1970) reported increases in the P content of pangolagrass (*Digitaria decumbens*) from 0.07–0.1% to 0.25–0.30% due to P fertilization.

Application of commercial N : P : K fertilizers to soils has generally had no consistent effect on plant uptake of trace elements (Reid and Horvath, 1980). Trace-element fertilization of soils is, however, frequently practiced as a measure of control of specific element deficiencies, and as indicated by Allaway (1968), the effectiveness of control varies greatly with the soil and the particular element involved. Certain impurities in commercial N : P : K fertilizers (e.g., Zn in superphosphate) do provide significant quantities of specific minerals (Ozanne et al., 1965). Likewise, a superphosphate fertilizer may contain 12% sulfur (S), while only insignificant quantities of this mineral are available from the more refined triple superphosphate. If purified fertilizers are used, these sources of other minerals are lost.

2. Soil Drainage and pH

The soil content of an element would seem the most important limitation. However, availability factors including soil pH, texture, moisture content, and organic matter are probably more often the limiting factors rather than soil mineral content (Williams, 1963).

The amount of most trace elements in herbage grown on freely drained soils is normally lower than on corresponding poorly drained soils (Mitchell et al., 1957; Swift, 1972). Poor drainage conditions increase extractable trace elements, thereby resulting in a corresponding increase in plant uptake. By increasing soil water content and/or reducing soil aeration, the availability of Fe, Mn, Co, and molybdenum (Mo) can be increased, whereas the availability of Se would be reduced (Reuter, 1975).

As soil pH increases, the availability and uptake of Fe, Mn, Zn, Cu, and Co decrease, whereas Mo and Se concentrations increase (Pfander, 1971; Williams, 1963; Latteur, 1962; Miller *et al.*, 1972). Forage grown on a 15-ppm Co soil with a neutral or slightly acid pH may contain more Co than that from a 40-ppm highly alkaline soil (Latteur, 1962). Williams (1963) considers that the availability of Mn is so strongly affected by soil pH that its amount in herbage bears no relation to its level in the soil. Liming a soil to raise the pH from 5.5 to 6.5 can reduce Mn uptake by one-half. On the contrary, a rise in pH from 6.0 to 6.5 can raise forage Mo content two- or threefold.

3. Forage Species and Varieties

Although mineral content of a soil ultimately depends on the parent rock from which the soil was derived, evidence indicates little relationship between soil chemistry and mineral composition of farm crops and vegetation growing on that soil (Hemphill, 1977). Consequently, mineral intake by animals depends more on the type of plant and level of consumption than on the parent rock from which the soil was derived and on which plants were grown.

Large variations in mineral content of different plant species growing on the same soil have been reported (Thompson, 1957; Underwood, 1981; Gomide *et al.*, 1969). Kayongo-Male *et al.* (1974) demonstrated a wide range of mineral concentrations in tropical forages grown in Puerto Rico. As an example, large variations in concentrations were found for Mg (0.11–0.78%) and Cu (13–97 ppm) in the various plant genera. *Brachiaria decumbens* nearly doubled its P content when soil P was doubled, but P content of other pasture species such as *Sporobulus* did not vary widely (Long *et al.*, 1969).

It is a generally accepted view that herbs and legumes are richer in a number of mineral elements than are grasses (Fleming, 1973). The concentrations of Cu, Zn, Mo, and Co are generally higher in legumes than in grasses, whereas the reverse is true for Se and, with the exception of lupins, for Mn. Various herbs such as capeweed (*Arctotheca calendula Levyns*) may contain higher concentrations of certain trace elements than do legumes, perhaps due to their greater depth of rooting (Reuter, 1975).

Butler and Johnson (1957) concluded that species was more important in determining forage iodine (I) content than either soil or season. From Rhodesia, Jones (1964) reported one variety of *Chloris gayana* to contain

300 ppm Na, while a second variety grown on the same site had a vastly different Na level of 3100 ppm. For most minerals, "accumulator" plants exist that contain extremely high levels of a specific mineral (Schütte, 1964). As an example, in certain regions accumulator species containing over 1000 ppm Se are found growing beside grasses containing less than 10 ppm (Beeson and Matrone, 1976). *Pandiaka methalorum,* in Zaire, analyzed 6260 ppm Co, whereas *Crassula vaginata* and *Euphorbia* species growing on the same soil analyzed 1190 ppm and 30 ppm Co, respectively. (Malaisse *et al.,* 1979)

4. Forage Maturity

Variations in mineral content of tropical grasses have been reported to occur with increasing age of the plant. Gomide (1978) studied the N, K, P, Ca, Mg, Mn, and Zn content of five tropical grasses of Brazil at different stages of maturity (Table 8.5). It was concluded that as forages neared maturity, mineral content declined for most minerals, and that K and P

TABLE 8.5

Variation of Mineral Composition with Forage Age[a]

Grass	Forage age (days)	Composition of forage dry matter						
		N (%)	P (%)	K (%)	Ca (%)	Mg (%)	Zn (ppm)	Mn (ppm)
Guineagrass	14	2.96	0.18	2.24	0.41	0.28	38	—
(Panicum maximum)	28	2.40	0.14	2.33	0.34	0.23	34	—
	42	1.81	0.13	2.80	0.34	0.20	36	—
	56	1.55	0.10	2.64	0.34	0.17	32	—
	70	1.26	0.08	2.53	0.31	0.14	32	—
Molassesgrass	14	2.88	0.20	2.3	0.28	0.25	116	178
(Melinis minutiflora)	28	2.54	0.18	2.3	0.27	0.25	135	106
	42	2.18	0.18	2.2	0.27	0.25	71	73
	56	1.47	0.06	2.0	0.20	0.20	73	62
	70	1.34	0.05	1.7	0.20	0.18	97	136
Elephantgrass	28	—	0.33	2.38	0.61	0.42	40	138
(Pennisetum purpureum)	84	—	0.15	1.20	0.38	0.28	28	111
	140	—	0.11	0.34	0.43	0.36	33	128
Pangolagrass	28	—	0.16	1.32	0.56	0.39	35	192
(Digitaria decumbens)	84	—	0.11	0.74	0.50	0.38	22	188
	140	—	0.12	0.37	0.66	0.39	31	317
Jaraguagrass	28	—	0.28	1.68	0.40	0.46	51	—
(Hyparrhenia rufa)	56	—	0.17	0.63	0.20	0.36	30	—
	84	—	0.11	0.57	0.23	0.58	37	—

[a] Modified from Gomide (1978).

could be deficient and Zn and protein could be borderline, for grazing cattle. In most circumstances, P, K, Mg, Na, chlorine (Cl), Cu, Co, Fe, Zn, and Mo decline as the plant matures (Underwood, 1981; Gomide *et al.*, 1969). Forage Ca concentration is less affected by advancing maturity (Gomide *et al.*, 1969; Gomide, 1978), thereby resulting in a detrimental widening of the ratio of this mineral with other elements (i.e., a wide Ca:P ratio).

As plants mature, mineral contents decline due to a natural dilution process (Fleming, 1973) and translocation of nutrients to the root system (Tergas and Blue, 1971). Fleming (1973) concluded that as photosynthetic areas increase, dry matter (DM) production outstrips mineral uptake, resulting in a decline in mineral concentration. The author states that this is generally true, but there are some exceptions. The process of translocation of nutrients is a more serious problem for tropical versus temperate regions since freezing conditions in temperate areas will stop translocation while the movement of nutrients to the root system is continuous throughout the dry season in tropical regions.

5. Pasture Management, Forage Yield, and Climate

Pasture management, forage yield, and climate influence the species of forage predominating and also change the leaf–stem ratio radically, thereby having a direct bearing on the mineral content of the sward. From India, Whyte (1962) observed that good perennial grasses are "grazed into the ground" late in the dry season, resulting in disastrous effects on their powers of regeneration and replacement with tougher species of inferior quality. From Africa, uncontrolled, heavy grazing pressure is also causing many palatable genuses such as *Brachiaria* and *Panicum* to disappear and be replaced by highly lignified *Sporobulus* species. Edorma (1981) reported increased dominance of *Sporobulus* species when native grasses were frequently clipped to emulate heavy-grazing pressure.

Increased crop yields remove trace elements from the soil at a fast rate, with the result that deficiencies are frequently found on the most progressive farms (Shütte, 1946). Climate also exerts an effect upon the mineral composition of forages. Reith (1965) notes that maximum content of major nutrients is obtained at soil temperatures that produce maximum growth, and low temperatures generally produce crops containing lower percentages of N, P, K, Ca, and Mg. A clearly demonstrated interaction between temperature and mineral uptake appears for P, which is absorbed more slowly at low temperatures, possibly by mechanisms of depressed root extension and membrane permeability (Nye and Tinker, 1977).

Climate limits the potential yield and the magnitude of yield response to trace element additions. Where climate varies markedly between seasons, large between-seasons differences in nutrient uptake are observed. Under the abundant rainfall and high temperatures of the wet tropics and semitropics, soils are often low in soluble minerals (Allman and Hamilton, 1949). Grass tetany predictably occurs primarily in the cool season (T' Hart, 1960). Also, the prevalence of Zn deficiency in cool, wet seasons has been associated with decreased solubility of soil Zn (Reuter, 1975).

Pasture management involving liming and fertilization can be extremely beneficial to increasing forage mineral concentrations. Overliming, however, can accentuate a Se or Mo toxicity in livestock by increasing plant concentrations of these elements and, at the same time, favor Co and Mn deficiencies due to lowered plant uptakes. The direct effect of fertilizer N on trace element levels in herbage is small, but by causing changes in soil pH and the botanical composition of a sward, the indirect effect may be large (Reith, 1970). Overuse of N and K fertilizers increases the incidence of grass tetany (Kemp et al., 1961), and K fertilization also dramatically reduces forage Na content in plants (Underwood, 1981).

III. SOILS AS SOURCES OF MINERALS

Soils are usually considered to influence animal nutrition only through the quantity and quality of the herbage they produce. However, the fact that soil is ingested by grazing animals indicates that not only the usual sequence of soil–plant–animal relationships occurs, but also a direct soil–animal effect should be considered (Healy, 1972), Healy (1972) considered that the bulk of ingested soil was probably taken in accidentally along with herbage and not because of depraved appetite. Healy (1968, 1972) summarized the following factors as the most important affecting soil ingestion by cattle and sheep: (a) soil type, (b) stocking rate, (c) earthworm population, (d) management, (e) seasonal variation, and (f) individual animal difference.

Seasonal pattern of soil intake, the quantities of soil ingested, and its relationship to soil type are closely related to excessive wear of teeth. Low stocking rate and supplementary feed reduced levels of soils in feces by one-half and wear of teeth by two-thirds as compared with controls (Healy, 1972).

In humid climates such as New Zealand, soil ingestion was highest during winter and was directly related to closeness of grazing and muddying of pasture (Mascola et al., 1974). Intake was inversely related to soil structure, since structural properties of topsoils influence puddling, drainage, and earthworm casting. Under range conditions, soil ingestion appeared to be related primarily to forage availability, closeness of grazing, and shallow-rooted plants (Mayland et al., 1975). Most of the soil appeared to be ingested with roots consumed with above ground parts, with dust on leaves and stems accounting for only a small portion.

Ingestion of soil by grazing ruminants has been studied in many regions of the world. Under certain circumstances, soil ingestion by the grazing animal can constitute a substantial fraction of the diet (Field and Purves, 1964). Miller et al. (1978) reviewed literature concerning soil consumption by cattle from widely separated areas of the world and found a range of 90 g to over 1.4 kg/cow/day.

In Scotland, Field and Purves (1964) reported a variation in soil ingestion of approximately 10–200 g/day (0.4–14% of the daily DM intake) for sheep and emphasized that during a period of scarcity of herbage, the intake of soil increased dramatically and constituted a substantial fraction of the sheep's DM intake. In New Zealand, Healy (1967,1968,1972) found that under conditions with animals wintered outdoors, sheep ingested up to about 75 kg of soil annually and dairy animals up to about 600 kg. These amounts represented daily intakes of approximately 200 and 1600 g, respectively, for sheep and cattle.

In Southwest England, Thornton (1974) indicated that soil ingestion by grazing cattle over the winter months varied from 140 to 1480 g/day, with an average of 620 g/day. Expressed as a percentage of DM intake, these values ranged from 1 to 11%, with an average of 4.5%. Mayland et al. (1975) determined soil ingestion for cattle on semiarid range and found daily soil ingestion levels to range from 100 to 1500 g, with a median of 500 g soil.

Studying wild ruminants and fecal ash concentrations, Kreulen and Jager (1984) reported that white-tailed deer and bighorn sheep would have a soil intake of 5–12% of total DM and, in absolute terms, an animal of approximately 150 kg, (i.e., wildebeest) would be capable of eating 500 g of soil daily.

A. Reasons for Deliberate Soil Consumption

Ruminants on pasture may inadvertently consume surprisingly large quantities of soil as a natural consequence of grazing (Healy, 1972). Deliberate soil consumption (geophagia), on the other hand, is classified as a form of "pica," which is defined as animals chewing on objects and eating materials not considered to be natural feedstuff. (Fig. 8.1).

What is the cause of earth eating and what psychological or behavioral complexes induce livestock to start eating soil or chewing abnormal materials and then to continue with these habits? The simplest and most common explanation is to postulate that the animals are in need of some

Fig. 8.1. The contents of the rumen of a cow that had exhibited a "pica" condition of eating soil, stones, and other objects. A large number of stones ($\simeq$ 19 kg) were found in the rumen of this animal in La Libertad, Chiapas, Mexico. (Courtesy of Carlos García Bojalil, I.T.E.S.M., Querétaro, Qro., Mexico.)

nutrient not available in sufficient quantities in their diet and that they eat the abnormal substances to correct these shortages (French, 1955). Additional reasons for soil eating could include a possible indication of "stress" or "psychological maladjustment" or merely the result of a "habit" of long standing (Ammerman et al., 1965).

Deliberate soil intake, in contrast to inadvertant soil intake with contaminated forage, is normally selective. It usually concerns exploitation of special local deposits, and sometimes even particular subsoil horizons. These sites are variably termed "natural lick," "mineral lick," or "salt lick" (Kreulen and Jager, 1984). Depending on the prevalence of the desired material and on the intensity and period of use, some licks are just scrapes in the topsoil, whereas others could be classified as pits or craters. From the llanos region of Venezuela, Fig. 8.2 illustrates an animal consuming soil, and Fig. 8.3 shows a pit formed from soil consumption.

B. Beneficial Effects of Soil Consumption

Ingested soil could affect animal health and productivity in a number of ways. Possible beneficial effects include providing a source of essential

Fig. 8.2. Cattle consuming soil at Hato El Frío in the llanos state of Apure, Venezuela. Note holes in soil from previous consumption. Excessive soil consumption is associated with pronounced mineral deficiencies. (Courtesy of Eliecer Alberto Velasco, Hato El Frío, Apure State, Venezuela.)

elements, supplying grit that may act as a physical conditioner, improving utilization of energy, and increasing availability of certain minerals (Miller et al., 1978).

The importance of soil minerals as dietary sources depends on the amount of soil ingested, the ratio of the mineral concentration in soil to that in herbage, and the ability of the ruminant to solubilize and absorb the soil-derived minerals (Mayland et al., 1975). The magnitude of release, absorption, and retention of mineral elements from soil depends on soil properties (i.e., element content, absorption power, particle size, etc.), forage quality and the related composition of the digestive liquors (i.e., element content, pH, complexing agents, etc.), and on the nutritional state of the animal (Kreulen and Jager, 1984).

Healy (1972) measured mineral contents of sheep ruminal, duodenal and ileal fluids from different soil types and found that soluble mineral concentrations were increased for some elements and reduced for others by ingested soil. Miller et al. (1977) fed 450 and 900 g of red clay subsoil to

Fig. 8.3. Evidence of soil consumption in the llanos state of Apure, Venezuela.

Holstein cows and concluded that amounts of soil that are likely to be consumed under normal and sparse grazing would not have an important effect on the utilization of most microminerals. From this study apparent absorption of Mg, Ca, and P was not altered; however, K was reduced 12 and 25% by ingesting 450 and 900 g, respectively, over controls.

Grace and Healy (1974) reported the effects of daily addition of 100 g of two types of soil to diets of sheep on apparent absorption and retention of macro-elements. Soil addition reduced fecal loss of Mg and Ca and increased apparent availability of both elements. The authors suggested some physiochemical process imparted by the soil itself, since the increased apparent absorption and retention of Mg and Ca could not be attributed directly to the content of those elements in the soils.

Studies in new Zealand have demonstrated that soils can make important contributions to the mineral nutrition of grazing animals. Table 8.6 illustrates mineral consumption by sheep consuming clean pasture compared with the same pasture contaminated with approximately 14% soil on a DM basis (Healy, 1973). For this particular contaminated pasture, only insignificant quantities of K, P, and Mo would result from contamination, while beneficial dietary intakes of other elements, particularly Fe,

Mn, Co, Se, and I, would be suggested. Underwood (1981) noted that with elements such as Co and I, which occur in soils in concentrations usually much higher than those of the plants growing on them, soil ingestion can be beneficial to the animal.

Both short-term radioisotope studies and longer-term feeding trials have shown that soil can be a source of mineral elements to animals. In a review by Kreulen and Jager (1984), balance trials with rats, pigs, and sheep fed clay soils have revealed substantial increases in absorption and retention of Na, Ca, Mg, P, Co, Cu, I, Mn, Se, and Zn and also the prevention or cure of negative balances and deficiency signs. Work at the universities of Washington State and Florida showed a beneficial effect for growing pigs fed 5% soil, but did cause the feces to be hard (Cunha, 1977).

TABLE 8.6

Comparison of Element Intake by Sheep per Day from Clean and Contaminated Pasture[a,b]

	Element intake/day	
	Clean pasture[c]	Contaminated pasture[c]
Macroelements (g)		
Calcium	4.2	5.7
Magnesium	1.8	2.6
Potassium	18	19
Sodium	1.2	2.7
Phosphorus	3.0	3.7
Trace elements (mg)		
Iron	150	4150
Manganese	60	160
Zinc	18	24
Copper	4.2	6.2
Molybdenum	0.6	0.7
Cobalt	0.18	0.68
Selenium	0.06	0.21
Iodine	0.3	0.8

[a] Modified from Healy (1973).

[b] Pasture and soil consumptions are an assumed "average."

[c] A DM intake of 600 g of clean pasture is assumed. In case of contaminated pasture, the same DM intake is assumed plus 100 g of soil—approximately 14% soil contamination on a DM basis or less than 2% on a fresh weight basis. Soil contamination is most often from wind blowing soil or from rain splashing soil on the forage.

Rigg and Askew (1934) noted that as little as 10 g of soil twice weekly prevented Co deficiency in sheep. In another similar study, Andrews (1956) noted an increase in Co in the livers of sheep grazing short pastures and attributed this to ingested soil. More recently, MacPherson *et al.* (1978) observed that even Co-poor soil was able to maintain growth rates of sheep fed a Co-deficient diet when soil was ingested at a rate of 100 g/day.

In New Zealand, postmortem and clinical examination of lambs showed that there was a high prevalence of enlarged thyroids in lambs from ewes on low-stocked pasture areas. This condition, however, was almost nonexistent in lambs on high-stocked pasture areas. Since I in herbage from high- and low-stocked areas did not differ, higher soil ingestion, from the high-stocked areas appeared to be an important source of I and prevented the development of goiter in lambs (Healy *et al.*, 1972).

Thornton and Kinniburgh (1978) found in Central England that soil contributed a source of Zn (3–36%) and Cu (6–16%) for cattle. Mayland *et al.* (1975) likewise considered ingested soil to be a dietary Zn source, because they did not find visual Zn deficiency signs in cattle whose diet did not otherwise appear to contain adequate Zn. Healy and Wilson (1971) reported that soil fed at a rate of 50 g/day resulted in significant blood Se increase by approximately 50% at the end of 6 weeks. The total Se contents of the two soils used ranged from 1.2 to 1.8 ppm.

C. Detrimental Effects of Soil Consumption

Among the potentially harmful effects of excess soil consumption is sand impaction in underfed livestock (Hunter, 1975). This causes abrasive action that accelerates tooth wear and erosion of digestive tract epithelium, decreases availability of certain dietary elements, and provides a source of undesirable materials, such as pesticide residues and heavy metals (Miller *et al.*, 1978).

In sheep, the most obvious effect of soil ingestion in large quantities is wear on incisor teeth. Healy (1967) cited studies directed at establishing the cause or causes of wear on sheep incisor teeth and noted the importance of soil type on the quantity of soil ingested, and hence wear on teeth. With excessive wear, ewes are reduced to "gummies" at a comparatively early age and are unable to graze in competition with younger animals (Healy, 1972).

The severity of worn teeth for cattle consuming excessive soil in the llanos region of Venezuela is illustrated in Fig. 8.4. Animals with severely worn teeth experience difficulty grazing and, when slaughtered, are discriminated against and mistakely considered much older and, consequently, net a smaller sales price. Table 8.7 illustrates the incidence of

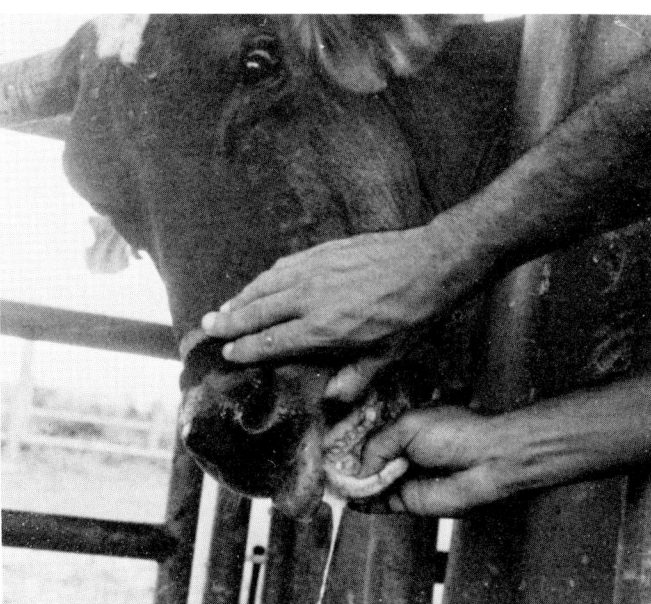

Fig. 8.4. Cattle from llanos of Venezuela that have been consuming large quantities of soil, resulting in abnormal wear on teeth. Teeth of the animal at top would be classified as moderate to severe, while the teeth of the animal at bottom are severely worn and close to the gum line. (Bottom photo courtesy of Eliecer Alberto Velasco, Hato El Frío, Apure State, Venezuela.)

TABLE 8.7

Dental Condition of Beef Cattle Culled Due to Poor Production in the Venezuelan Llanos[a,b]

Ranch name[b]	No. of animals	Teeth wear classification (% of total teeth)[c]			
		Good	Typical	Severe	Very severe
La Apontera	41	34.1	19.5	14.6	31.7
El Hato	62	30.6	27.4	12.9	29.0
Chupadero	166	23.5	31.3	15.1	30.1
El Hato	94	47.9	21.3	6.4	24.5
Chupadero	62	27.4	17.7	24.2	30.6
Total	425	31.5	25.4	14.1	28.9

[a] E. Velasco (personal communication).
[b] Different animals from the ranches El Hato and Chupadero were evaluated at two different times during the year.
[c] All teeth were worn to some degree; however, classification of severe and very severe would be worn to a degree where forage consumption would be affected.

worn teeth on ranches in the llanos region of Venezuela. A total of 43.2% of the culled cattle had tooth wear classified as severe and very severe (E. Velasco, 1982, unpublished data). When teeth were examined from the general cow herds (695 animals) of three ranches, an average of 29.5% of the cows were classified as having severe or very severe tooth wear.

Excess of certain minerals obtained from soil consumption may be antagonistic to required elements that are borderline to deficient in grazing ruminant diets. Among the minerals present in high quantities in tropical acid soils are Al and Fe, with recognized interference of P utilization by animals (Standish et al., 1969; Standish and Ammerman, 1971; Valdivia et al., 1982; Garcia-Bojalil, 1984). Iron is likewise antagonistic to Se, forming complexes less available to plants under acid conditions.

Rosa (1980) studied the effect of soil ingestion in sheep and reported that inclusion of 10% Costa Rican soil decreased body weight, increased unabsorbed P, and decreased apparent and true P absorption. Valdivia et al. (1982) investigated the effect of the addition of 10% sand and two types of soil of different composition to the diet of sheep. The authors suggested that soils, through their mineral content, especially Al and Fe, reduced dietary P utilization by lambs.

Langlands et al. (1982) studied the effect of stocking rate and soil ingestion on Cu and Se status of grazing animals. It was concluded that both Se and Cu concentrations in the tissues of grazing sheep may sometimes be reduced when stocking rate is increased, and that the reduction may be sufficient to induce clinical signs of deficiency when the elements are marginally deficient. Suttle *et al.* (1975) concluded that the Cu antagonists Mo and Zn are biologically available in soils, and their ingestion from soil contamination of herbage may be a factor in the etiology of hypocuprosis in cattle and "swayback" (Cu deficiency) in sheep.

Soil ingestion, in relation to consuming insecticides and toxic metals, needs to be considered. The fact that insecticides can remain concentrated in the upper inch of soil and that sheep can ingest soil suggests potential insecticide intake (Healy, 1967). Thornton (1974) reported that areas in southwestern England have soils that contain up to 2500 ppm arsenic (As), 1000 ppm tin (Sn), 1500 ppm Cu, 1700 ppm lead (Pb), and 1200 ppm Zn. According to data on soil ingestion by grazing cattle in that region, the author suggested that animals may ingest up to 10 times the amount of Cu, Pb, and As in the form of soil to that in herbage. Calculations based on soil, plant, and fecal analyses show that from 9 to 80% of the Pb and 34 to 90% of the As intake in cattle on contaminated land is due to ingested soil (Thornton and Abrahams, 1983).

In conclusion, soil ingestion could affect animal health by (a) providing a source of essential elements; (b) acting as a physical conditioner, improving performance from various rations; (c) affecting availability of dietary elements in the digestive tract; (d) causing sand impaction in underfed cattle; (e) having an abrasive effect causing erosion of ruminal epithelium and tooth wear; and (f) supplying undesirable materials, such as pesticide residues or heavy metals that may be absorbed in surface soil particles.

REFERENCES

Allaway, W. H. (1968). *Adv. Agron.* **20**, 235–274.
Allman, R. T., and Hamilton, T. S. (1949). "Nutritional Deficiencies in Livestock." FAO Agricultural Studies No. 5., Washington, D.C.
Ammerman, C. B., Loggins, P. E., and Wing, J. M. (1965). *Feedstuffs* **37**(9), 26–27.
Andrew, C. S., and Robins, M. F. (1969). *Aust. J. Agric. Res.* **20**, 665–674.
Andrews, E. D. (1956). *N. Z. J. Agric.* **92**, 239–244.
Arnold, G. E., McManus, W. R., and Bush, I. G. (1966). *Aust. J. Exp. Agric. Anim. Husb.* **6**, 101.
Aubert, H., and Pinta, M. (1977). "Trace Elements in Soils." Elsevier, Amsterdam.
Beeson, K. C., and Matrone, G. (1976). "The Soil Factor in Nutrition: Animal and Human." Dekker, New York.
Butler, G. W., and Johnson, J. M. (1957). *Nature (London)* **179**, 216–217.
Conrad, J. H., and McDowell, L. R. (1978). *In* "Proceedings Latin American Symposium on

Mineral Nutrition Research with Grazing Ruminants" (J. H. Conrad and L. R. McDowell, eds.) Univ. of Florida Press, Gainesville.
Cunha, T. J. (1977). "Swine Feeding and Nutrition," p. 80. Academic Press, New York.
Edroma, E. L. (1981). *Afr. J. Ecol.* **19**, 313–326.
Egan, A. R. (1975). *In* "Trace Elements in Soil-Plant-Animal Systems." (D. J. Nicholas and A. R. Egan, eds.), pp. 371–384. Academic Press, New York.
Field, A. C., and Purves, D. (1964). *Proc. Nutr. Soc.* **23**, 24–25.
Fleming, G. A. (1973). *In* "Chemistry and Biochemistry of Herbage" (G. W. Butler and R. W. Bailey, eds.), pp. 529–563. Academic Press, London.
French, M. H. (1955). *East Afr. Agric. J.* **20**, 168–175.
French, M. H. (1957). *Herb. Abstr.* **27**, 1–9.
Garcia-Bojalil, C. (1984). M.S. thesis, Univ. of Florida, Gainesville.
Gomide, J. A. (1978). *In* "Proceedings Latin American Symposium on Mineral Nutrition Research with Grazing Ruminants" (J. H. Conrad and L. R. McDowell, eds.), pp. 32–40. Univ. of Florida, Gainesville.
Gomide, J. A., Noller, C. H., Mott, G. O., Conrad, J. H., and Hill, D. L. (1969). *Agron. J.* **61**, 120–123.
Grace, N. D. (1972). *Proc. N. Z. Soc. Anim. Prod.* **32**, 77–84.
Grace, N. D., and Healey, W. B. (1974). *N. Z. J. Agric. Res.* **17**, 73–78.
Hartmans, J. (1970). *Trace Elem. Metab. Anim. Proc. WAAP/IBP Int. Symp. 1969*, pp. 441–445.
Healy, W. B. (1967). *Proc. N. Z. Soc. Anim. Prod.* **27**, 109–120.
Healy, W. B. (1968). *N. Z. J. Agric. Res.* **11**, 487–499.
Healy, W. B. (1972). *N. Z. J. Agric. Res.* **15**, 289–305.
Healy, W. B. (1973). *In* "Chemistry and Biochemistry of Herbage" (G. W. Butler and R. W. Bailey, eds.), pp. 567–588. Academic Press, New York.
Healy, W. B., and Wilson, G. F. (1971). *J. Agric. Res.* **14**, 122–131.
Healy, W. B., Crouchley, G., Gillett, R. L., Rankin, P. C., and Watts, H. M. (1972). *J. Agric. Res.* **15**, 778–782.
Hemphill, D. D. (1977). *In* "Geochemistry and the Environment," pp. 124–131. Natl. Acad. Sci., Washington, D. C.
Hunter, R. (1975). *J. Am. Vet. Med. Assoc.* **166**, 1179.
Jerez, M., McDowell, L. R., Martin, F. G., Hargus, W. A., and Conrad, J. H. (1984). *Trop. Anim. Prod.* **9**, 12–21.
Jones, D. I. H. (1964). *Rhod. J. Agric. Res.* **2**, 57–59.
Kayongo-Male, H., Thomas, J. W., and Ullrey, D. W. (1974). *Trace Elem. Metab. Anim. Proc. Int. Symp. 2nd*, pp. 455–457.
Kemp, A., Deijs, W. B., Hemkes, O. J., and Van Es. A. J. H. (1961). *Neth. J. Agric. Sci.* **9**, 134–149.
Kincaid, R. L., and Cronrath, J. D. (1983). *J. Dairy Sci.* **16**, 821–824.
Kirk, W. G., Shirley, R. L., Hodges, E. M., Davis, G. K., Peacock, F. M., Easley, J. F., and Martin, F. G. (1970). *Bull—Fla. Agric. Exp. Stn.* **735**.
Kreulen, D. A., and Jager, T. (1984). *In* "Symposium on Herbivore Nutrition in the Sub-Tropics and Tropics—Problems and Prospects." (F. M. C. Gilchrist and R. I. Mackie, eds.), pp. 204–221. Craighall, South Africa.
Langlands, J. P., Bowles, J. E., Donald, G. E., and Smith, A. J. (1982). *Aust. J. Agric. Res.* **33**(2), 313–320.
Latteur, J. P. (1962). "Cobalt Deficiencies and Sub-deficiencies in Ruminants." Center D'Information du Cobalt, Brussels.
Loneragon, J. F. (1975). *In* "Trace Elements in Soil-Plant-Animal Systems" (D. J. Nicholas and A. R. Egan, eds.), pp. 109–134. Academic Press, New York.

Long, M. I. F., Ndyanabo, W. K., Marshall, B., and Thornton, D. D. (1969). *Trop. Agric. (Trinidad)* **46**, 201–209.
McDowell, L. R. (1976). *In* "Beef Cattle Production in Developing Countries" (A. J. Smith, ed.), pp. 216–241. Centre for Tropical Veterinary Medicine, Univ. of Edinburgh, Edinburgh.
McDowell, L. R., and Conrad, J. H. (1977). *World Anim. Rev.* **24**, 24–33.
McDowell, L. R., Conrad, J. H., Thomas, J. E., Harris, L. E., and Fick, K. R. (1977). *Trop. Anim. Prod.* **2**, 273–279.
MacPherson, A., Voss, R. C., and Dixon, J. (1978). *Trace Elem. Metab. Man Anim. Proc. Int. Symp. 3rd* pp. 496–498.
Malaisse, F., Grégoire, J., Morrison, R. S., Brooks, R. R., and Reeves, R. D. (1979). *Oikos* **33**, 472–478.
Mascola, J. H., Barth, K. M., and McLaren, J. B. (1974). *J. Anim. Sci.* **38**, 1298–1303.
Mayland, H. F., Florence, A. R., Rosenau, R. C., Lazar, V. A., and Turner, H. A. (1975). *J. Range Manage.* **28**(6), 448–452.
Milford, R., and Haydock, K. P. H. (1965). *Aust. J. Exp. Agric. Anim. Husb.* **5**, 13–17.
Miller, J. K., Madsen, F. C., and Swanson, E. W. (1977). *J. Dairy Sci.* **60**, 618–622.
Miller, J. K., Swanson, E. W., and Madsen, F. C. (1978). *Feedstuffs* **50**, 31–32, 55.
Miller, W. J., Lassiter, J. W., and Jones, J. B. (1972). *In* "Proceedings Georgia Nutrition Conference for Feed Industry," pp. 94–106. University of Georgia Press, Atlanta.
Mitchell, R. L., Reith, J. W. S., and Johnston, I. M. (1957). *J. Sci. Food Agric.* **8** (*Suppl.*), 51–59.
Montalvo, M. I., Veiga, J. B., McDowell, L. R., Ocumpaugh, W. R., and Mott, G. O. (1983). *Agron. Abstr.*, p. 135.
Moore, J. E., and Mott, G. O. (1973). *In* "Anti-Quality Components of Forages" (A. G. Matches, ed.), pp. 53–96. Crop Science Society of America, Special Publication No. 4, Madison, Wisconsin.
Mtimuni, J. P. (1982). Ph.D. dissertation, Univ. of Florida, Gainesville.
Nye, P. H., and Tinker, P. B. (1977). "Solute Movement in the Soil-Root System." Univ. of California Press, Berkeley.
Ozanne, P. G., Shaw, T. C., and Kirton, D. J. (1965). *Aust. J. Exp. Agric. Anim. Husb.* **5**, 29–33.
Peducassé, C. A., McDowell, L. R., Parra, L. A., Wilkins, J. V., Martin, F. G., Loosli, J. K., and Conrad, J. H. (1983). *Trop. Anim. Prod.* **8**, 118–130.
Perdomo, J. T., Shirley, R. L., and Chicco, C. F. (1977). *J. Anim. Sci.* **45**, 1114–1119.
Pfander, W. H. (1971). *J. Anim. Sci.* **33**, 843–849.
Reid, R. L., and Horvath, D. J. (1980). *Anim. Feed Sci. Technol.* **5**, 95–167.
Reith, J. W. S. (1965). *NAAS Q. Rev.* **68**, 150–156.
Reith, J. W. S. (1970). *Trace Elem. Metab. Anim. Proc. WAAP/IBP Int. Symp.* 1969, pp. 410–411.
Reith, J. W. S. (1973). *In* "Hill Pasture Improvement and Its Economic Utilization" (P. A. Gething, P. Newbould, and J. B. E. Patterson, eds.), pp. 5–12. Potassium Institute, Henley-on-Thames.
Reuter, D. J. (1975). *In* "Trace Elements in Soil-Plant-Animal Systems." (D. J. Nicholas and A. R. Egan. eds.), pp. 291–324. Academic Press, New York.
Rigg, T., and Askew, H. O. (1934). *Emp. J. Exp. Agric.* **2**(5), 1–8.
Rosa, I. V. (1980). Ph.D. thesis, Univ. of Florida, Gainesville.
Schütte, K. H. (1964). "Biology of the Trace Elements. Their Role in Nutrition." Lippincott, Philadelphia.
Standish, J. F., and Ammerman, C. B. (1971). *J. Anim. Sci.* **33**, 481–484.

Standish, J. F., Ammerman, C. B., Simpson, C. F., Neal, F. C., and Palmer, A. Z. (1969). *J. Anim. Sci.* **29,** 496–503.
Suttle, N. F., Alloway, J., and Thornton, I. (1975). *J. Agric. Sci.* **84,** 249–254.
Swift, G. (1972). "A Review of Factors Affecting the Mineral Content of Herbage in Relation to Animal Requirements with Particular Reference to the North of Scotland." Technical Report No. 1. North of Scotland College of Agriculture, Scotland.
Tejada, R. (1984). Ph.D. dissertation, Univ. of Florida, Gainesville.
Tergas, L. E., and Blue, W. G. (1971). *Agronomy* **63,** 6–9.
T'Hart, M. (1960). *In* "Conference on Hypomagnesemia" pp. 88–101. British Veterinary Association, London.
Thompson, A. (1957). *J. Sci. Food Agric.* **8** (Suppl.), 72–81.
Thornton, I. (1974). *Trace Elem. Metab. Anim. Proc. Int. Symp. 2nd,* pp. 451–454.
Thornton, I., and Abrahams, P. (1983). *Sci. Total Environ.* **28,** 287–294.
Thornton, I., and Kinniburgh, D. G. (1978). *Trace Elem. Metab. Man Anim. Proc. Int. Symp. 3rd.,* p. 499 (abstr).
Underwood, E. J. (1981). "the Mineral Nutrition of Livestock," 2nd ed. Commonwealth Agricultural Bureaux, Slough, England.
Valdivia, R., Ammerman, C. B., Henry, P. R., Feaster, J. P., and Wilcox, C. J. (1982). *J. Anim. Sci.* **55,** 402–410.
Vargas, R., McDowell, L. R., Conrad, J. H., Martin, F. G., Buergelt, C., and Ellis, G. L. (1984). *Trop. Anim. Prod.* **9,** 103–113.
Whitehead, D. C. (1966). "Commonw. Bur. Pastures Field Crops." Mimeographed Publication No. 1/1966. Commonwealth Agricultural Bureaux, Bucks, England.
Whitehead, D. C. (1970). *Bull. Commonw. Bur. Pastures Field Crops* **48.**
Whyte, R. O. (1962). *Trop. Agric. (Trinidad)* **39,** 1–11.
Williams, R. (1963). "Minor Elements and Their Effects on the Growth and Chemical Composition of Herbage Plants." Mimeographed Publication No. 1/1959. Commonwealth Agricultural Bureaux, Bucks, England.

9

Calcium, Phosphorus, and Fluorine

L. R. McDOWELL

Department of Animal Science
University of Florida
Gainesville, Florida

I.	General.	189
II.	Calcium and Phosphorus.	190
	A. Introduction.	190
	B. Metabolism.	190
	C. Requirements.	192
	D. Deficiency.	193
	E. Prevention and Control.	203
	F. Toxicity.	204
III.	Fluorine.	204
	A. Introduction.	204
	B. Essentiality.	204
	C. Toxicity.	205
	D. Chemical Forms.	208
	E. Prevention and Control.	209
	References.	210

I. GENERAL

Calcium (Ca) and phosphorus (P) have long been recognized as important essential mineral elements, while adverse effects result from fluorine (F) toxicity. Calcium and P are considered together in this review because they are closely interrelated along with vitamin D in bone metabolism, and the requirements of these nutrients are dependent on the concentrations of each other. These minerals are concerned not only with bone development, growth, productivity, and reproduction but also with essentially all important metabolic processes. Calcium and P are very closely related to the extent that a deficiency or an excess of one will likely interfere with the proper utilization of the other. For grazing livestock, particularly cattle, P deficiency is the most severe mineral limitation in tropical countries. Next to salt, P is the nutrient most frequently given as a supplement to grazing ruminants. Phosphorus supplements if not defluorinated contain toxic levels of F, which over a period of time result in the development of bone lesions.

II. CALCIUM AND PHOSPHORUS

A. Introduction

Calcium and P are the two most abundant mineral elements in the animal body. Naturally occurring deficiencies of Ca and P in domestic animals usually develop in quite different circumstances, and a dual deficiency, in which the two minerals are equally limiting, is less common (Underwood, 1981).

Phosphorus deficiency is predominantly a condition of grazing ruminants, especially cattle, whereas Ca deficiency is more a problem of animals fed predominantly concentrates, especially pigs and poultry and also feedlot cattle finished on high-grain diets. Either one or both elements may be deficient under grazing conditions in certain geographical areas, although P is the element most likely to be deficient. Forages tend to be higher in Ca and lower in P, but grains are generally lower in Ca and higher in P. Phosphorus deficiency is the most widespread and is of the highest economical importance of all the mineral deficiencies for grazing livestock (McDowell, 1976; Underwood, 1981).

B. Metabolism

Calcium and P have a vital function in almost all tissues in the body and must be available to livestock in the proper quantities and ratio. These elements make up over 70% of the total mineral elements in the body. Ninety-nine percent of the Ca and 80% of the P in the entire body are found in bones and teeth.

Calcium is essential for skeletal formation, normal blood clotting, rhythmic heart action, neuromuscular excitability, enzyme activation, and permeability of membranes. Approximately 20% of body P is distributed throughout the soft tissues, being especially concentrated in red blood cells, muscle, and nerve tissues. In addition to skeletal formation, P is also essential for proper functioning of rumen microorganisms, especially those that digest plant cellulose, utilization of energy from feeds, buffering of blood and other fluids, many enzyme systems, and protein metabolism. Phosphorus is intimately associated with the normal function of all animal tissues by virtue of its role in the processes of energy exchange. Thus, it may be expected that any limitation to the P supply is reflected in a generalized impairment of body function. Phospholipid formation requires P, as do amino acid metabolism and protein formation. Phosphorus plays an important role in blood and rumen buffer systems.

The utilization and metabolism of both Ca and P are influenced by many of the same factors, including adequate levels of one to the other. An excess or deficiency of one will interfere with the proper utilization of the other, making the ratio between them critical. A Ca:P ratio of 1:1 to 2:1 is usually recommended, with a close ratio most critical if P intake is marginal or inadequate.

The main site of both Ca and P absorption is the small intestine. The solubility of Ca compounds, and hence the absorption of Ca, is favored by acid conditions and hindered by alkaline conditions in the small intestine. Thus, most Ca is absorbed in the proximal portion of the duodenum [National Research Council (NRC), 1984]. The percentage of absorption of Ca decreases with age, high F intakes, and high Ca intakes, or low vitamin D intakes (Ammerman and Goodrich, 1983). Phosphorus absorption is influenced by the P source, intestinal pH, animal age, and dietary intakes of several other minerals including Ca, iron (Fe), aluminum (Al), manganese (Mn), potassium (K), and magnesium (Mg) (Irving, 1964). Calcium absorption, bone deposition and removal, maintainence of adequate serum Ca, and excretion are regulated by parathyroid hormone, calcitonin, and metabolically active vitamin D forms including 1,25-dihydroxycholecalciferol and 24,25-dihydroxycholecalciferol. Vitamin D is probably less important in the absorption of P, but it is reported as a factor.

The daily turnover of P in ruminant saliva is similar to or greater than daily P intake (Cohen, 1980). Phosphorus output in saliva is dependent on P intake, and salivary P concentration is directly related to plasma P concentration. It is significant that although grazing ruminants would often be consuming forages with a wide Ca:P ratio, due to high P salivary content, the ratio of these minerals would be narrow at the absorption site. The major route of excretion for both Ca and P occurs through feces. Urinary loss is minimal, due to reabsorption by the kidneys. Phosphorus reabsorption is more efficient for ruminants than for nonruminants and thus is a protective mechanism for grazing ruminants consuming low-P diets. A large loss of Ca and P would result from lactating animals, as, for example, cow's milk (3.5% fat) averages 1.17% Ca and 1.05% P.

C. Requirements

The NRC publications on nutrient requirements of beef cattle (NRC, 1984), dairy cattle (NRC, 1978), sheep (NRC, 1975), and goats (NRC, 1981), in addition to the British nutrient requirements of ruminants publication [Agricultural Research Council (ARC), 1980], serve as the main basis for mineral requirements. Recommended mineral requirements and factors modifying requirements are discussed in Chapter 2 of this volume. The requirements of beef cattle indicate that 0.18–1.04% Ca and 0.18–0.70% P are adequate for growing and fattening steers and heifers, 0.43–0.60% Ca and 0.31–0.40% P for lactating dairy cows, 0.21–0.52% Ca and 0.16–0.37% P for sheep, and 0.21–0.52% Ca and 0.16–0.37% P for goats.

Requirements for mineral elements have been established by the factorial method and the feeding experiment method (ARC, 1980; Miller, 1983). With the factorial approach, the net quantity of Ca or P required for maintenance, deposition into tissues, and secretions is determined. The total represents the quantity of the available mineral needed. In the feeding-experiment method, response of animals is studied when various amounts of the mineral are fed. There is lack of agreement in regard to the Ca and P requirements of ruminants and appropriate methods for estimating these requirements (Cohen, 1980; Braithwaite, 1983). A large difference in recommendations results in determining the percentage of Ca or P that is available from a feedstuff (Miller, 1983). As an example, the ARC (1980) concluded that 68% of the Ca in most cattle feeds was available for absorption, while the NRC (1978) used a figure of 45%. An additional error arises from making the assumption that total endogenous loss is constant irrespective of feed intake, age, or physiological state (Braithwaite, 1984). Different Ca and P requirements may include safety margins in some NRC publications, but ARC values are minimums with no added safety margins.

Adequate Ca and P nutrition depends not only on sufficient total dietary supplies, but also on the chemical forms in which they occur in the diet and on the vitamin D status of the diet on the animal. The dietary Ca:P ratio also can be important. The balance of the minerals is often upset when legumes with a Ca:P ratio of 6 to 10:1 are fed, or likewise a wide ratio exists when only overly mature tropical forages, particularly those low in P, are available to grazing livestock during extended dry seasons. However, Ca in alfalfa is only 50–70% as available to dairy cattle as that from inorganic sources (Cunha, 1984). A dietary Ca:P ratio between 1:1 and 2:1 is assumed to be ideal for growth and bone formation since this is approximately the ratio of the two minerals in bone. Actually, ruminants can tolerate a wider range of Ca:P, particularly, when their vitamin D status is high. Nine Ca:P ratios ranging from 0.41:1 to 14.3:1 were tested by Wise *et al.* (1963) with dietary ratios below 1:1 and over 7:1 adversely affecting growth and feed efficiency.

D. Deficiency

The most well known deficiency sign of Ca and P is abnormal metabolism or calcification of bones. An inadequate intake of Ca may cause weakened bones (Fig. 9.1), slow growth, low milk production, and tetany (convulsions) in severe deficiencies. The increased demand for Ca at parturition may result in milk fever (parturient paresis) in lactating animals. Milk fever, which is an impaired metabolic condition that is related to Ca status, to previous Ca intake, and to parathyroid function, can be quite effectively prevented by maintaining a narrow Ca:P ratio (Stott, 1965). Animals that develop milk fever are unable to meet the sudden demand for Ca that is brought about by the initiation of lactation. Advanc-

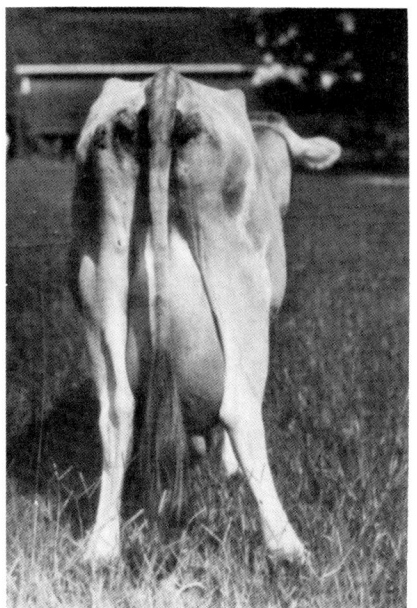
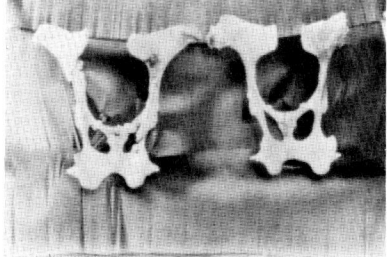

Fig. 9.1. Both hips of the cow shown above were broken (knocked down) as a result of a low-calcium ration. Her skeletal reserve of calcium was depleted to the point that her weakened bones were broken easily. The right photo shows the pelvic bones of a herd mate broken in three places. (Courtesy of R. B. Becker, University of Florida, Gainesville, Florida.)

ing age is also associated with a high incidence of milk fever, with animals in their first and second lactation able to more easily mobilize Ca from bone.

Signs of P deficiencies are not easily recognized except in severe cases characterized by fragile bones (Figs. 9.2 and 9.3), general weakness, weight loss (Figs. 9.4 and 9.5), emaciation, stiffness, reduced milk production, and chewing of wood, rocks, bones (Fig. 9.6), and other objects. Abnormal chewing of objects may also occur, however, with other dietary deficiencies. Reduction of appetite will have the result of reduced energy and protein intakes and, consequently, loss of weight. Conversely, despite normal P intake, bone mineralization may be restricted by inadequate intakes of energy and protein. With growing sheep high-energy-protein increased retention of Ca (49.9 versus 26.8%) and P (72.5 versus 54.4%) when compared to low-energy protein controls (Rosero et al., 1983).

In the young animals, lack of Ca, P, or vitamin D results in rickets (Fig. 9.7), and in the adult or more mature animal, osteomalacia. Rickets is characterized by an insufficient deposition of Ca and P in the cartilaginous matrix for development of strong bones.

In addition to the balance between Ca:P ratio and vitamin D, deficiencies of these minerals are influenced by availability of different sources of these minerals and interrelationships with additional mineral elements or nutrients. In India, Ca deficiency is reported for cattle fed straw due to the large amounts of oxalates in this feed (Ray, 1963). Additional countries reporting deleterious effects to cattle from oxalates are Costa Rica (Kiatoko et al., 1978) and Brazil (Pimentel and Thiago, 1982). In many tropical countries, high amounts of soil Fe, Ca, and Al accentuate a P deficiency by forming insoluble phosphate complexes. Soils in humid tropics are characteristically acid with high percentages of Fe and exchangeable Al, which complexes with P, making it unavailable to plants. Also, direct soil consumption by grazing animals can result in adverse effects due to high concentrations of Fe and Al (Rosa et al., 1982).

Calcium deficiency is much less of an area problem for grazing livestock than is P deficiency. Todd (1967) states that, with the exception of cereals and their by-products, most animal feeds contain between 0.2 and 1.0% Ca in the dry matter and that uncomplicated Ca deficiency is rare in grazing livestock unless the pasture contains less than 0.2% Ca. Compared to P inadequacies, Ca deficiency is rare in grazing cattle, with the exception of cows lactating large quantities of milk or those grazing on acid, sandy, or organic soils in humid areas where the herbage consists mainly of quick growing grasses and is devoid of legume species (Underwood, 1981). Although Ca deficiency can easily be produced in young,

9. Calcium, Phosphorus, and Fluorine 195

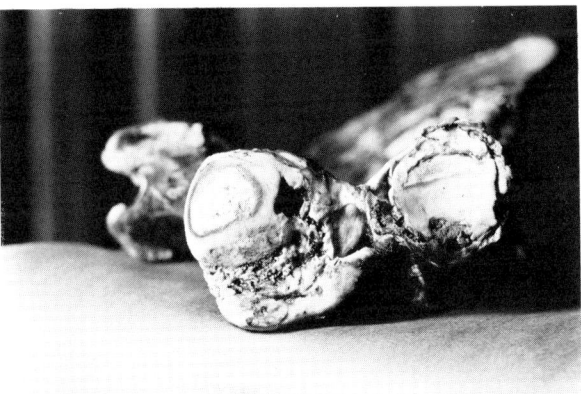

Fig. 9.2. Native cow (top) in a P-deficient, Florida range cow area. Note the thin condition and crippled right shoulder (sweeny) as result of inadequate phosphorus in her ration. This cow exhibited a depraved appetite and chewed objects continually; her rumen contained rags, pieces of inner tube, shoe nails, metal, oyster shells, etc. Her blood phosphorus level was 2.0 mg/100 ml. The bottom photo from the same cow shows the severely eroded and weakened scapula and humerous bones. Note the hole worn into the articular surface to the bone marrow cavity. (Courtesy of R. B. Becker, University of Florida, Gainesville, Florida.)

Fig. 9.3. Lamb fed a P-deficient ration and showing a typical knock-kneed condition. (From NRC, 1975.)

growing animals and lactating dairy cows fed native forages supplemented with concentrates, the deficiency is infrequently reported in grazing beef cattle even during lactation (Loosli, 1978). A lower incidence of Ca than of P disorders is attributable to three major factors: (1) a higher concentration of Ca than of P in the leaves and stems of most plant species; (2) a wider distribution of P-deficient than of Ca-deficient soils; and (3) a lesser decline in the concentration of Ca than of P with advancing maturation of the plant (Underwood, 1981).

For grazing livestock, the most prevalent mineral element deficiency throughout the world is lack of P (McDowell, 1976; Underwood, 1981). In a review (McDowell *et al.*, 1984), P deficiency was reported in 46 tropical countries of Latin America, Southeast Asia, and Africa (see Chapter 16 of this volume). In most livestock grazing areas of tropical countries, soils and plants are low in P. Many grass species containing over 0.3% P during early stages of growth are available to grazing livestock for only short periods. For the greater part of the year, mature forages contain less than 0.15% P. In the veld country of South Africa, where the classical studies of bovine aphosphorosis were made, herbage concentrations fall typically from between 0.13 and 0.18% P in the wet summer to as low as 0.05–0.07% P in the dry winter and can remain low for 6–8 months of the year

9. Calcium, Phosphorus, and Fluorine   197

Fig. 9.4. This cow (top) had a depraved appetite and had produced an average of only 2.9 kg of milk daily and gained only 0.07 kg/day for 285 days on a P-deficient ration. After P (monobasic sodium phosphate) was added to the same ration (below), she had gained 1.0 kg/day for a 90-day period. (Courtesy of R. B. Becker, University of Florida, Gainesville, Florida.)

(Underwood, 1981). Pastures from northern Australia will drop from 0.25% in young plants to as low as 0.05% when the grass is mature (Davies *et al.*, 1938).

In the 1974 edition of *Latin American Tables of Feed Composition*, the number of forages containing average Ca and P values are 1123 and 1129, respectively. Of the forages from these feed tables, 72.8% of the P values (<0.30%) and 31.1% of the Ca values (<0.30%) were borderline to deficient in these elements for most classes of grazing livestock (McDowell *et al.*, 1977). Table 9.1 summarizes forage Ca and P concentrations

Fig. 9.5. An almost adult ox, raised on the P-deficient "lavrado" fields of Roraima, northern Brazil, showing severely stunted growth. (Courtesy of Jürgen Döbereiner and Carlos H. Tokarnia, EMBRAPA, Rio de Janeiro, Brazil.)

TABLE 9.1

Forage Calcium and Phosphorus for Selected Tropical Regions

Location	Season	Sample no	Calcium Mean	Percentage of samples below 0.30%	Phosphorus Mean	Percentage of samples below 0.25%
Malawi[a]	Dry	21	0.63	13	0.19	75
	Wet	48	0.25	81	0.25	56
Bolivia[b]	Dry	8	0.49	8	0.20	75
	Wet	16	0.44	17	0.18	89
Bolivia[c]	Dry	20	0.21	90	0.15	100
	Dry	84	0.25	57	0.12	100
Dominican Republic[d]	Dry	69	0.48	24	0.17	83
Colombia[e]	Dry	36	0.16	95	0.16	92
	Wet	35	0.16	100	0.17	83
Guatemala[f]	Dry	84	0.29	71	0.23	57
	Wet	84	0.34	32	0.26	60

[a] Mtimuni (1982).
[b] McDowell et al. (1982).
[c] Peducassé et al. (1983).
[d] Jerez et al. (1984).
[e] Vargas et al. (1984).
[f] Tejada (1984).

Fig. 9.6. Bone chewing is often associated with a phosphorus deficiency. A cow (top) chewing bone in the llanos region of Santa Maria de Ipire, state of Guarico, Venezuela. (Courtesy of David Morillo, Centro de Investigaciones Agronómicas, Maracay, Venezuela.) A bone (below) that was being consumed in Argentina. Saliva could still be seen on the bone when the photo was taken. (Courtesy of Bernardo Jorge Carrillo, C.I.C.V., INTA, Castelar, Argentina.)

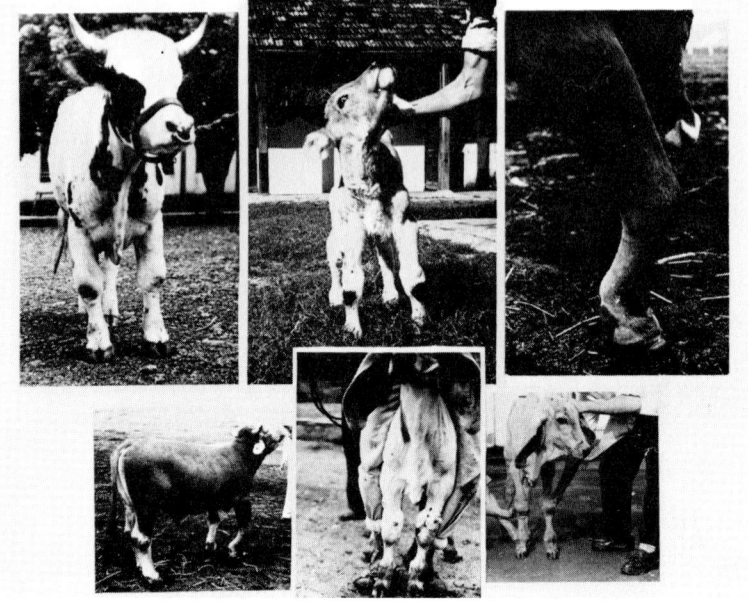

Fig. 9.7. Calves and young bulls with rickets. (Courtesy of Francisco Megale, Universidade Federal de Minas Gerais, Escola de Veterinaria, Belo Horizonte, MG, Brazil.)

from five countries of Latin America and Africa. The majority of all samples were deficient in P ($< 0.25\%$), with many forages also low in Ca, depending on the season of the year. From six ranches in Northern Mato Grosso, Brazil, forage P averaged 0.08 and 0.20 for the dry and wet season, respectively (Sousa *et al.*, 1979). For forage Ca, the values in the dry and wet seasons were 0.67 and 0.34%, respectively. From Costa Rica, 1468 forage samples contained 0.18% P in the wet season, and 1335 samples averaged 0.11% P in the dry season (J. Sanchez, personal communication). Table 9.2 illustrates extremely low Ca and P (particularly P) forage concentrations for 11 ranches in Venezuela during the wet and dry seasons.

Many reports from tropical regions of the world, dating back to the early part of the century, have revealed the beneficial effects of P supplementation on overall performance. Reports of improved weight gains by P-supplemented cattle have been summarized for various world regions (Tokarnia and Döbereiner, 1973; McDowell, 1976; Cohen, 1975; Fick *et al.*, 1978). From Brazil, Moraes *et al.* (1982) related supplemental P to daily gains of growing cattle. Treatments of 0, 4.3, 6.7, and 8.4 g of P per day resulted in daily gains of 216, 379, 465, and 564 g, respectively. From

TABLE 9.2
Mean Forage Calcium and Phosphorus Concentration (%) from Cattle Ranches in the States of Monagas and Guarico, Venezuela[a,b]

	Monagas				Guarico			
	Calcium		Phosphorus		Calcium		Phosphorus	
Ranch no	Dry season	Wet season	Dry season	Wet season	Dry season	Wet season	Dry season	Wet season
1	0.26	0.31	0.05	0.10	0.19	0.30	0.01	0.16
2	0.19	0.18	0.08	0.16	0.19	0.17	0.05	0.16
3	0.14	0.12	0.12	0.05	0.24	0.35	0.02	0.13
4	0.17	0.23	0.07	0.08	0.16	0.29	0.18	0.02
5	0.17	0.22	0.07	0.08	0.15	0.33	0.26	0.22
6	0.06	0.16	0.09	0.07	0.12	0.10	0.01	0.02
7	0.16	0.15	0.08	0.10	0.12	0.18	0.01	0.02
8	—	—	—	—	0.21	0.12	0.01	0.04
9	—	—	—	—	0.32	0.14	0.04	0.04
10	—	—	—	—	0.49	0.28	0.03	0.07
11	—	—	—	—	0.38	0.36	0.01	0.03
Overall mean	0.16	0.19	0.08	0.09	0.23	0.24	0.06	0.08
Standard deviation	0.06	0.06	0.02	0.03	0.12	0.10	0.08	0.07

[a] Adapted from Velasquez (1978) and Faria et al. (1981).
[b] Each mean is based on 4 pastures/farm and 10 subsamples/pasture.

Peru, Echevarria et al. (1973) obtained weight gains of 0.59 kg in steers supplemented with dicalcium phosphate and 0.27 kg for controls. Bolivian cattle gained 96.4 kg when receiving a source of P versus 79.4 kg for controls (McDowell et al., 1982). Estevez-Cancino (1960) reported increased milk production (up to 24%) with bone-meal supplementation in farms that had P-deficient forages.

The most devastating economic result of P deficiency is reproductive failure, with P supplementation dramatically increasing fertility levels in grazing cattle from many parts of the world. In 2-year observations of 200 South African cattle, calf crops increased from 51% in control cattle to approximately 80% for cattle supplemented with bone meal or other P sources (Theiler et al., 1928). Latin American countries reporting increased fertility rates resulting from P administration include Bolivia, Brazil, Colombia, Panama, Peru, and Uruguay, and from Asia, the Philippines and Thailand (McDowell et al., 1983). From this review, averaging 16 cattle reports on the effect of supplementation on calving percentage,

Fig. 9.8. A cow in Argentina suffers from botulism as a result of eating bones. The animal is weak and has difficulty rising. (Courtesy of Bernardo Jorge Carrillo, C.I.C.V., INTA, Castelar, Argentina.)

those receiving P supplement averaged a 75.6% calf crop versus 52.9% for controls.

From Zimbabwe (Ward, 1968) and northern Australia (Lamond, 1970), the effects of undernutrition on fertility were apparent. With few exceptions, lactating cows were not calving 2 years in succession. In P-deficient areas, if a calf is produced, cows may not come into a regular estrus again until body P levels are restored, either by feeding supplementary P or by cessation of lactation.

In South Africa in the early 1900s, pioneer P-supplementation studies (Van Niekerk, 1978) revealed the cause of bovine botulism and aphosphorosis. Both conditions were the result of a severe P deficiency, with cattle exhibiting subnormal growth and reproduction and a depraved appetite or "pica," as illustrated by bone chewing. Besides South Africa, other countries reporting death from botulism as a result of bone chewing include Argentina (Fig. 9.8), Brazil, and Senegal. In the area of Piaui, Brazil, an estimated 2–3% of approximately 10,000 cattle die annually of botulism (Tokarnia *et al.*, 1970).

Signs of borderline Ca and P deficiencies are not easily distinguishable from other deficiencies, and, therefore, detection must rely on response from supplementation or chemical analysis. Due to limitation of serum Ca and P as an indicator of status (see Chapter 15 of this volume), analysis of the ration, bone composition, and breaking strength are the best ways of assessing a deficiency of Ca and P.

E. Prevention and Control

The P requirement of a grazing ruminant is rarely met by forages; therefore, supplementation is necessary. Calcium and P deficiencies can be prevented or overcome by direct treatment of the animals through supplementation of the diet or the water supply or, indirectly, by appropriate fertilizer treatment of the soils on which the pastures to be consumed are grown.

The choice of supplementation procedure depends on the conditions of husbandry. On sparse P-deficient grazing, the direct method is preferred because the use of phosphate fertilizers involves high transport and application cost, and herbage productivity is usually limited by climate or soil problems. In more climatically favored and intensively farmed areas, phosphate applications designed primarily to increase pasture yields also increase P concentrations. In extensive range conditions where fertilizer applications are uneconomical, as in many areas of Latin America, Asia, and Africa, direct provision of additional P can be achieved by the use of phosphate containing supplements as part of free-choice mineral mixtures (see chapter 17 of this volume). The easiest and cheapest procedure is to provide a phosphatic mineral supplement in troughs or boxes protected from the rain. A biological availability comparison of Ca and P sources is found in Chapter 17 of this volume. Good sources of mineral P are dicalcium phosphate and superphosphate; ground raw rock phosphate is relatively unpalatable, and most sources are too high in F for safe use. Procedures that require the use of water-soluble phosphates, Na_2HPO_4 or ammonium polyphosphate, are good but more expensive than those that require the less soluble phosphates. Many of the materials used as P supplements supply significant amounts of Ca.

The protein content of forages often declines with P, so that protein deficiency, and frequently also a deficiency of available energy, are exacerbating factors in the malnutrition of livestock in P-deficient areas, and, therefore, P supplementation programs also need to consider the potential benefit, in relation to cost, of energy–protein supplements. Reduction of appetite and food consumption restricts energy intake so that a response to P supplementation can be anticipated only up to a level where other nutrients, for example protein, restrict energy intake (Cohen, 1980). Since P and protein contents are correlated, supplementation of low-protein pastures with P alone may not be successful. Responses from feeding low-

P/high-protein supplements (i.e., oilseed meals) or the use of low-P-tolerant legumes (i.e., *Stylosanthes humilis*) to increase portein intakes of grazing livestock may be maximized only if supplements of P are also provided (Cohen, 1980). Nevertheless, a number of cattle supplementation trials from tropical countries in Latin America, Southeast Asia, and Africa have illustrated that P is more limiting than either energy or protein, with P supplementation studies resulting in dramatic weight gains and increases in calving percentages (McDowell *et al.*, 1983).

F. Toxicity

The maximum tolerable level of Ca is 2% of diet dry matter, and for P is 0.6% for sheep and 2% for cattle (NRC, 1984). Excess of either of these minerals may cause bone disorders and reduce feed consumption and gain. Excess P has also been reported to be involved in urinary calculi, reduced serum Ca, and increased serum P (NRC, 1980). The greatest detrimental result from large amounts of Ca and P is the antagonistic effect toward the metabolism of other elements. Calcium is particularly antagonistic toward P, Mg, Fe, iodine (I), zinc (Zn), and Mn.

III. FLUORINE

A. Introduction

The three elements most frequently causing toxicity problems for grazing livestock are F, molybdenum (Mo), and selenium (Se), with Mo and Se discussed in subsequent chapters in this volume. Although apparently essential for most species, only toxic effects of F are likely to be of importance to grazing livestock. Industrial contamination of forages and the growing use of F-containing phosphates as mineral supplements has increased the F hazards to ruminants.

B. Essentiality

Research with laboratory animals suggests that F is an essential element for all farm animals (Underwood, 1981). In limited amounts, F has been demonstrated to increase resistance of teeth to cavities in human children and experimental animals. Likewise, F is reported to decrease osteoporosis in bones in adult humans (Underwood, 1977). It also has been reported to benefit mice by reducing anemia and improving fertility, as well as increasing growth of rats. However, if it is an essential element for most animals, the requirements are exceedingly low. No one at this date has produced an environment so low in F that animal survival has been vitally threatened (NRC, 1980).

C. Toxicity

Fluorine is a very toxic element, with fluorosis found in many parts of the world (see Chapter 16 of this volume). Chronic fluorosis is generally observed under three conditions: (1) continuous consumption of high F mineral supplements, (2) drinking water high in F (3–15 ppm or more), and (3) grazing F-contaminated forages adjacent to industrial plants that emit F fumes or dust. With notable exceptions, the F content of plants is seldom more than 1 to 2 ppm, since most plants have a limited capacity to absorb this element (Underwood, 1981). However, highly toxic concentrations, mainly as fluoracetate, have been reported in several South African plants (Marais, 1943).

Water high in F (3–15 ppm or more) is usually from deep wells originating from deep rock formations rather than from surface water supplies. Allowing water to stand in troughs will result in a higher concentration due to evaporation. The more common incidence of fluorosis results when grazing ruminants consume contaminated forages near industrial plants that have emitted F compounds into the environment, or when they have consumed mineral supplements high in this mineral. The principal F-containing mineral supplement is rock phosphate.

The major clinical signs of F toxicity are found in teeth and bone. If animals are young, the teeth may become modified in shape, size, and color. The incisors may become pitted, and the molars may show cavities due to fracture or wear, especially if excess F has been consumed prior to development of the permanent teeth. Jaw and long bones develop exostosis (Fig. 9.9), and joints may become thickened, causing the animal to become stiff and lame. A level of 20–30 ppm of total F in the diet will cause dental mottling; above 50 ppm, F will cause a significant incidence of lameness and decreased milk production in lactating cows. Cattle frequently have decreased feed intake when F is greater than 50 ppm in the diet. Figs. 9.10–9.13 show teeth and metatarsal bones of cattle suffering from fluorosis. Table 9.3 contains suggested F tolerances for animals (Thompson, 1978).

Intake of F in water containing 1 ppm by cattle has been estimated to provide 0.01–1.0 mg of F per kg of body weight per day, which is not of health significance. However, mottled teeth have been observed in cows drinking water that contained 4–5 ppm of the element. In total intake calculations, the F present in water should be taken into account.

Fig. 9.9. Cow with exostosis (bony growth) of the bones caused by ingesting large amounts of fluorine in Polk County, Florida. (Courtesy of R. L. Shirley, University of Florida, Gainesville, Florida.)

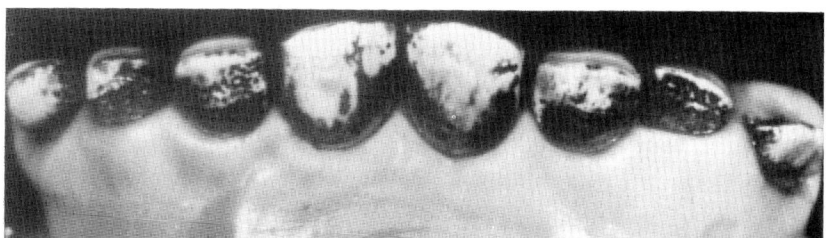

Fig. 9.10. Incisors from 4-year-old bovine with severe dental fluorosis. Lesions include enamel hypoplasia, hypocalcification, staining, and abnormal wear and reflect a nearly constant elevated fluoride intake during the period of tooth formation. (Courtesy of J. L. Shupe and A. E. Olson, Utah State University, Logan, Utah.)

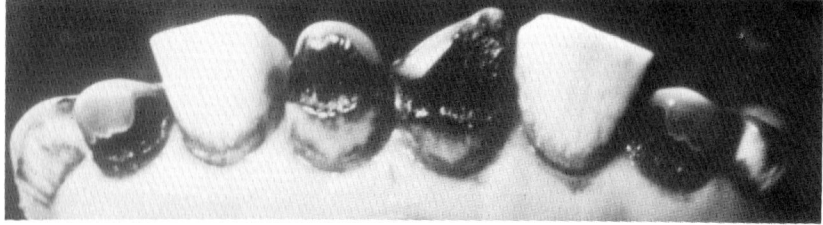

Fig. 9.11. Incisors from 5-year-old bovine with severe dental fluorosis reflect intermittent periods of elevated fluoride ingestion during the period of tooth formation. (Courtesy of J. L. Shupe and A. E. Olson, Utah State University, Logan, Utah.)

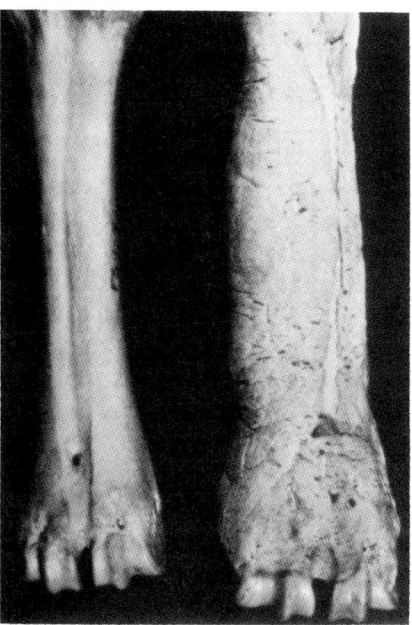

Fig. 9.12. Left side of photo shows normal bovine metatarsals. Right side of photo shows severe periosteal hyperostosis with roughened irregular surface consisting of disorganized and poorly mineralized bone. Note that the articulating surfaces are unaffected. (Courtesy of J. L. Shupe and A. E. Olson, Utah State University, Logan, Utah.)

Livestock are protected against fluorosis by increased F urinary excretion and by deposition in the skeletal tissue. Nevertheless, F is a cumulative poison, and once bone tissue is saturated, continued intakes are deposited in soft tissues, with the result of metabolic disturbances and death. Due to deposition in bones, F content of osteous tissue is an accurate indicator of toxicity of this element (see Chapter 15 of this volume). Toxicity of F is a reflection of amount and duration of ingestion, solubility of fluorides ingested, age of the animal, nutrition, stress factors, and individual animal differences (Underwood, 1981). Long-term experiments with beef cattle indicated that 30 ppm of dietary F resulted in excessive wearing and staining of teeth (Hobbs and Merriman, 1962). Suttie *et al.* (1957) observed that lactating cows could tolerate 30 ppm with no apparent difficulty, that 40 ppm was a marginal tolerance, and that 50 ppm would result in fluorosis within 3 to 5 years.

Cattle are less tolerant to F toxicity than other grazing livestock (Phillips *et al.*, 1960). Cattle in tropical countries that are provided with inadequate supplies of energy and protein during extended dry seasons are

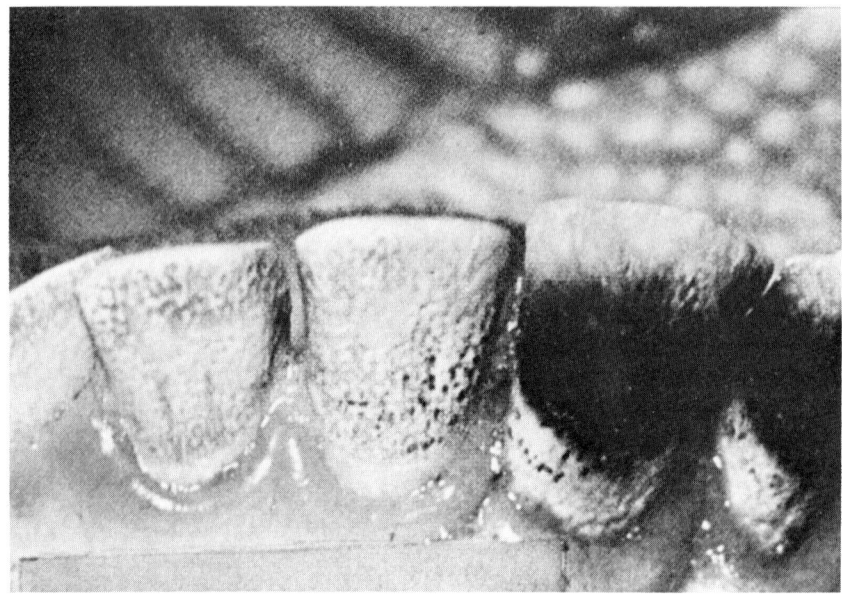

Fig. 9.13. Heifer that received 107 ppm of F (100 ppm as NaF). Note wear on new incisors following F ingestion and stain on central pair that had erupted prior to treatment. (Courtesy of S. L. Hansard, University of Tennessee, Knoxville, Tennessee.)

particularly susceptible to F toxicity. General undernutrition apparently enhances the deleterious effects of F toxicosis.

D. Chemical Forms

The chemical form of F is significant, since F in sodium fluoride is much more available than in calcium fluoride or rock phosphate. Fluorine from dicalcium phosphate and raw rock phosphate was reported to be 50% as available as F from sodium fluoride (Clay and Suttie, 1981). It is apparent that P sources manufactured by the furnace process are relatively free of fluorides. Those manufactured from defluorinated phosphoric acid will contain safe levels when added to feed supplements and salt mixtures for livestock and meet the recommendation of having not more than 1 part of F per 100 parts of P. Soft rock phosphate and ground raw rock phosphate generally exceed this ratio by about tenfold and for this reason should be closely monitored as to both the quantity of F provided and the length of time provided to livestock. Bone meal, fish meal, and poultry by-product meals may at times have considerable F present, and this should be taken into account in total F intake.

9. Calcium, Phosphorus, and Fluorine

TABLE 9.3

Suggested Fluorine Tolerances for Animals[a,b]

Animal	Tolerance level (NRC) (F ppm in diet)	Definitely unsafe (F ppm in diet)
Breeding animals		
Beef or dairy calves[c]	—	40 and above
Beef or dairy heifers[d]	30–40	50 and above
Mature beef or dairy cattle[e]	40–50	60 and above
Ewes	60	70 and above
Horses	40–60	80 and above
Sows	100–150	160 and above
Laying hens	400	440 and above
Animals to be slaughtered		
Beef or dairy calves[c]	35	65 and above
Beef or dairy heifers[d]	50	80 and above
Growing chickens (broilers)	300	340 and above
Mature beef or dairy cows[e]	100	120 and above
Feeder lambs	150	170 and above
Finishing pigs	150	200 and above

[a] Modified from Thompson (1978).
[b] These levels are based on the assumption that the animals receive an otherwise adequate diet. The values presented assume the ingestion of a soluble fluoride, such as NaF. When the fluoride in the ration is present in the form of dicalcium phosphate, these tolerances are increased by approximately 50%.
[c] Calves up to 4 months of age.
[d] Heifers 4 months to 2 years of age.
[e] Cattle 3 years of age or older.

E. Prevention and Control

For the prevention of fluorosis, F content of water and supplemental phosphates should be determined along with visual observations to detect early signs of fluorosis. Many nutritionists from developed, industrialized countries make the statement that "only defluorinated phosphates should be fed to ruminants." This is a generally acceptable recommendation since defluorinated phosphates are available and relatively low in cost for developed versus developing countries. Under some circumstances, however, exceeding the upper recommendation of 30–50 ppm of F for grazing livestock may be justified. When defluorinated phosphates are unavailable or prohibitively expensive, as occurs in most developing tropical countries, fertilizer or untreated phosphates would be recommended, but only for short periods of time. As an example, phosphates containing higher levels of F would be more appropriately provided to feedlot cattle than to animals retained in the breeding herd. Likewise, successful mineral supplements have been formulated using a mixture of fertilizer phosphates (40%) with defluorinated phosphates.

REFERENCES

Ammerman, C. B., and Goodrich, R. D. (1983). *J. Anim. Sci.* 57 (Suppl. 2), 519–533.
Agricultural Research Council (ARC) 1980. "The Nutrient Requirements of Farm Livestock, No. 2, Ruminants. Agricultural Research Council, London.
Braithwaite, G. D. (1983). *Br. J. Nutr.* **50,** 711–722.
Clay, A. B., and Suttie, J. W. (1981). *Am. Soc. Anim. Sci., Proc. Annu. Meet. 72nd.,* p. 352 (abstr.).
Cohen, R. D. H. (1975). *World Rev. Anim. Prod.* **16**(2), 27–43.
Cohen, R. D. H. (1980). *Livest. Prod. Sci.* **7,** 25–37.
Cunha, T. J. (1984). *Agric. Dig.* **18**(1), 2.
Davies, J. G., Scott, A. E., and Kennedy, J. F. (1938). *J. Counc. Sci. Ind. Res. (Aust.)* **11,** 127–139.
Echevarria, M., Valdivia, R., Barúa, J., and Campos, L. (1973). *In* "3rd Reunion de Especialistas e Investigadores Forrajeros del Peru," pp. 83–92. San Maros University, Lima, Peru.
Estevez-Cancino, J. A. (1960). *Acta Agron. (Palmira)* **10,** 169–183.
Faria, J. A., Velásquez, J., and López, M. G. (1981). *In* "Situacion de la Nutricion Mineral En Fincas de las Sabanas Orientales del Estado Guarico." Boletin No. 6. Fondo de Investigación Agropecuaria, Guarico, Venezuela.
Fick, K. R., McDowell, L. R., and Houser, R. H. (1978). *In* "Proceedings Latin American Symposium on Mineral Nutrition Research with Grazing Ruminants" (J. H. Conrad and L. R. McDowell), pp. 149–162. Univ. Florida Press, Gainesville.
Hobbs, C. S., and Merriman, G. M. (1962). *Tenn. Agric. Exp. St. Bull.* **351.**
Irving, J. T. (1964). *In* "Mineral Metabolism" (C. L. Comer and F. Bronner, eds.) Vol. 2, p. 249, Academic Press, New York.
Jerez, M., McDowell, L. R., Martin, F. G., Hargus, W. A., and Conrad, J. H. (1984). *Trop. Anim. Prod.* **9,** 12–21.
Kiatoko, M., McDowell, L. R., Fick, K. R., Fonseca, H., Camacho, J., Loosli, J. K., and Conrad, J. H. (1978). *J. Dairy Sci.* **61,** 324–330.
Lamond, D. R. (1970). *Anim. Breed. Abstr.* **38,** 354–372.
Loosli, J. K. (1978). *In* "Proceedings Latin American Symposium on Mineral Nutrition Research with Grazing Ruminants" (J. H. Conrad and L. R. McDowell, eds.), pp. 5–8. Univ. of Florida, Gainesville.
McDowell, L. R. (1976). *In* "Beef Cattle Production in Developing Countries" (A. J. Smith, ed.), pp. 216–244. Univ. of Edinburgh Press, Edinburgh.
McDowell, L. R., Conrad, J. H., Thomas, J. E., Harris, L. E., and Fick, K. R. (1977). *Trop. Anim. Prod.* **2,** 273–279.
McDowell, L. R., Bauer, B., Galdo, E., Koger, M., Loosli, J. K., and Conrad, J. H. (1982). *J. Anim. Sci.* **55,** 964–970.
McDowell, L. R., Conrad, J. H., Ellis, G. L., and Loosli, J. K. (1983). Minerals for Grazing

Ruminants in Tropical Regions." Bull. Univ. Fl. Coop. Ext. Serv. Univ. of Florida, Gainesville.
McDowell, L. R., Conrad, J. H., and Ellis, G. L. (1984). *In* "Symposium on Herbivore Nutrition in Sub-Tropics and Tropics—Problems and Prospects" (F. M. C. Gilchrist and R. I. Mackie, eds.), pp. 67–88. Pretoria, S. Africa.
Marais, J. (1943). *Onderstepoort J. Vet. Sci. Anim. Ind.* **18**, 203–206.
Miller, W. J. (1983). *Feedstuffs* **55**(42), 19–22.
Moraes, E., Italiano, E. C., and Pieniz, C. (1982). *In* "Proceedings Sociedade Brasileria de Zootecnia," p. 156 (abstr.). Piracicaba, São Paulo, Brazil.
Mtimuni, J. P. (1982). Ph.D. dissertation, Univ. of Florida, Gainesville.
National Research Council (NRC) (1974). "Effects of Fluorides in Animals," Natl. Acad. Sci., Washington, D.C.
National Research Council (NRC) (1975). "Nutrient Requirements of Sheep," 5th rev. ed. Nutrient Requirements of Domestic Animals, No. 3. Natl. Acad. Sci., Washington, D.C.
National Research Council (NRC) (1978). "Nutrient Requirements of Dairy Cattle," 5th rev. ed. Nutrient Requirements of Domestic Animals, No. 3. Natl. Acad. Sci., Washington, D.C.
National Research Council (NRC) (1980). "Mineral Tolerance of Domestic Animals." Natl. Acad. Sci., Washington, D.C.
National Research Council (NRC) (1981). "Nutrient Requirements of Goats," 1st ed. Nutrient Requirements of Domestic Animals, No. 4. Natl. Acad. Sci., Washington, D.C.
National Research Council (NRC) (1984). "Nutrient Requirements of Beef Cattle," 6th rev. ed. Nutrient Requirements of Domestic Animals, No. 4. Natl. Acad. Sci., Washington, D.C.
Peducassé, C. A., McDowell, L. R., Parra, L. A., Wilkins, J. V., Martin, F. G., Loosli, J. K., and Conrad, J. H. (1983). *Trop. Anim. Prod.* **8**, 118–130.
Phillips, P. H., Greenwood, D. A., Hobbs, C. S., Huffman, C. F., and Spencer, G. R. (1960). "The Fluorosis Problem in Livestock Production,"National Research Council Publication 824. Natl. Acad. Sci., Washington, D.C.
Pimentel, D. M., and Thiago, R. L. S. *In* "Proceedings Sociedade (1982). Brasileira de Zootecnia," p. 384, (abstr.) Piracicaba, São Paulo, Brazil.
Ray, S. M. (1963). *In* "Proceedings 1st World Conference on Animal Production," pp. 109–114, European Association of Animal Production, Rome.
Rosa, I. V., Henry, P. R., and Ammerman, C. B. (1982). *J. Anim. Sci.* **55**, 1231–1240.
Rosero, O. R., McDowell, L. R., Martin, F. G., Conrad, J. H., and Ellis, G. L. (1983). *Nutr. Rep. Int.* **28**, 1179–1188.
Sousa, J. C., Conrad, J. H., Blue, W. G., and McDowell, L. R. (1979). *Pesqui. Agropecu. Bras.* **14**(4), 387–395.
Stott, G. H. (1965). *J. Dairy Sci.* **48**, 1485–1489.
Suttie, J. W., Miller, R. F., and Phillips, P. H. (1957). *J. Dairy Sci.* **40**, 1485–1491.
Tejada, R., (1984). Ph.D. dissertation, Univ. of Florida, Gainesville.
Theiler, A., Green, H. H., and deToit, P. J. (1928). *J. Agric. Sci.* **18**, 369–371.
Thompson, D. J. (1978). *In* "Proceedings Latin American Symposium on Mineral Nutrition Research with Grazing Ruminants" (J. H. Conrad and L. R. McDowell, eds.), pp. 47–53. Univ. of Florida, Gainesville.
Todd, J. R. (1967). *Vet. Rec.* **81**, 6–8.
Tokarnia, C., and Döbereiner, J. (1973). *Pesqui. Agropecu. Bras. Ser. Vet.* **8**, 1–6.
Tokarnia, C. H., Langenegger, J., Langenegger, C. H., and Carvalho, E. V. (1970). *Pesqui. Agropecu. Bras.* **5**, 465–472.

Underwood, E. J. (1977). "Trace Elements in Human and Animal Nutrition" 3rd ed. Academic Press, New York.
Underwood, E. J. (1981). "The Mineral Nutrition of Livestock." Commonwealth Agricultural Bureaux, London.
Van Niekerk, B. D. H. (1978). In "Proceedings Latin American Symposium on Mineral Nutrition Research with Grazing Ruminants" (J. H. Conrad and L. R. McDowell, eds), pp. 194–200. Univ. of Florida, Gainesville.
Vargas D., R., McDowell, L. R., Conrad, J. H., Martin, F. G., Buergelt, C., and Ellis, G. L. (1984). *Trop. Animal Prod.* **9**, 103–113.
Velasquez, J. A. (1978). In "Situación de la Nutición Mineral del Ganado Bovino en el Estado Monagas." Univ. de Oriente, Juspein, Venezuela.
Ward, H. K. (1968). *Rhod. J. Agric. Res.* **6**, 93–101.
Wise, M. B., Ordoveza, A. L., and Barrick, E. R. (1963). *J. Nutr.* **79**, 79–84.

10

Common Salt (Sodium and Chlorine), Magnesium, and Potassium

L. R. McDOWELL
Department of Animal Science
University of Florida
Gainesville, Florida

I.	General.	214
II.	Common Salt (Sodium and Chlorine).	214
	A. Introduction.	214
	B. Metabolism.	214
	C. Requirements.	215
	D. Deficiency.	216
	E. Prevention and Control.	218
	F. Toxicity.	220
III.	Magnesium.	220
	A. Introduction.	220
	B. Metabolism.	221
	C. Requirements.	222
	D. Deficiency.	222
	E. Prevention and Control.	226
	F. Toxicity.	228
IV.	Potassium.	228
	A. Introduction.	228
	B. Metabolism.	228
	C. Requirements.	229
	D. Deficiency.	229
	E. Prevention and Control.	233
	F. Toxicity.	233
	References.	233

I. GENERAL

The strong craving for salt exhibited by grazing animals under most natural conditions was undoubtedly recognized before recorded history. Common salt is the basis for free-choice, ad libitum mineral supplements and often in tropical regions is the only mineral provided to grazing ruminants. Magnesium (Mg) and potassium (K) supplements, unlike salt, are unpalatable and generally are not required for grazing ruminants, except under special circumstances. It is convenient to consider sodium (Na), chlorine (Cl), K, and Mg together because of the broad similarities in their functions and requirements and their interactions with each other.

II. COMMON SALT (SODIUM AND CHLORINE)

A. Introduction

Sodium, in conjunction with Cl as common salt, has been known since ancient times as a required part of diets for livestock. The craving of livestock for common salt is well established. Tropical forages normally do not contain sufficient quantities of Na to meet the requirements of grazing ruminants throughout the year. This inadequacy is easily overcome by the practice of providing common salt ad libitum. Nevertheless, most grazing livestock in tropical countries either receive insufficient salt or have only very limited access to salt at certain times of the year.

B. Metabolism

Sodium and Cl, in addition to K, all function in maintaining osmotic pressure and regulating acid–base equilibrium. These two mineral elements function as electrolytes in body fluids and are specifically involved at the cellular level in water metabolism, nutrient uptake, and transmission of nerve impulses. Chlorine is necessary for activation of amylase and is essential for formation of gastric hydrochloric acid.

Sodium and Cl ions are absorbed by animals principally from the upper small intestine. Absorption of Na and Cl may occur from the rumen, with Neathery (1981) reporting that Cl is almost completely absorbed and that absorption occurs throughout the digestive tract. Approximately 80% of the Na and Cl entering the gastrointestinal tract arises from internal secretions such as saliva, gastric fluids, bile, and pancreatic juice [National Research Council (NRC), 1980]. Small quantities of Na and Cl are lost in feces and perspiration, with the majority excreted in urine. Regulation of body concentrations is controlled by hormones, including aldosterone and an antidiuretic hormone of the posterior pituitary. Both hormones act to maintain a constant ratio of Na to K in the extracellular fluid. Chloride metabolism is controlled in relation to Na so that excess kidney excretion of Na is accompanied by Cl. Chloride excretion is also influenced by bicarbonate ion, with a rise in plasma bicarbonate resulting in the excretion of a comparable amount of Cl (NRC, 1980).

C. Requirements

The essential need for Na and Cl by livestock has been demonstrated for thousands of years by a natural craving for common salt. Sodium is the critical nutrient in salt, and evidence of a naturally occurring dietary deficiency of Cl, as distinct from Na, has not been established; some recent studies indicate a need to reevaluate Cl needs in animals, since it also may be needed in certain situations. Therefore, the Na and Cl requirement is often expressed as salt (NaCl).

However, the reason for Na as the important element in salt is that Na is present in most natural feed ingredients at much lower levels than is Cl. The recommended Na requirement for grazing ruminants is between 0.04 and 0.18%, with the higher level recommended for lactating dairy cows (see Chapter 2 of this volume.) The Cl requirement for ruminants is generally unknown. From limited data, the Cl requirement for lactating dairy cows is estimated to be between 0.10 and 0.18% and would not exceed 0.27% (Fettman et al., 1984). However, Underwood (1981) suggested that the Cl requirement for lactating dairy cows should be substantially higher than the 0.15% Na requirement since cow's milk contains more than twice as much Cl as Na. Lowering dietary Cl in young growing cattle from 0.5 to 0.038% Cl caused significant changes associated with acid–base balance, but it did not adversely affect health and growth for 7 weeks (Burkhalter et al., 1980). The minimum Cl requirements of sheep and goats, as well as other ruminant species besides cattle, have apparently not been studied (Underwood, 1981). Sodium requirements are highest and deficiency is most likely to occur: (1) during lactation, due to secretion of Na in milk, (2) in rapidly growing animals, (3) under tropical or hot, semiarid conditions where large losses of water and Na occur in the sweat and where pastures are low in Na, (4) in animals grazing pastures heavily fertilized with K, which depresses herbage Na levels. Even after prolonged severe deficiency, NaCl (salt) levels secreted in milk remain high. Thus, lactating animals suffer most from lack of salt in the diet (Loosli, 1978). Cow's milk contains 630 ppm Na and 1150 ppm Cl (Cunha, 1983). Heavy milk producers need more salt than do others. With a borderline Na deficiency, clinical signs may not appear until lactation is initiated.

Various researchers, including Dirven (1963), suggest that under tropical or hot, semiarid conditions where large losses of water and Na occur in sweat, the Na requirement may be higher. Cattle deprived of salt may be so voracious that they often injure each other in attempting to reach salt.

D. Deficiency

The initial sign of Na and Cl deficiency is a craving for salt, demonstrated by the avid licking of wood, soil, and sweat from other animals and by the drinking of water. Even untamed cattle are known to approach horseback riders in order to lick the sweat of horses. A prolonged deficiency causes loss of appetite, decreased growth, unthrifty appearance, reduced milk production, and loss of weight (Fig. 10.1). When these clinical signs occur shortly after calving with a high-producing cow, her breakdown and death can be very sudden (Cunha, 1983). If salt is supplied before collapse, there is a rapid and almost dramatic recovery.

Experimentally produced Cl deficiency, independent of Na deficiency, results in clinical signs in dairy cows that include decreased body weight and milk production, depraved appetite, lethargy, anorexia, emaciation, constipation, cardiovascular depression, and milk dehydration (Fettman et al., 1984). Neathery (1981) reported clinical signs for young dairy calves to include anorexia, weight loss, lethargy, mild polydipsia, and milk polyuria (Fig. 10.2). Severe eye defects (scleral infection, sunken eyes, scaliness around eyes) and reduced respiration rate were observed as the deficiency progressed.

Natural forages low in Na have been reported in numerous tropical countries throughout the world (see Chapter 16 of this volume). The University of Florida summarized the available analyses of Latin American feeds, and of 1615 forage samples, only 146 had been analyzed for sodium. Of these, 18% contained 0.05% Na or less, 41% had between 0.05 and 0.10%, 18% ranged from 0.11 to 0.20%, and 22% were above 0.2% (McDowell et al., 1974). In western Panama, *Hyparrhenia rufa* (Jaragua grass) from 11 different locations contained $0.05 \pm 0.02\%$ Na in April compared with $0.036 \pm 0.006\%$ in November (Chicco, 1972).

Forage Na analyzed from cattle ranches in northern Mato Grosso, Brazil, during both the wet and dry seasons, were extremely deficient, being able to meet only between 14 and 30% of the animals requirements (Sousa et al., 1982). Table 10.1 lists five Latin American and one African country, illustrating the great extent of deficient forage Na concentrations. Several experiments with grazing animals in tropical areas have demonstrated no benefits from salt supplementation. As the Na deficiency de-

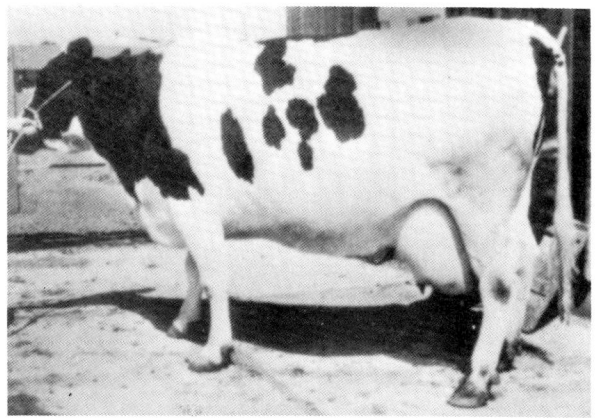

Fig. 10.1. Salt deficiency in a dairy cow. Illustrates before (top) and after (bottom) 1 year of salt deprivation. (Courtesy of S. E. Smith, Cornell University, Ithaca, New York.)

velops, Na content of urine, plasma, saliva, and other fluids is decreased. However, for livestock that are growing, lactating, or working, Na-conserving mechanisms will eventually be inadequate, particularly in hot environments where low-Na forages are consumed.

French (1955) concluded that salt is the most needed nutrient for livestock throughout East Africa. According to Sutmöller et al. (1966), the insufficiency of Na is the most widespread mineral deficiency in the Amazon Valley of Brazil ($\simeq$3.5 million square kilometers). From Bolivia, steers receiving salt had significantly higher average weights than unsupplemented controls had, 385 versus 370 kg, respectively (McDowell et al., 1984). In Zambia, Walker (1957) reported that the daily feeding of 28 g of

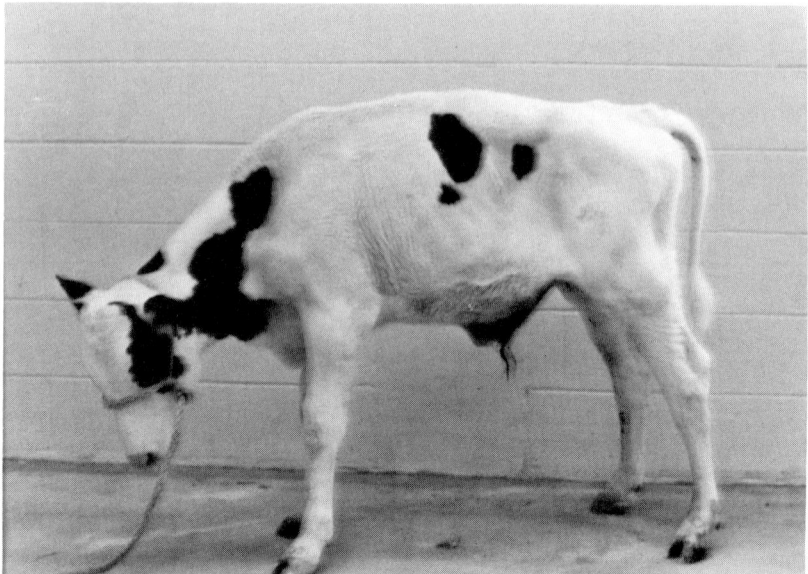

Fig. 10.2. This calf had been on the low-Cl diet and abomasal contents removed for 22 days when photo was made. It died 2 days later. Notice the emaciated appearance and the spraddle-leg stance. The animal had difficulty in walking and maintaining its balance. (Courtesy of M. W. Neathery, University of Georgia, Athens, Georgia.)

NaCl resulted in a marked increase in weight gains for grazing steers. Australian workers (Murphy and Plasto, 1973) noted dramatic responses in beef cows supplemented with salt. Thailand cattle supplemented with Na had higher weight gains, reduced calf mortality, and better reproductive performance than unsupplemented controls had (Falvey, 1980). Evaluation of salt deficiency in ruminants is discussed in Chapter 15 of this volume. Craving for salt is the earliest and most obvious criterion for Na deficiency. However, Na concentration in the urine and saliva are more accurate indicators of Na status. Because of animals' rapid reaction to deficiency long before clinical signs appear, the best criterion for assessment of Na status is concentration of Na and K in saliva [Netherlands Committee on Mineral Nutrition (NCMN), 1973]. A deficiency causes a fall in Na and a rise in K.

E. Prevention and Control

The amount of supplemental salt needed varies with amounts consumed in feeds and in water. Tropical forages normally do not contain

TABLE 10.1
Forage Sodium Concentrations (%) for Selected Tropical Regions

Location	Season	Number of samples	Mean	Percentage of samples below 0.06% of Na
Malawi[a]	dry	21	0.05	96.9
	wet	48	0.05	95.7
Bolivia[b]	dry	8	0.03	91.7
	wet	16	0.03	83.3
Bolivia[c]	dry	20	0.01	100
	dry	84	0.01	100
Dominican Republic[d]	dry	33	0.07	78.0
Colombia[e]	dry	36	0.02	100
	wet	35	0.03	100
Guatemala[f]	dry	84	0.03	55.0
	wet	84	0.09	88.0

[a] Mtimuni (1982).
[b] McDowell et al. (1982).
[c] Peducassé et al. (1983).
[d] Jerez et al. (1984).
[e] Vargas et al. (1984).
[f] Tejada (1984).

sufficient quantities of Na to meet the requirements of grazing livestock throughout the year. This inadequacy is overcome by the practice of providing common salt ad libitum. The salt needs of grazing cattle, for example, can easily be met with mineral mixtures containing 20–35% salt and consumed at a rate of 45 g/head daily. It is recommended that feedlot rations contain 0.25% added salt, one-half of the 0.5% level recommended a few years past. An advantage of the lower salt level in modern feedlot rations is the prevention of salt buildup in feedlot waste and lessening problems in waste treatment and utilization as a fertilizer.

Supplementary Na is invariably supplied in practical husbandry conditions as common salt because of its palatability, relatively low cost, and ready availability. Unfortunately neither the cost nor the availability of this supplement is always satisfactory in tropical, developing countries where its need may be the greatest (Underwood, 1981). If the cost is too high for delivering common salt to the areas where livestock are produced in tropical countries, the expense may not be justified unless there is a substantial response. There are additional advantages for providing salt to grazing livestock other than as a nutrient requirement. Salt is used as a management tool to control animal location in different pastures of a ranch. The use of salt blocks or free-choice supplements in scattered localities is a method to distribute grazing animals throughout a range area (Cunha, 1983). Also, salt can be placed in less frequently utilized forage areas to increase forage use.

F. Toxicity

Most animals can tolerate large quantities of dietary salt when an adequate supply of water is available. Clinical signs following consumption of water with more than 7000 ppm of dissolved salt include low consumption of feed and water, mild digestive disturbances, low rates of gain, and diarrhea.

Salt toxicity for grazing livestock can occur when drinking water contains high levels of salt or when animals are deprived of salt and then have access to the mineral without access to sufficient water. In large areas of the arid or semiarid regions of Africa, Asia, America, and Australia, millions of livestock subsist for many months of the year upon water supplies of high saline content (Underwood, 1981). Salt concentrations and tolerance of animals for salt in water are discussed in Chapter 3 of this volume. Water containing salt concentrations up to 5 gm/liter is safe for lactating cattle and up to 7 gm/liter for nonlactating cattle and sheep, but cattle and sheep can adapt to concentrations considerably higher than these levels, at least in temperate climates. In such an environment, where the winters were cool to mild and the forage at this time was high in moisture, sheep were found to tolerate water containing 13 gm/liter.

III. MAGNESIUM

A. Introduction

The practical importance of Mg is its relationship to the serious metabolic disorder grass tetany. The disease is also referred to as lactation tetany, grass staggers, and wheat pasture poisoning. Because of the low serum Mg of afflicted animals, it is also known as "hypomagnesemic tetany."

Grass tetany is a serious problem of grazing ruminants in many parts of the world. Clinical tetany is endemic in some countries, affecting only a small proportion of cattle (1–2%). However, individual herds may report tetany as high as 20%. Females are mainly affected, with cattle more susceptible than are sheep or goats. This is especially a problem with grazing beef cattle, where a first indication of grass tetany may be finding one or more dead animals. Excellent and extensive reviews concerning hypomagnesemic tetany are cited by Rendig and Grunes (1979) and Underwood (1981).

B. Metabolism

Magnesium is present in the body of the animal in a proportion of approximately 0.05%, with about 60–70% found in the skeleton and the remainder in the soft tissue. Besides being an essential constituent of the bones and teeth, it participates directly or indirectly in the function of a wide array of enzymes that require Mg for optimum activity. Magnesium is involved in the metabolism of carbohydrates and lipids, is required for cellular oxidation, and exerts a potential influence on neuromuscular activity, low concentrations inducing tetany.

The rumen–reticulum is the major site of Mg absorption in ruminants, with the large intestine playing a major role under some circumstances. In ruminants, the absorption rate of Mg varies from 5 to 30%. Low absorption is caused by high K content (due to a higher negative electrical potential in rumen fluid against blood), low fiber content in feed (due to a more rapid feed passage along the forestomaches), and high pH in rumen fluid (due to intake of protein-rich feeds) (Meyer, 1976). Young, lush pastures would be detrimental to Mg absorption by having high concentrations of protein and K and being low in fiber, which would increase the passage rate and thereby reduce absorption.

It is well established that high dietary K, experienced by ruminants grazing immature forages, will decrease Mg absorption from the gut and thereby induce hypomagnesemia and tetany. Preintestinal Mg absorption was decreased 39% when 4.8% was fed to sheep (Greene et al., 1983). Other workers have emphasized the importance of the Na:K ratio, ammonia concentrations in rumen contents, and the binding of Mg by fatty acids or bacterial proteins to Mg absorption. Endogenous Mg is primarily excreted via feces, but urine is suggested to be the major excretion route for Mg absorbed in excess of requirements (Rook et al., 1958). Milk is the main source of Mg losses from the body in milking cows (Meyer, 1976).

The Mg reserves of the body (mainly in the skeleton) for compensating an Mg deficiency are quite low. The Mg content in bone ash varies from 0.7% in young animals to 0.5% in older ones (Meyer, 1976). Also, the amount that can be mobilized from bone decreases with the age of the cow. Blaxter and McGill (1956) calculated that older cows with more than six calves were 14 times more likely to develop grass tetany than were first-lactation heifers, which relates to available bone reserves of Mg.

C. Requirements

Magnesium requirements for various ruminant animals according to NRC requirements have been presented in Chapter 2 of this volume. The dietary Mg requirements of ruminants vary with the species and breed of animals, age and rate of growth or production, and biological availability in the diet.

Minimum needs of sheep and cattle for growth can generally be met by pastures or rations containing 0.10%. A higher proportion, 0.18–0.20%, is considered necessary for lactating cows. Magnesium requirements for gestating beef cows have been estimated to be between 7 and 9 g/day, and between 18 and 22 g/day during lactation (Fontenot, 1980). In order to meet the lactation requirement, forages would need to contain between 0.16 and 0.19% Mg. High dietary levels of protein, K, calcium (Ca), P, and possibly aluminum (Al) increase requirements for Mg due to their depression of Mg absorption in ruminants. Other factors, such as high levels of dietary organic acids or higher fatty acids, have been implicated in the incidence of grass tetany (Ammerman and Goodrich, 1983). For high-yielding dairy cows on pastures heavily fertilized with N and K, which lowers availability of Mg, Dutch workers state the Mg requirements as 0.30%–0.38% of dry matter (NCMN, 1973).

Biological availability of different sources of Mg for ruminants is considerable. Peeler (1972) found that the availability of Mg ranged from 10 to 25% in forages and from 30 to 40% in grains and concentrates. Greater Mg availability in concentrates, along with the substantial amounts of grain usually fed to lactating dairy cows, is one of the major reasons why grass tetany is less prevalent among dairy cows than among beef cows in the United States. This would not be the situation in many developing tropical countries, since dairy cows often do not receive concentrates in substantial quantities.

Kemp et al. (1961) reported that Mg availability improves with increasing maturity of grasses and may be decreased by heavy K and N fertilization. Likewise, usually Mg in preserved forages is more available than in pastures. The availability of Mg may range from 5 to 33% in succulent feeds and from 10 to 40% in hay and concentrates (Wilkinson and Stuedemann, 1979).

D. Deficiency

Grass tetany (or hypomagnesemia) is a complex ruminant metabolic disorder that is affected by forage species and mineral composition, soil properties, fertilizer practices, season of the year, temperature, animal

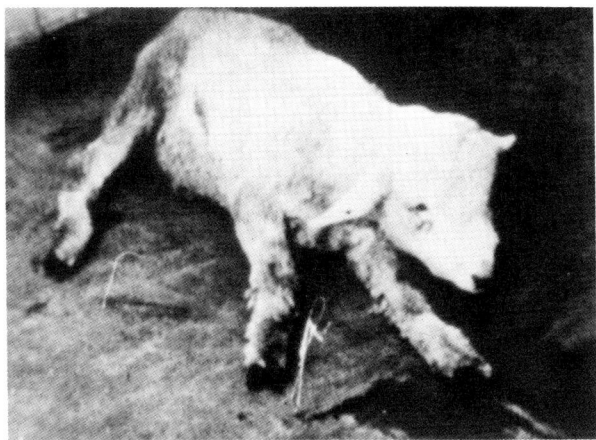

Fig. 10.3. Magnesium deficiency in a lamb. Note the stiff legs. (Courtesy of U. S. Garrigus, University of Illinois, Urbana, Illinois.)

species, breed, and age. The name of grass tetany is not entirely appropriate because the disease is not limited to animals receiving grass and is characterized by convulsions rather than tetany.

Magnesium deficiency, expressed by clinical signs of hypomagnesemia tetany, is encountered in both grazing ruminants and young ruminants reared too long on milk, without access to other feeds. Clinical signs in young ruminants include reduced appetite, increased excitability, profuse salivation, and convulsions (Fig. 10.3). Signs of grass tetany in older animals are similar to those reported for deficient young animals.

Boling (1982) has described the external or visible signs of grass tetany syndrome in cows. Initially, the cow may have a depressed appetite and exhibit a dull, lethargic appearance. As the condition progresses, signs of stiffness may be apparent as she walks, and ultimately, a staggering gait may become apparent. As severity of the condition progresses, the cow becomes highly excitable and nervous and has readily visible muscular tremors. Chewing, hypersalivation, and blinking of the third eyelid are particularly characteristic of hypomagnesemic tetany. In the most severe stage, the animal collapses to the ground with continuation of the tetanic muscular spasms. The legs will usually thrash the ground around the cow, uprooting forage. Death occurs after the collapse stage of tetany if the animal does not receive medical treatment (Fig. 10.4). Crookshank and Sims (1955) note that 6–10 hr are usually required from the time of the first clinical signs until the animal passes into a comatose condition. If treatment is not initiated before coma, there is little chance of recovery.

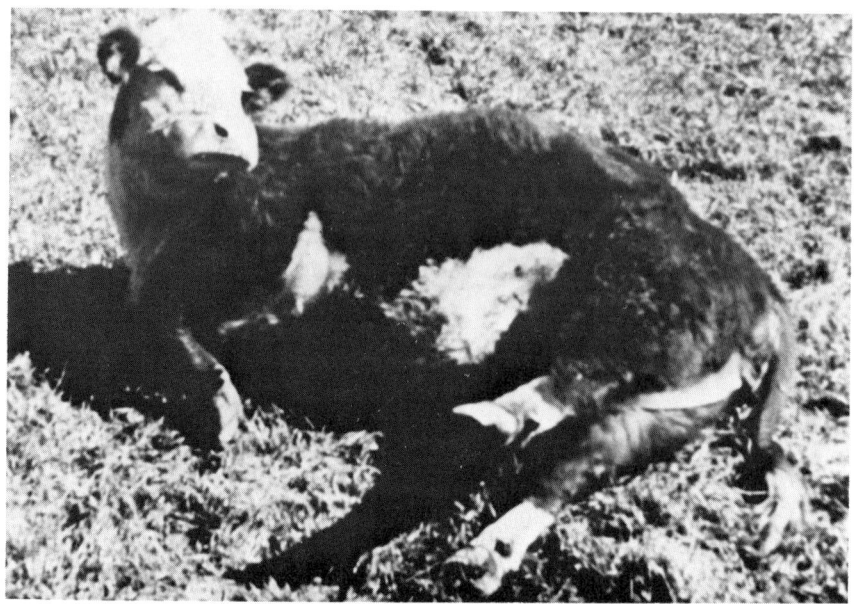
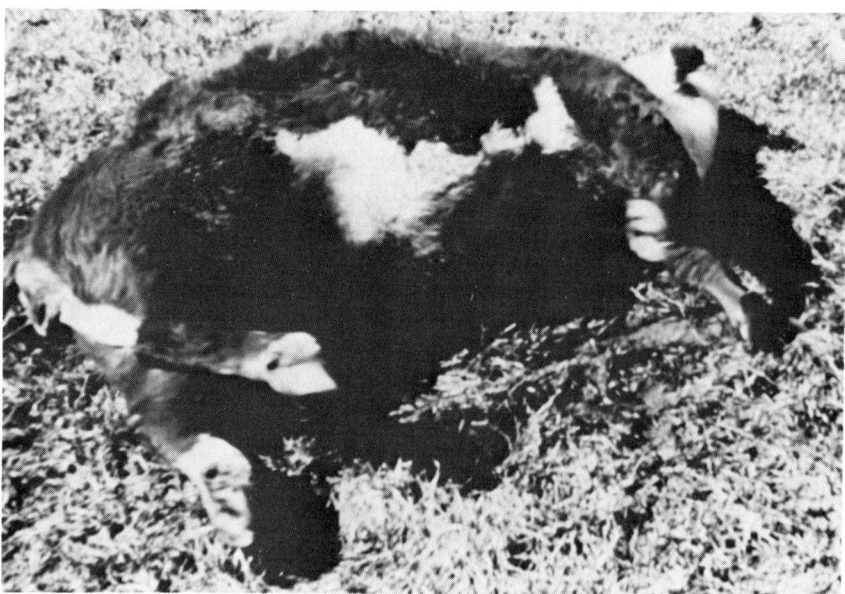

Fig. 10.4. Cow in collapse stage of tetany and death from tetany. Note area around forelegs where ground has been thrashed during convulsions. (Courtesy of J. A. Boling, C. W. Absher, and D. E. Miksch, University of Kentucky, Lexington, Kentucky.)

Confirmation of grass tetany is only justified when blood or urine samples are low in Mg (see Chapter 15 of this volume.) Likewise, a reduction in Mg content of cerebrospinal fluid (<1.6 mg %) is an accurate indicator of deficiency (Meyer, 1976). For diagnosis the following ranges in serum plasma Mg level for cattle and sheep are as follows: normal values, 1.8–3.2 mg %; slight hypomagnesemia, 1.2–1.8 mg %; and severe hypomagnesemia, 1.2 mg % or less.

Magnesium concentration in blood plasma does not fall until there is a severe deficiency. In contrast, an excess or a lack of Mg is immediately reflected in daily excretion of Mg in urine. Hence, daily urinary excretion is a better criterion of Mg supply than the level of plasma concentration. Tentative criteria for Mg in urine are as follows: more than 10.0 mg/100 ml, adequate to liberal; 2.0–10.0 mg/100 ml, inadequate; less than 2.0 mg/100 ml, severe deficiency and danger of tetany. A rough assessment of supply can be obtained from the content of Mg, N, and K in pasture. This approach is more accurate when the pasture is sampled close to the date of grazing. If the dates are more than a week apart, the assessment is unreliable. This method can be used only for grazing cattle, whereas the urine method is reliable on indoor rations as well as pasture (NCMN, 1973).

Grass tetany generally occurs during early spring, or a particularly wet autumn, among older animals grazing grass or small-grain forages in cool weather. Occurrence of tetany is seen where ruminant production is highly developed, high-quality pastures are available, and high-yielding, quick-maturing stock is raised. Voisin (1963) suggests that present-day farming methods have given rise to imbalance in the soil and herbage, which is upsetting the Mg metabolism of grazing ruminants. It is apparent that the incidence of a Mg deficiency is influenced as much or more by management considerations as by geographical location.

According to Underwood (1981), grass tetany, in its acute form, results from the insufficiency of Mg or from unavailable Mg in the diet, associated with inability of the animal to mobilize Mg from its skeleton in sufficient quantities to maintain levels of Mg in the serum, which is influenced by known as well as unknown dietetic factors. Tetany generally does not occur without tetany clinical signs. Although not characterized by death, incidence of nonclinical hypomagnesemia is far greater than that of clinical tetany, and economic consequences of lowered production are substantial (NCMN, 1973). Chronic marginal deficiency of Mg will result in reduced feed intake and performance. Suboptimal performance can be related to a decline in cellulose digestion. Studies with lambs and steers suggest that feed intake and cellulose digestion decline faster than does serum Mg when Mg-free diets are fed (Chicco *et al.*, 1973; Emery, 1976).

Hypomagnesemia tetany in sheep is almost exclusively a disease of the first 8 weeks of lacation; incidence is highest 1–4 weeks after lambing (Herd, 1966). However, in Australia a high incidence of hypomagnesemia tetany in breeding ewes has been correlated with periods of rapid winter growth of pastures (Underwood, 1981).

Clinical tetanies in cattle, milk-fed calves, and sheep can combine hypomagnesemia and hypocalcemia (low blood Ca). A high incidence of hypocalcemia has been reported with wheat-pasture poisoning (Bohman et al., 1983). It is apparent that both Ca and Mg metabolism are interrelated in this malady. Contreras et al. (1982) demonstrated that hypomagnesemia reduced Ca mobilization rate in experimental ruminants; therefore, hypomagnesemia may make livestock prone to hypocalcemia.

Incidence of hypomagnesemic tetany occurs in most European countries, North America, Australia, South Africa, and New Zealand. Developing tropical countries where Mg tetany is encountered or highly suspected include: Argentina, Brazil, Chile, Colombia, Costa Rica, Guatemala, Guyana, Haiti, Honduras, Jamaica, Kenya, Malawi, Peru, Surinam, Trinidad, Uruguay, and Venezuela (see Chapter 16 of this volume.) Viana (1978) reviewed the literature on Mg deficiency in Latin America and reported a scarcity of data. Reports of Mg deficiency are much more prevalent in temperate than in tropical regions. Some of the reasons for fewer reports of grass tetany in tropical regions include the prevalence of the condition during cooler temperatures (8–14°C), and for many tropical countries, the generally low productivity of pastures and of herds, and lack of fertilization of pasture with N and K. Nevertheless, grass tetany can be a problem for grazing ruminants in tropical countries since forages are often low in Mg. Slightly over one-third of 288 forages included in the 1974 "Latin American Tables of Feed Composition" contained 0.2% Mg or less (McDowell et al., 1977).

E. Prevention and Control

Animals that develop tetany should receive medical treatment immediately. Treatment can include subcutaneous injection of a single dose of 400 ml of a 25% solution of $MgSO_4$ or intravenous injection of a similar dose of Mg lactate (Underwood, 1981). These treatments will restore serum Mg of an affected cow to near normal within about 10 min. Serum Mg concentrations will fall again unless the cow is immediately removed from the tetany-producing pasture and fed Mg adequate diets.

Several safe and practical means of raising the Mg intakes of animals enough to prevent losses from tetany have been devised. Fertilizing pastures with relatively high levels of Mg increases Mg in the forage. However, applications of large amounts of Mg are required. This method of control has limitations on many soil types and usually has to be accompanied by other means of supplying additional Mg. Dusting or spraying pastures with Mg compounds increases Mg of the forage, but the Mg is easily washed off by rain.

For calves and cows that are being fed concentrates, provision of 50 gm of MgO per day in 300–400 gm of concentrate mixture is adequate. Other nutritionists have recommended from 10 to 20 gm of Mg daily per head of mature cattle. Unlike common salt (NaCl), most Mg salts are quite unpalatable. An important practical aspect in feeding supplemental Mg is combining it with other palatable ingredients. Various combinations of MgO with salt, protein supplements, molasses, other concentrate ingredients, and other feeds have been used or studied extensively (Miller, 1979).

The success of cobalt (Co) bullets in providing supplemental Co led to the development of Mg alloy bullets. These bullets are presumably retained by the ruminoreticular fold and release a limited amount of Mg. The amount of Mg released daily is rather low. Some of the bullets are regurgitated and the rate of decomposition of the bullets is variable (Fontenot, 1980).

Ammerman *et al.* (1972) found that the biological availabilities of Mg supplements in reagent-grade $MgCO_3$, MgO, and $MgSO_4$ were 43.8, 50.9, and 57.6%, respectively (see Chapter 17 of this volume.) If cost per unit of Mg and its biological availability are considered, Mg oxide generally is the best form of Mg for supplementation purposes (Wilkinson and Stuedemann, 1979). However, if supplementation is via water, more soluble forms of Mg including sulfate, chloride, and acetate must be used.

Supplemental Mg, as part of a concentrate mixture, is the best way of ensuring adequate intake by ruminants. When concentrates are not fed, which is more often the situation for grazing livestock in tropical countries, free-choice Mg feeding is a commonly recommended practice. There is a diversity of free-choice methods to provide Mg. Several relatively successful formulas are as follows: (1) MgO plus molasses at a ratio of 1:1; (2) 97% molasses + 3% $MgCl_2$ (often with urea and a source of P); (3) equal parts of MgO, salt, bonemeal, and grain; and (4) a 1:1 ratio of salt and MgO. In the southeastern United States, a complete mineral mixture with 25% MgO (14% Mg) has been effective in preventing grass tetany in beef cattle (Cunha, 1973).

In the case of Mg, an oral supplement would only be of value during the seasonal occurrences of grass tetany (Allcroft, 1961). Unfortunately, many commercial Mg-containing, free-choice mineral supplements are often of little value because: (1) they contain inadequate quantities of Mg to protect against tetany during susceptible periods, and (2) provision of such supplements to normal animals during nonsusceptible periods is useless as a prophylactic measure, since additional Mg will not provide a depot of readily available Mg for emergency use. In some cases, however, some producers will feed Mg supplements about a month before the Mg tetany season, believing it decreases the amount of Mg needed daily during the susceptible period.

F. Toxicity

Toxicosis due to ingestion of natural feedstuffs high in Mg has not been reported and does not appear likely. NRC (1984) established the maximum tolerable level of Mg as 0.4% of the diet. Diets commonly fed to ruminants are well below toxic levels. Holstein calves were fed toxic levels of 1, 2, and 4% supplemental Mg (oxide), with diarrhea the most obvious effect of excess Mg (Gentry *et al.*, 1978). Intensity of diarrhea was closely related to dietary Mg, with reduced feed consumption and gains in evidence for the higher (2 and 4%) Mg concentrations.

IV. POTASSIUM

A. Introduction

While K is the third most abundant mineral in the body, this element has been generally ignored for ruminant diets. Several changes in recent years have tended to lower the K contents of rations and create a more critical situation for this element. Reduced use of natural proteins (urea substitution) and roughages have lowered K intake. High-forage diets typically contain several times the amount of K present in high-grain diets (Ammerman and Goodrich, 1983). Various types of stress in different species, including that derived from shipping animals and as a result of heat, have resulted in higher K requirements.

B. Metabolism

Potassium is chemically very much like Na and is associated with Na in many biological systems. In contrast with Na, which is the main electrolyte in the plasma and extracellular fluids, K is present primarily inside the cells (Thompson, 1978). Potassium is essential for life, being required for a variety of body functions including osmotic balance, acid–base equilibrium, several enzyme systems, and water balance. An ionic balance exists between K, Na, Ca, and Mg. Potassium is not readily stored and must be supplied daily in the diet. There are no appreciable reserves other than that in the muscle and nerve cells.

Potassium is absorbed mainly from the small intestine and, to some extent, in the large intestine. The majority of K excretion is in the urine and also via sweat and secretion in milk. There is more K in milk than any other element. Aldosterone and Na intake affect K excretion. The hormone increases Na reabsorption in the kidney, and there is usually an inverse relationship between Na and K excretion.

C. Requirements

The requirement for K is higher for ruminants than for nonruminants. For ruminant species the requirement is estimated to be between 0.5 and 0.8% (see Chapter 2 of this volume). The K requirement appears to be increased for livestock under stress. Excitement tends to increase urinary loss of K, and diseases with fever or diarrhea further increase K loss. A study from Texas revealed increased weight gains for steers that had been stressed by shipping when fed feedlot rations containing 1.0–1.5% K (Hutcheson, 1979). Other results substantiate advantages of higher dietary K levels for beef cattle grazing winter range. Karn and Clanton (1977) concluded that weaning calves should receive supplements containing at least 2% K.

Florida studies indicated that 0.8% K is not adequate under heat stress, particularly with high-producing dairy cows (Beede et al., 1983). These researchers conclude that lactating dairy cows may have a higher dietary requirement for K compared with other domestic animals because of lactational stress, associated with higher milk production and high K content of milk, and because of heat stress, due to an increased loss of endogenous K through sweating and decreased daily K intake. Increasing K from 0.66 to 1.08% increased feed intake and milk yield in heat-stressed lactating cows (Beede et al., 1983). Increasing the K level in the diet also increased the requirements for Na.

D. Deficiency

Potassium deficiency for ruminants results in nonspecific signs such as slow growth (Fig. 10.5), reduced feed and water intake, lowered feed efficiency, muscular weakness, nervous disorders, stiffness, decreased pliability of hide, emaciation, intracellular acidosis, and degeneration of vital organs (Thompson, 1978). Beede et al. (1983) reported that K deficiency for lactating dairy cows resulted in dramatic reductions in feed and water intake, milk yield, and blood plasma K concentrations within 3–5

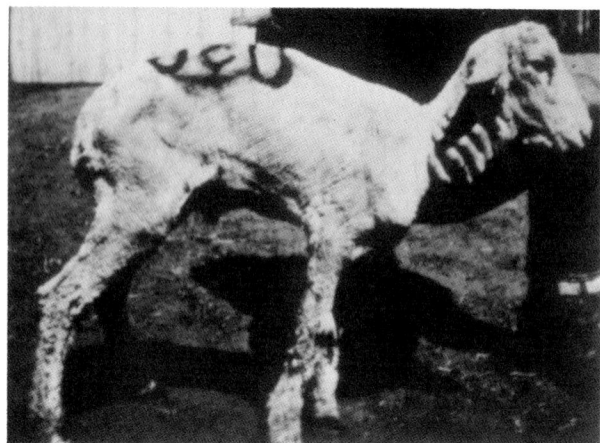

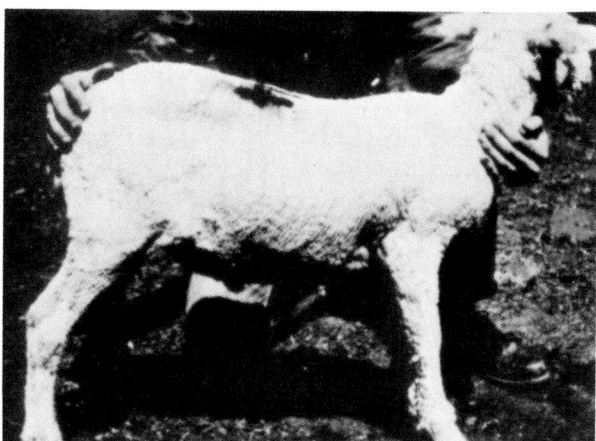

Fig. 10.5. The lamb at top received a K-deficient ration (0.1% K), whereas the lamb at bottom received sufficient dietary K (0.6% K). (Courtesy of R. L. Preston, University of Missouri, Columbia, Missouri.)

days after administration of a K-deficient diet. Near complete inanition, pica, and death of three cows occurred (Fig. 10.6). Potassium therapy of deficient cows reversed the condition within 12–24 hr.

The K content of many concentrates, which would be basic ingredients for feedlot cattle and high-producing dairy cows, is below the requirement. The estimated requirement of 0.6–0.8% K for beef and dairy cattle receiving high grain diets would not be provided, since grains often contain less than 0.5% K. The likelihood of K deficiency for ruminants will increase as high-forage diets are replaced with greater quantities of low-K-containing grains and nonprotein nitrogen supplements (i.e., urea).

Fig. 10.6. Potassium deficiency in a dairy cow. During 4 weeks while fed a K-adequate diet (1.1% K), this cow (top) consumed an average 23.6 kg-ration of dry matter and yielded 26.4 kg of milk per day. Abruptly switched to a K-deficient diet (0.12% K; same basal diet as adequate diet, equal in Cl but without KCl), the same cow's feed intake dropped 60%, milk yield declined 54%, and water intake dropped 43% within 4 days. By the eighth day of K restriction, the cow had lost 109 kg of body weight (presumably much was a loss of gut contents). Potassium restriction resulted in pica (by day 3) and severe inanition. On the eighth day of restriction (12 hr after the bottom picture was taken) the cow died exhibiting tetany-like symptoms. (Courtesy of D. K. Beede and P. G. Mallonée, Department of Dairy Science, University of Florida, Gainesville, Florida.)

Research directed toward determining K adequacy for ruminants consuming roughage rations is limited. Potassium in growing forage is usually quite high (1–4%); thus, it had been assumed that grazing livestock consuming primarily a forage diet would receive adequate K. Present information indicates that mature winter pastures that have weathered or hay that has been exposed to rain and sun or was overly mature when harvested can have K levels that are less than adequate for good nutrition (Karn and Clanton, 1977; NRC, 1980).

In tropical regions, it is possible that K deficiencies could arise in view of the decreasing content of this mineral with increasing forage maturity during the extended dry season and the use of urea that supplies none of this element. Low K values have been reported for forages in Florida, Brazil, Panama, Nigeria, Swaziland, Uganda, and Venezuela, (McDowell et al., 1983). In Brazil, average K content of six grasses at 4 weeks was 1.42% versus 0.30% at 36 weeks of age (Gomide et al., 1969). In Florida, a 1979 forage analysis from four regions indicated adequate K in September–October, but five of seven ranches sampled in February–March contained forage K concentrations considerably less than 0.6% (Kiatoko et al., 1982). Kalmbacher and Martin (1981) report creeping bluestem (*Schizachyrium stoloniferum*) affected by seasonal patterns and low K concentrations, ranging from 0.33 to 0.54%, with an average of 0.42%.

There are very few confirmed reports of K deficiency for ruminants grazing exclusively forages. One report from Nigeria indicated clinical manifestations of K deficiency in 27 cattle, raised in the traditional semi-nomadic herding system, that consumed forages analyzing 0.20% K (Smith et al., 1980). Clinical signs quickly disappeared after administration of K. The main reason for lack of widespread K deficiency, even when forages contain less than the requirement, is likely due to other nutrient deficiencies of forages. Mature tropical forages are often deficient in energy, protein, P, Na, Ca, and a number of trace elements. It is likely that a K deficiency will not be expressed as long as there are other nutrients that are even more deficient. Research is needed to evaluate any benefits derived from K supplementation to grazing livestock, since there is almost no information on this subject.

Evaluation of a K deficiency is difficult. Low serum K analyses have some diagnostic value for establishing a forage deficiency, but low serum K may be caused also by malnutrition, negative N balance, gastrointestinal losses, and endocrine malfunction. Reduced feed consumption appears to be an early sign of inadequate dietary K. Because reliable evaluations of K deficiency based on tissue analyses are not available, dietary K concentration appears to be the best indicator of K status (see Chapter 15 of this volume).

E. Prevention and Control

Potassium supplementation should always be considered when substantial portions of dietary concentrates are substituted for roughages. Depending upon K levels in forages and other ingredients used, it may be necessary to add supplemental K. This is particularly true when grazing winter or dry-season range or tropical pastures and when urea is substituted for plant proteins. Several chemical forms of K, including the chloride, carbonate, bicarbonate, and orthophosphate sources, are approximately equal in value (see Chapter 17 of this volume), and K from forages also appears to be efficiently utilized.

F. Toxicity

High K content in forages during critical times of the year can be antagonistic to Mg absorption and/or utilization, and thus can influence the incidence of grass tetany (see Section III). Potassium toxicity can be mitigated by Na salts and increased intake of Mg (NRC, 1980).

NRC (1984) gives the maximum tolerable level of K as 3% of dry matter. During the early part of the growing season, grazing ruminants consume immature forages that often contain 3–4% K. The estimated tolerable level of 3% would seem illogical for these ruminants since no detectable detrimental effects have been reported for naturally high K levels (3%+) found in forages at a young stage of maturity. Also, because ingested K beyond the requirement is quickly excreted, K toxicity is not a practical problem.

REFERENCES

Allcroft, R. (1961). *Vet. Rec.* **73**, 1255–1266.
Ammerman, C. B., and Goodrich, R. D. (1983). *J. Anim. Sci.* **57**(Suppl. 2), 519–533.
Ammerman, C. B., Chicco, C. F., Loggins, P. E., and Arrington, L. R. (1972). *J. Anim. Sci.* **34**, 122–126.
Beede, D. K., Schneider, P. L., Mallonee, P. G., Collier, R. J., and Wilcox, C. J. (1983). *In* "6th Annual International Mineral Conference (IMC)," pp. 5–26. St. Petersburg, Florida.
Blaxter, K. L., and McGill, R. F. (1956). *Vet. Rev. Annot.* **2**, 35–55.
Bohman, V. R., Horn, F. P., Littledike, E. T., Hurst, J. G., and Griffin, D. (1983). *J. Anim. Sci.* **57**, 1364–1373.
Boling, J. A. (1982). *Anim. Nutr. Health* **37**, 20–24.
Burkhalter, D. L., Neathery, M. W., Miller, W. J., Whitlock, R. H., Allen, J. C., and Gentry, R. P. (1980). *J. Dairy Sci.* **63**, 269–276.
Chicco, C. F. (1972). *In* "Program de Pastos y Forrajes," United Nations Development Project, Food and Agriculture Organization, Rome. David, Panama.

Chicco, C. F., Ammerman, C. B., and Loggins, P. E. (1973). *J. Dairy Sci.* **56,** 822–824.
Contreras, P. A., Manston, R., and Sansom, B. F. (1982). *Res. Vet. Sci.* **33,** 10–16.
Crookshank, H. R., and Sims, F. H. (1955). *J. Anim. Sci.* **14,** 964–969.
Cunha, T. J. (1973). *Feedstuffs* **45,**(20) 27–28.
Cunha, T. J. (1983). "Salt and Trace Minerals for Livestock, Poultry and Other Animals." Salt Institute, Alexandria, Virginia.
Dirven, J. G. P. (1963). *Neth. J. Agric. Sci.* **2,**(4), 295–307.
Emery, R. S. (1976). *In* "Symposium Magnesium in Ruminant Nutrition," pp. 1–16. Israel Chemicals Ltd., Tel Aviv.
Falvey, L. (1980). *In* "Proceedings First Seminar on Mineral Nutrition in Thailand," pp. 8–17. Kasetsart Univ., Bangkok.
Fettman, M. J., Chase, L. E., Bentinck-Smith, J., Coppock, C. E., and Zinn, S. A. (1984). *J. Dairy Sci.* **67,** 2321–2335.
French, M. H. (1955). *East Afr. Agric. J.* **20,** 168–175.
Fontenot, J. P. (1979). *In* "Symposium Grass Tetany" (V. V. Rendig and D. L. Grunes, eds.), pp. 51–59. Am. Soc. Agron. Madison, Wisconsin.
Fontenot, J. P. (1980). *Anim. Nutr. Health* **35,** 38–40.
Gentry, R. P., Miller, W. J., Pugh, D. G., Neathery, M. W., and Bynum, J. B. (1978). *J. Dairy Sci.* **61,** 1750–1755.
Gomide, J. A., Noller, C. H., Mott, G. O., Conrad, J. H., and Hill, D. L. (1969). *Agron. J.* **61,** 120–123.
Greene, L. W., Fontenot, J. P., and Webb, K. E., Jr. (1983). *J. Anim. Sci.* **57,** 503–510.
Herd, R. P. (1966). *Aust. Vet. J.* **42,** 160–164.
Hutcheson, D. (1979). *Anim. Nutr. Health* **34,** 11.
Jerez, M., McDowell, L. R., Martin, F. G., Hargus, W. A., and Conrad, J. H. (1984). *Trop. Anim. Prod.* **9,** 12–21.
Kalmbacher, R. S., and Martin, F. G. (1981). *J. Range Manage.* **34,** 406–408.
Karn, J. F., and Clanton, D. C. (1977). *J. Anim. Sci.* **45,** 1426–1434.
Kemp, A., Deijs, W. B., Hemkes, O. J., and Van Es, A. J. H. (1961). *Neth. J. Agric. Sci.* **9,** 134–149.
Kiatoko, M., McDowell, L. R., Bertrand, J. E., Chapman, H. L., Pate, F. M., Martin, F. G., and Conrad, J. H. (1982). *J. Anim. Sci.* **55,** 28–37.
Loosli, J. K. (1978). *In* "Proceedings Latin American Symposium on Mineral Nutrition Research with Grazing Ruminants" (J. H. Conrad and L. R. McDowell, eds.), pp. 54–58. Univ. of Florida, Gainesville.
McDowell, L. R., Conrad, J. H., Thomas, J. E., and Harris, L. E. (1974). "Latin American Tables of Feed Composition." Univ. of Florida, Gainesville.
McDowell, L. R., Conrad, J. H., Thomas, J. E., Harris, L. E., and Fick, K. R. (1977). *Trop. Anim. Prod.* **2,** 173–179.
McDowell, L. R., Bauer, B., Galdo, E., Koger, M., Loosli, J. K., and Conrad, J. H. (1982). *J. Anim. Sci.* **55,** 964–970.
McDowell, L. R., Conrad, J. H., Ellis, G. L., and Loosli, J. K. (1983). "Minerals for Grazing Ruminants in Tropical Regions." Dept. Anim. Sci., Univ. of Florida, Gainesville.
McDowell, L. R., Koger, M., Peducassé, A., Loosli, J. K., Conrad, J. H., Bauer, B., and Galdo, E. (1984). *Trop. Agric. (Trinidad)* **61,** 29–34.
Meyer, H. (1976). *In* "Symposium Magnesium in Ruminant Nutrition," pp. 35–49. Israel Chemicals Ltd., Tel Aviv.
Miller, W. Y. (1979). "Dairy Cattle Feeding and Nutrition." Academic Press, New York.
Mtimuni, J. P. (1982). Ph.D. dissertation, Univ. of Florida, Gainesville.

Murphy, G. M., and Plasto, A. W. (1973). *Aust. J. Exp. Agric. Anim. Husb.* **13,** 369–374.
National Research Council (NRC) (1980). "Mineral Tolerances of Domestic Animals." Natl. Acad. Sci., Washington, D.C.
National Research Council (NRC) (1984). "Nutrient Requirements of Domestic Animals, No. 4. Nutrient Requirements of Beef Cattle," 6th rev. ed. Natl. Acad. Sci. Washington, D.C.
Neathery, M. W. (1981). *In* "Proceedings Georgia Nutrition Conference for the Feed Industry," pp. 78–81. Univ. Georgia Press, Athens.
Netherlands Committee on Mineral Nutrition (NCMN) (1973). "Tracing and Treating Mineral Disorders in Dairy Cattle." Centre for Agricultural Publishing and Documentation, Wageningen, Netherlands.
Peducassé, C. A., McDowell, L. R., Parra, L. A., Wilkins, J. V., Martin, F. G., Loosli, J. K., and Conrad, J. H. (1983). *Trop. Anim. Prod.* **8,** 118–130.
Peeler, H. T. (1972). *J. Anim. Sci.* **35,** 695–712.
Rendig, V. V., and Grunes, D. L. eds. (1979). "Symposium Grass Tetany." Am. Soc. Agron. Madison, Wisconsin.
Rook, J. A. F., Balch, C. C., and Line, C. (1958). *J. Agric. Sci.* **51,** 189–198.
Smith, O. B., Kasali, O. B., Adeyanju, S. A., and Adegbola, A. A. (1980). *J. Anim. Sci.* **51**(Suppl. 1), 396 (abstr.).
Sousa, J. C. de, Conrad, J. H., Mott, G. O., McDowell, L. R., Ammerman, C. B., and Blue, W. G. (1982). *Pesqui. Agropecu. Bras.* **17**(1), 11–20.
Sutmöller, P., Vahia de Abreu, A., Van der Grift, J., and Sombroek, W. G. (1966). "Mineral Imbalances in Cattle in the Amazon Valley," Communication No. 53. Department of Agriculture Research, Royal Tropical Institute, Amsterdam, Netherlands.
Tejada, R. (1984). Ph.D. dissertation, Univ. Florida, Gainesville.
Thompson, D. J. (1978). *In* "Proceedings Latin American Symposium on Mineral Nutrition Research with Grazing Ruminants" (J. H. Conrad and L. R. McDowell, eds.), pp. 73–79, Univ. of Florida, Gainesville.
Underwood, E. J. (1981). "The Mineral Nutrition of Livestock." Commonwealth Agricultural Bureaux, London.
Vargas, R., McDowell, L. R., Conrad, J. H., Martin, F. G., Buergelt, C., and Ellis, G. L. (1984). *Trop. Anim. Prod.* **9,** 103–113.
Viana, J. A. C. (1978). *In* "Proceedings Latin American Symposium on Mineral Nutrition Research with Grazing Ruminants" (J. H. Conrad and L. R. McDowell, eds.), pp. 59–66. Univ. of Florida, Gainesville.
Voisin, A. (1963). "Grass Tetany." Charles C. Thomas, Publ., Springfield, Illinois.
Walker, C. A. (1957). *J. Agric. Sci.* **49,** 394–400.
Wilkinson, S. R., and Stuedemann, J. A. (1979). *In* "Symposium Grass Tetany." (V. V. Rending and D. L. Grunes, eds.), pp. 93–117. Am. Soc. Agron. Madison, Wisconsin.

11

Copper, Molybdenum, and Sulfur

L. R. McDOWELL
Department of Animal Science
University of Florida
Gainesville, Florida

I.	Introduction	237
II.	Metabolism of Copper and Molybdenum	238
III.	Copper and Molybdenum Requirements	239
IV.	Copper Deficiency	240
V.	Prevention and Control of Copper Deficiency	248
VI.	Toxicity of Copper and Molybdenum	249
VII.	Sulfur	251
	A. Sulfur Metabolism	251
	B. Sulfur Requirements	252
	C. Sulfur Deficiency	252
	D. Prevention and Control of Deficiency	254
	E. Sulfur Toxicity and Interrelationships	254
	References	255

I. INTRODUCTION

With the exception of phosphorus (P) deficiency, copper (Cu) deficiency is the most severe mineral limitation to grazing livestock throughout extensive regions of the tropics (McDowell *et al.*, 1984). Copper deficiencies in ruminants occur mainly under grazing conditions, with gross signs of the deficiency rare when concentrates are fed.

Copper, molybdenum (Mo), and sulfur (S) are discussed together due to the important nutritional and biochemical interactions of these elements, especially from the standpoint of grazing ruminants. The essentiality of Cu for ruminants was first established in the 1930s when a Cu deficiency in grazing cattle was demonstrated in Florida (Neal *et al.*, 1931). The antagonism between Cu and Mo was first reported in England, when it was discovered that a severe scouring disease of cattle was caused by the ingestion of forage containing high levels of Mo (Underwood, 1981). This toxicity of Mo was overcome when cattle were treated with Cu. The effect of Mo in limiting Cu retention was demonstrated in sheep when Dick (1952) reported that this effect was only exerted in the presence of adequate amounts of inorganic sulfate. Findings indicating a three-way interaction between Cu, Mo, and S gave a great stimulus to further studies of dietary mineral interactions in livestock. They also drew attention to the significance of dietary S status in ruminants (Underwood, 1981).

II. METABOLISM OF COPPER AND MOLYBDENUM

Copper is necessary in hemoglobin production, iron (Fe) absorption from the small intestine, and Fe mobilization from tissue stores. Ceruloplasmin, which is synthesized in the liver and contains Cu, is necessary for the oxidation of Fe, permitting it to bind with the Fe transport protein, transferrin. Copper is also a component of various body pigments and is involved in the central nervous system, bone metabolism, and heart function. With a Cu deficiency, a primary defect of the organic matrix of bone is related to the failure of collagen to undergo cross-linking and maturation (Harris and O'Dell, 1974). The activity of the Cu metalloenzyme, lysyloxidase, decreases severely in Cu deficiency, and the mature collagen and elastin are not oxidized. The observed aortic aneurysms and rupture, with decreased elastin content and tensile strength in these fibers, are accredited to the failure in conversion of lysine to desmosine, the cross-linking residue in elastin.

Copper absorption occurs primarily from the upper gastrointestinal tract, with only 1–3% absorbed in ruminants. In sheep, considerable absorption takes place from the large intestine (Underwood, 1977). Dietary phytates, high levels of calcium (Ca) carbonate, Fe, zinc (Zn), cadmium (Cd), or Mo also reduce absorption. Matrone (1970) suggests Mo interferes with absorption by the formation of a Cu:Mo complex (Cu × Mo × SO_4), resulting in depression of Cu absorption. Sulfur, in the absence of Mo, may also cause a Cu deficiency due to the formation of insoluble, unabsorbed Cu sulfide in the gut. The formation of both Cu sulfide and Cu thiomolybdate would greatly reduce Cu absorption, since both render Cu unavailable [National Research Council (1984)]. High or prolonged intakes of Mo or S produce marked changes in blood Cu and subsequent tissue distribution. The inhibitory effects of Fe-rich soils on Cu availability in S-rich diets may be a carryover of sulfide as FeS from the rumen to the abomasum, where it forms insoluble CuS (Suttle and Peter, 1984). Copper excretion is an active process in which Cu is released into bile and ultimately into feces (Underwood, 1977).

In contrast to the many key roles played by Cu, the only established function of Mo is as an indispensable component of the flavoprotein enzyme, xanthine oxidase, as well as aldehyde oxidase and sulfide oxidase. These enzymes are involved with oxidation of purines and in Fe metabolism (Underwood, 1977). Absorption of Mo is from the intestine, and excretion is primarily via urine, with small amounts excreted in bile and milk.

III. COPPER AND MOLYBDENUM REQUIREMENTS

The precise Cu requirement of the ruminant awaits evaluation of the various interfering factors for absorption and metabolism. General recommendations for the minimum Cu requirement of grazing livestock cannot reasonably be made without reference to pasture Cu, Mo, and S concentrations. Where soils have been limed and Ca carbonate is high, the Cu requirement of merino sheep has been placed at 10 ppm, whereas in western Australia where soil liming does not occur and the Mo contents of the pastures are generally below 1.5 ppm, 6 ppm of Cu was found adequate (Underwood, 1981). Increasing soil pH favors plant uptake of Mo, while reducing concentrations of Cu. When Mo and S intakes are normal, a minimum of 5 ppm is suggested for British breeds of sheep and 6–7 ppm for merino sheep (Miller, 1979).

The Cu requirements of cattle for growth and health are higher than those of sheep. The suggested minimum Cu requirement for cattle is about 8–10 ppm (NRC, 1978, 1984). Even when Mo intakes were low in studies with Holstein calves, 8 ppm Cu (dry diet) did not entirely meet their needs, and 10 ppm was suggested as the minimum requirement (Mills et al., 1976).

Molybdenum and S can either increase or decrease the Cu status of an animal. From Argentina, Bingley and Carrillo (1966) reported that when the Cu:Mo ratio of forages in the presence of adequate S is less than 2.8:1, the incidence of hypocuprosis in the animals is evident. Miltimore and Mason (1971) reported that the critical Cu:Mo ratio in feeds appears to be 2.0:1, and feeds or pastures with lower ratios result in a "conditioned" Cu deficiency.

The Cu requirements of cattle on pastures have been shown to be higher than those of cattle fed the same forage as hay. The Cu content of forage declines with increased maturity. However, the Cu status of cattle grazing more mature forage is better than in those grazing immature forage (Hartmans, 1969). This suggests low availability of Cu in immature forage, or a change in the relationship of Cu to interfering factors.

The Mo requirements of ruminants are extremely low, with no Mo deficiencies having been reported or identified in grazing ruminants. Requirements for Mo are not established, but are probably less than 2 ppm (NRC, 1984). An exact estimate of the Mo requirement is impossible since Cu and S alter Mo metabolism. Furthermore, the majority of research with Mo has concentrated on this interrelationship with Cu and S, rather than on a specific requirement (NRC, 1984). Work by Ellis *et al.* (1958) with growing lambs showed faster gains and improvement in cellulose digestiblity when dietary Mo was increased from 0.36 to 2.37 ppm. However, many pastures grazed by ruminants contain less than 0.36 ppm of Mo, with no evidence of a deficiency.

IV. COPPER DEFICIENCY

Copper deficiency is a severe limitation to grazing ruminants and has been observed in many parts of the world. A total of 34 tropical countries of Latin America, Africa, and Asia (see Chapter 16 of this volume) have reported deficiencies, more than any other mineral with the exception of P. It is noteworthy that many developing countries have as yet not reported Cu deficiencies due to lack of detection methods, including chemical analysis and, therefore, are unable to realize the substantial production losses as a result of an inadequacy.

Of the numerous world reports of Cu deficiency in ruminants, only a few are concerned with a deficiency induced by the presence of unusually low concentrations of Cu (<3 ppm) in the feed. The majority of world reports are concerned with a "conditioned" Cu deficiency, where normal amounts of Cu (6–16 ppm) are inadequate due to other forage constituents such as Mo, S, and other factors that block utilization of Cu (Russell and Duncan, 1956). Copper deficiencies usually occur when forage Mo exceeds 3 ppm and the Cu level is below 5 ppm (Cunha, 1973).

Ward (1977) categorized Cu deficiencies into four groups where the feed contained (1) higher levels of Mo (more than 20 ppm), (2) low Cu but significant amounts of Mo (i.e., ratio <2:1), (3) deficient Cu (<5 ppm), and (4) normal Cu and low Mo, with high levels of soluble protein. It is suggested that the last situation is the result of high intakes of soluble protein from fresh pasture, which increases the amounts of sulfide produced in the rumen, thus resulting in unavailable Cu sulfide.

Studies with cattle (Campbell *et al.*, 1974) and sheep (Suttle and Peter,

1984) have suggested that the intake of Fe may depress Cu status of ruminants. An elevated Fe intake could result from consumption of groundwater, from soil ingestion, or from an increase in plant Fe concentrations due to waterlogging of the soil. A large number of studies from tropical regions has indicated that forages grown on acid soils are extremely high in Fe (see Chapter 13 of this volume), which may be aggravating the low Cu status of grazing livestock in many regions.

A wide variety of disorders in ruminants are associated with a simple or induced (due to high Mo and S) Cu deficiency including anemia, severe diarrhea, depressed growth, change of hair color, neonatal ataxia, temporary infertility, heart failure, and weak, fragile long bones that break easily (Underwood, 1981). Dietary Mo (5 ppm) significantly delayed in cattle the occurrence of first ovulation (26 to >46 weeks) compared with controls (Phillippo et al., 1984). Anemia is a general clinical sign for most species, while other signs may be observed in one or more species. The first world report of Cu essentiality for ruminants was with a Jersey heifer in Florida (Fig. 11.1) in 1931. The anemic animal had a hemoglobin reading of 5.9 g/100 ml versus 13.7 g/100 ml after Cu administration.

A very sensitive index of a Cu deficiency is achromatrichia, or loss of pigment in hair (Fig. 11.2) and wool. Loss of hair around the eye of a cow is illustrated in Fig. 11.3 as a result of Cu deficiency. Sheep deficient in Cu also fail to impart a crimp in the wool fibers, which results in an almost straight hairlike fiber called "stringy" or "steely" wool.

Another Cu deficiency sign is the development of fragile bones (Fig. 11.4), particularly the long bones, which break easily, sometimes without apparent cause. Cattle that show these skeletal abnormalities move like a pacing horse rather than like normal cattle (Fig. 11.5). Copper-deficient cattle may die suddenly when exerted, and postmortem examination may reveal small lesions of the heart. Demyelination of the central nervous system, or neonatal ataxia, is a nervous disorder of lambs and goats. This is characterized by incoordination of movement and high mortality. Fig. 11.6 illustrates a type of mild paralysis in Saltillo, Mexico, that responded to Cu therapy. Not all of these signs necessarily occur in every Cu-deficient animal, and some may be due to a combination of causes.

Young and growing animals have higher Cu requirements and, therefore, a higher deficiency incidence than do older livestock (Mills, 1966). Munro (1957) reports that Cu therapy increased the conception rate in heifers following one or, at most, two inseminations from 52 to 95%, but no improvement was noted for cows.

Subclinical Cu deficiencies are thought to be very widespread and are likely to be of more economic significance than are easily recognized cases. Thornton et al. (1972) reported Cu supplementation during a 6-

Fig. 11.1. Jersey Heifer E-15 had a hemoglobin reading of 5.9 g per 100 ml of whole-blood on October 21, 1930 (top). Administration of ferric ammonium citrate and copper sulfate (50 Fe:1 Cu) was started on December 15, 1930; it was 13.74 g per 100 ml. E-15 was fully recovered by March 27, 1931, and became a normal cow in the dairy herd (bottom). This was the first bovine to confirm that Cu was essential for ruminants. Meanwhile, two herd mates, receiving Fe as the sole supplement to the basal ration died of anemia. (Courtesy of R. B. Becker, W. M. Neal, and A. L. Shealy, University of Florida, Gainesville, Florida.)

month period increased liveweight gains in cattle by 10–70% over controls, even though, with few exceptions, control stock showed no clinical signs of hypocuprosis. Slow growth and loss of body weight due to Cu deficiency is illustrated in cattle from Piaui, Brazil (Fig. 11.7).

Evaluation of Cu status in livestock by determination of Cu in the diet or pasture has limited diagnostic value unless other elements with which Cu interacts, particularly Mo and S, are determined also. The criterion

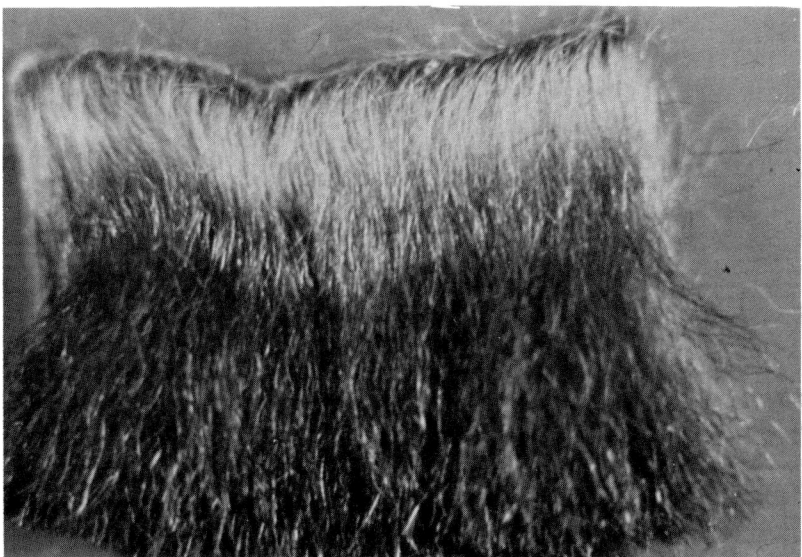

Fig. 11.2. Both illustrate hair color changes as a result of Cu deficiency. The dark color is normal when cattle receive adequate Cu. (Courtesy of B. J. Carrillo, C.I.C.V., INTA, Castelar, Argentina.)

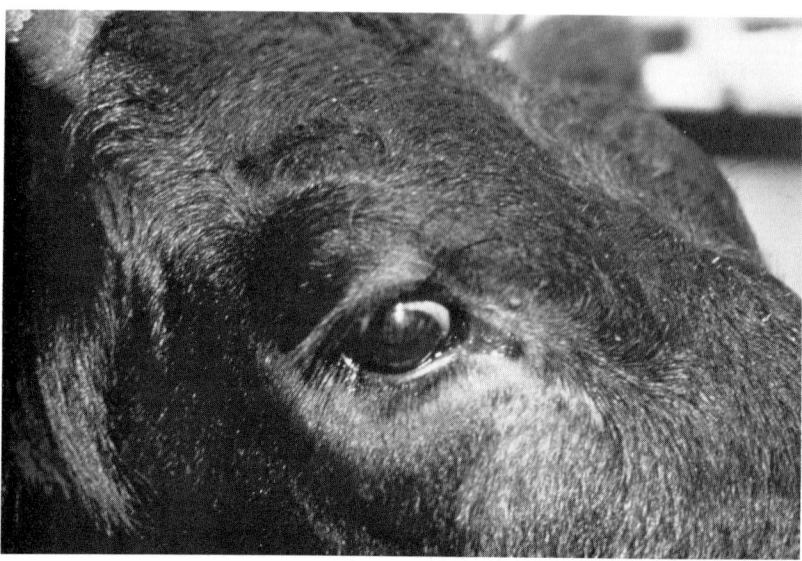

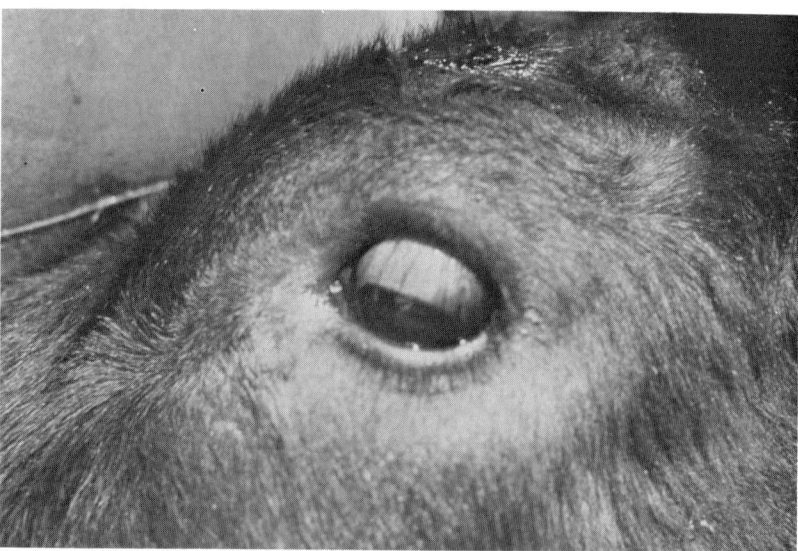

Fig. 11.3. The top photo illustrates normal hair and color around the eye of a cow in Argentina. The animal with a copper deficiency as illustrated by loss of hair ring around the eye is seen in the bottom photo. (Courtesy of Bernardo Jorge Carrillo, C.I.C.V., INTA, Castelar, Argentina.)

Fig. 11.4. Broken bones on the leg of this calf are the result of Cu deficiency. Swelling can be seen on one leg. (Courtesy of B. J. Carrillo, C.I.C.V., INTA, Castelar, Argentina.)

most widely used for Cu deficiency is the concentration of Cu in the liver. The liver is the main storage organ of the body for Cu, so liver Cu concentrations would be expected to provide a useful index of the Cu status of the animal (see Chapter 15 of this volume). Among domestic livestock, liver Cu values in healthy sheep and cattle have a normal range of 100–400 ppm on a dry basis. In sheep and cattle, liver concentrations vary slightly from birth to maturity. Liver Cu concentrations reflect the dietary status (Mills *et al.*, 1976), but they are influenced by the dietary proportions of Mo and S and by high intakes of Zn and Ca carbonate and other dietary compounds. They must, therefore, be used with caution as diagnostic aids. Evidence suggests that the Cu values below 25–75 ppm of liver dry matter should be used to differentiate deficient from normal animals.

Fig. 11.5. A dry cow showing a pacing gait characterized by an apparent stiffness in the hocks and springiness in the pasterns, typical of cows grazing certain Florida muck soil (marsh ranges) containing high Mo levels. In advanced cases, animals have difficulty breathing after minor exertion. The condition can be prevented or corrected with proper amounts of Cu to offset the high Mo in the forage (Becker *et al.*, 1965). (Courtesy of R. B. Becker, University of Florida, Gainesville, Florida.)

Fig. 11.6. There is a high incidence of paralysis in lambs found in the regions of Saltillo, Mexico. Animals with this condition respond to Cu supplementation.

Fig. 11.7. Copper-deficient cattle in the sandy coastal region of Piaui, Brazil. (Courtesy of J. Döbereiner and C. H. Tokarnia, EMBRAPA, Rio de Janeiro, Brazil.)

Whole-blood or plasma concentrations also reflect the dietary Cu status, although the normal range is wide. For sheep, cattle, and goats, the normal range is 0.6–1.5 μg/ml of Cu. It is widely accepted that whole blood or plasma Cu values consistently below 0.6 μg/ml are indicative of deficiency in sheep and cattle [Netherlands Committee on Mineral Nutrition (NCMN), 1973].

Changes in the activities of a number of Cu metalloenzymes in the blood and tissues occur during the development of Cu deficiency in livestock and offer diagnostic possibilities. It has been shown by Todd (1970) that ceruloplasmin (ferroxidase I) estimations on blood seum provide advantages over whole-blood or plasma Cu determinations because of the relative stability of the enzyme, the small size of the serum samples required, and the technical convenience of the assay.

Clinical signs, growth response, and analysis of forages and tissues have all been used to establish the extensive world areas of Cu deficiency. A total of 47% of forages included in the 1974 "Latin American Tables of Feed Composition" contained low concentrations (<10 ppm) of this element (McDowell *et al.,* 1977). Table 11.1 illustrates recent forage Cu analyses from five tropical countries, the vast majority of which were deficient in Cu. Each of these investigations also examined liver Cu concentrations and found corresponding low levels of Cu. Native forage samples were analyzed from three regions of the Colombian llanos during the early rainy, late rainy, and dry seasons, with mean Cu concentrations extremely low, ranging from 1.6 to 2.0 ppm (Lebdosoekojo *et al.,* 1980). Hill *et al.* (1962) reported the extent of Cu inadequacy in Malaysia by noting that 80% of the surveyed cattle and buffalo were Cu deficient. Sheep from herds with outbreaks of neonatal ataxia and raised under the

TABLE 11.1

Mean Forage Copper Concentrations (ppm) by Season and Percentage of Samples below Critical Concentrations for Six Tropical Countries[a]

Location	Wet season				Dry season			
	No of samples	Mean	cc	Percentage below cc	No of samples	Mean	cc	Percentage below cc
Bolivia[b]	16	5.8	10	94.4	84	1.3	10	100
Brazil[c]	120	1.5	—	—	192	2.1	—	—
Colombia[d]	35	3.6	10	100	36	2.8	10	100
Dominican Republic[e]	—	—	—	—	69	9.0	10	64.0
Guatemala[f]	84	21	10	58	84	8	8	92
Malawi[g]	48	9.5	8	52.6	21	3.1	8	89.5

[a] Critical concentrations (cc) (McDowell et al., 1984).
[b] Peducasse et al. (1983) (wet season); McDowell et al. (1982) (dry season).
[c] Sousa et al. (1980).
[d] Vargas et al. (1984).
[e] Jerez et al. (1984).
[f] Tejada (1984).
[g] Mtimuni (1982).

seminomadic conditions of the Sudan were evaluated for Cu status (Idris et al., 1976); deficient serum concentrations (<0.6 µg/ml) were found in 50% of 500 sheep analyzed.

V. PREVENTION AND CONTROL OF COPPER DEFICIENCY

Under range conditions in the tropics, Cu deficiency can be prevented by the provision of Cu-containing supplements, by dosing or drenching animals at intervals with Cu compounds, or by injection of organic complexes of Cu. Mineral supplements containing 0.1–0.2% Cu sulfate are generally consumed voluntarily by grazing animals in amounts sufficient to maintain adequate and safe total Cu intakes. Supplemental sources of Cu, from highest to lowest biological availability, include $CuSO_4$, $CuCO_3$, and CuO (see Chapter 17 of this volume).

Subcutaneous or intramuscular injection of some safe and slowly absorbed forms of Cu (i.e., glycinate and EDTA) constitute satisfactory means of treating animals in Cu-deficient areas where the pasture Mo contents are moderate, even at intervals as long as 4–7 months (Bohman et al., 1984). Copper oxide needles and Cu-containing controlled release glass are forms of supplemental Cu that have been used successfully. The application of Cu-containing fertilizers can be an effective means of raising the Cu content of pasture to levels adequate for grazing livestock and increasing pasture yields. The amounts required vary with the soil type and climatic conditions. Australian experience indicates that a single dressing of 5–7 kg/ha of $CuSO_4$ or its Cu equivalent in the form of cheaper Cu ores, is usually sufficient for 3 or 4 years, except on calcareous soils.

VI. TOXICITY OF COPPER AND MOLYBDENUM

In acute Cu toxicity, animals may experience nausea, vomiting, salivation, abdominal pain, convulsions, paralysis, collapse, and death. Hypercupremic conditions may also predispose the animal to anemia, muscular dystrophy, decreased growth, and impaired reproduction (NRC, 1980).

Chronic copper toxicity occurs under grazing conditions as a result of high Cu intake or very low Mo and S intakes. As an example, Cu toxicity in sheep under grazing conditions can occur where soil and pasture Cu concentrations are normal but Mo concentrations are very low (0.1–0.2 ppm). A Cu content in feed over 20 ppm can cause chronic poisoning in sheep (NCMN, 1973). It is estimated that cattle can tolerate 70–100 ppm of Cu over an extended period, and higher levels for shorter periods. Todd (1969), in reviewing Cu toxicity, concluded that chronic Cu toxicity in ruminants is almost entirely confined to sheep. Calves are affected much more than older cattle, but a comparable resistance to the toxicity does not occur with sheep as they mature. Feeding of manure from swine or poultry fed high-Cu diets can cause Cu toxicity in ruminants, especially in sheep (Fontenot and Webb, 1974). Likewise, fertilizing with high Cu wastes results in forages containing toxic levels of Cu for sheep.

Clinical signs of Mo toxicity are similar to Cu deficiency. An excessive intake of Mo will seriously deplete Cu reserves in cattle, quickly leading to scouring (Fig. 11.8), anorexia, anemia, loss of condition, and other signs associated with Cu deficiency. Fig. 11.9 illustrates faded haircoat in cattle consuming high forage Mo from Saltillo, Mexico. Depigmentation of skin and hair, with a loss of crimp in wool, are the clinical signs of Cu toxicity exhibited by sheep. Molybdenum levels of 5–6 ppm inhibit Cu storage and produce signs of molybdenosis (NRC, 1980). Likewise, even 2 ppm or less Mo can be toxic if forage Cu levels are sufficiently low. The chemical form of Mo may have an important effect on its toxicity. For

Fig. 11.8. The animals pictured exhibit severe diarrhea as a result of excess dietary Mo and too little Cu. (Courtesy of B. J. Carrillo, C.I.C.V., INTA, Castelar, Argentina.)

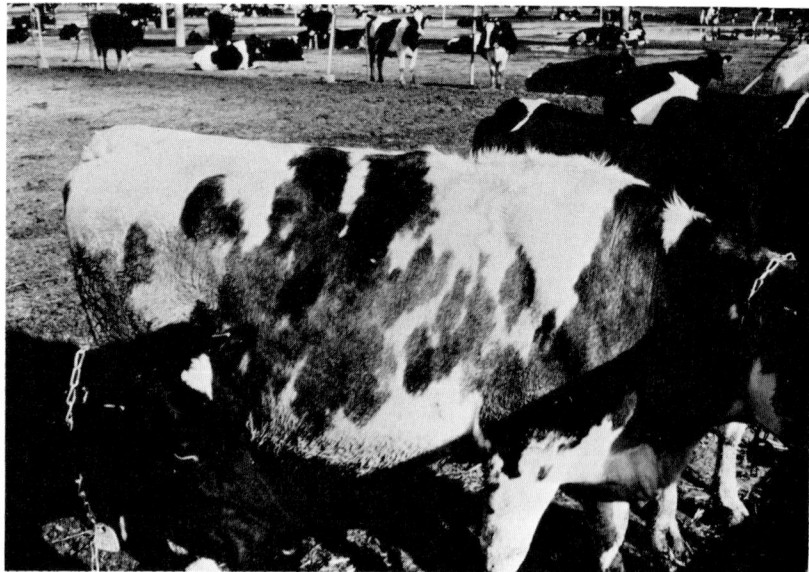

Fig. 11.9. Faded haircoat of animal in Saltillo, Mexico. Animals received diets high in Mo and low in Cu.

example, Mo in pasture may be much more toxic than a similar amount experimentally fed (Cunningham, 1950).

Molybdenum-toxic areas characteristically occur on poorly drained neutral or alkaline soils that favor Mo uptake by plants and reduce the availability of Cu. Both Mo toxicity and Cu deficiency are generally corrected by providing additional Cu to the animal diets. From severe Mo-toxic areas, injections of Cu compounds are often the preferred method of administration, since the primary site for Cu and Mo interaction is the gut (Suttle and Field, 1974).

VII. SULFUR

A. Sulfur Metabolism

Sulfur is an important element in the synthesis of protein, since two important amino acids, methionine and cysteine, contain S. Likewise, S is a part of the vitamins, thiamin and biotin, and of sulfate polysaccharides, including chondroitin. Chondroitin is a key component of cartilage, bone, tendons, and blood vessel walls. Body functions that involve S include protein synthesis and metabolism, fat and carbohydrate metabolism, blood clotting, endocrine function, and intra- and extracellular fluid acid–base balance.

Most forms of S must be either oxidized to sulfate or reduced to sulfide before they can be utilized by ruminants. This is largely a microbial process. Plants and most bacteria can reduce sulfate to sulfide as indicated by their growth with sulfate as the only source of S. The S content of plants depends largely on the amount of S in the plant protein. Methionine and cysteine usually comprise over 90% of the organic S in plants and represent the essential nature of plant proteins for animals. Sulfur absorption occurs in the rumen and small intestine, with substantial amounts absorbed through the rumen wall (Bray, 1969). Both organic and inorganic S compounds are readily and efficiently absorbed from the intestine. Approximately 75% of Na sulfate was excreted via urine in sheep, with cattle excreting a slightly larger percentage via urine (Hansard and Mohammed, 1969). Sulfur can be recycled to the rumen, with similarities to the recycling system for the urea–nitrogen system. The amount of S recycled in sheep is much less than for cattle (Kennedy *et al.,* 1975). Under grazing conditions, particularly with mature grasses rather than with legumes, and possibly under more intensive conditions, recycled S becomes nutritionally very significant (Moir, 1979).

B. Sulfur Requirements

Sulfur requirements of ruminants are not well defined. Between 0.1 and 0.32% S is the estimated requirement for grazing ruminants (see Chapter 2 of this volume). Since the S requirement for optimum microbial action appears to be the highest need for S of ruminants, the effect of S on rumen function is studied. The optimum S level for cellulose digestion *in vitro* has been reported to be 0.16–0.24% of dry matter.

It has been suggested (Kennedy *et al.*, 1968) that the S requirement for cellulose digestion is greater than that for starch. Unfortunately, beef cattle requirements have been determined with high-starch diets, resulting in unreasonably low recommended S requirements (Bull, 1979). Therefore, grazing ruminants that are consuming high-cellulose diets require a total S consumption of 0.20% or more.

The S requirement of ruminants can be approached from a consideration of the dietary N:S ratio. Tissues of cattle contain a N:S ratio of 15:1, and it has been shown that dietary ratios of 12 to 15:1 are excellent for cattle. Due to requirements of S for wool, a ratio of 10 to 12:1 is commonly recommended for sheep rations.

C. Sulfur Deficiency

Signs of S deficiency have been described as loss of weight (Fig. 11.10), weakness, lacrimation, dullness, and death. Rakes and Clark (1984) suggested that lameness in dairy cattle may be associated with S deficiency, as characterized by slower-growing, less flexible hooves. In a S deficiency, microbial protein synthesis is reduced, and the animal shows signs of protein malnutrition. A lack of S also results in a microbial population that does not utilize lactate; therefore, lactate accumulates in the rumen, blood, and urine. It is difficult to diagnose a deficiency, especially a borderline one. Serum sulfate levels have been suggested as an indicator of S deficiency, but blood lactate and dietary S levels may be the most reliable indicators of S status.

Soils of the tropics generally have low levels of S compared to those of temperate regions. Responses to fertilizer S are widespread in the tropics and have been recorded in 40 tropical countries with 23 different crops [International Fertilizer Development Center (IFDC), 1979]. The incidence of S deficiency in tropical countries is increasing due to the increasing use of high-analysis, low-S-containing fertilizers and the increased need for S brought about by increases in yields (Coleman, 1966). Shifting from the use of ordinary superphosphate (12% S) to triple superphosphate (1% S) eliminates a source of available S (IFDC, 1979). Likewise for

11. Copper, Molybdenum, and Sulfur

Fig. 11.10. Lambs fed a low-S diet. The lamb on the left received 3 g of S per pound of ration, whereas the other lamb received none. The deficient lamb exhibited excessive salivation, lacrimation, and shedding of wool. (Courtesy of U. S. Garrigus, University of Illinois, Urbana, Illinois.)

animal feeds, natural proteins contain appreciable S, but urea and other nonprotein nitrogen (NPN) compounds do not.

Animal nutritionists at one time erroneously believed that, if plants were able to grow, they contained enough S for livestock. In addition, S analysis in biological materials is comparatively difficult. For these reasons, there is a scarcity of S analyses of tropical forages. A limited amount of analysis has indicated that many tropical forages contained less than an optimum S concentration of 0.20%. Sulfur analyses of 10 forage samples from the llanos rangelands of both Colombia and Venezuela were low, ranging from 0.032 to 0.088% (Miles and McDowell, 1983). From these regions, severe leaching of soils and frequent burning of grasslands led to the assumption that many, if not most, llanos rangeland forages will be found deficient in S. Ruminants consuming large quantities of corn silage will likely be receiving inadequate S. The amount of S in corn silage samples analyzed in the Georgia Forage Testing Laboratory in 1973–1974 averaged 0.07% in dry matter, with a range of 0.005–0.19%. More than 80% of the corn silages had less than one-half the estimated 0.20% S needed for lactating cows (Miller, 1979).

A recent review (Miles and McDowell, 1983) summarized four cattle S-supplementation trials in which control diets contained between 0.04 and 0.10% S. From these studies, intake by supplemented cattle increased between 7 and 260%, and production of milk and meat increased anywhere from 6 to more than 400%. Herbage with a low S concentration has been reported in Australia, Ecuador, and Uganda, with improved ruminant animal production following S supplementation (McDowell, 1976; Stobbs and Minson, 1980).

D. Prevention and Control of Deficiency

Sulfur supplementation will most likely be needed to meet the requirements of ruminants when poor-quality roughages grown on S-deficient soils or feeds combined with NPN, such as urea, are fed. Since there is no S in urea, the element may need to be added when high levels of urea are fed. Sulfur may be provided in the diet by both organic and inorganic sources. Ruminants may utilize the S in methionine, methionine hydroxy analog, sodium (Na), Ca, magnesium (Mg), potassium (K) or ammonium sulfate salts, or elemental S. Sulfur as the highly insoluble elemental S or lignin sulfonate is much less available, and it is suggested that elemental S (flowers of S) is utilized about one-third as efficiently as in the sulfate or methionine forms. Sulfur in corn and corn silage has been found to be less available than that in Na sulfate, methionine, and methionine hydroxy analog.

Although economically prohibitive in many tropical regions, S fertilization is an effective way of increasing forage S, as well as yields. An additional value of S fertilization is that some reports from tropical regions have indicated that S fertilization may increase forage intake by improving the palatability of less palatable species [Centro Internacional de Agricultura Tropical (CIAT), 1981; Rees *et al.*, 1974]. With S-deficient Pangolagrass, applying fertilizer S increased intake by 44%, compared with only 28% when the sheep were drenched with Na sulfate (Rees *et al.*, 1974). Sulfur fertilization to *Desmonium ovalifolium* in the Colombian llanos increased foliar S, nitrogen (N), and biomass production, decreased tannin content in leaves and, most importantly, increased intake of the forage (CIAT, 1981).

E. Sulfur Toxicity and Interrelationships

The interrelationship of S with Cu and Mo has been discussed, with Cu requirements increased by both S and Mo. Likewise, the interrelationship between selenium (Se) and S is due in part to their similar structures. Selenium can replace S in some organic compounds but the metabolic activity of the seleno-compound is less than that of the normal S-containing compound. Sulfur has been used to counteract the feeding of Se when fed in toxic concentrations.

The maximum tolerable level for S is reported to be 0.40 ppm (NRC,

1980, 1984), but is less well defined than the maximum requirement. Kandylis (1984) reports that excessive quantities of dietary S (above 0.3–0.4%) as sulfate or elemental S may cause toxic effects and, in extreme cases, can be fatal. Cattle can tolerate more S from natural feed ingredients than from added sulfate. Excessive dietary S levels may cause acute toxicity, resulting in clinical signs of abdominal pain, muscle twitching, diarrhea, severe dehydration, strong odor of sulfide on the breath, congested lungs, and acute enteritis (Miller, 1979).

REFERENCES

Bingley, J. B., and Carrillo, B. J. (1966). *Nature (London)* **209**, 834–835.
Bohman, V. R., Drake, E. L., and Behrens, W. C. (1984). *J. Dairy Sci.* **67**, 1468–1473.
Bray, A. C. (1969). *Aust. J. Agric. Res.* **20**, 734–748.
Bull, L. S. (1979). In "The 2nd Annual International Minerals Conference," pp. 111–130. International Minerals & Chemical Corporation, Mundelein, Illinois.
Campbell, A. G., Coup, M. R., Bishop, W. H., and Wright, D. E. (1974). *N. Z. J. Agric. Res.* **17**, 393–399.
Centro International de Agricultura Tropical (CIAT) (1981). "Centro Internacional de Agricultura Tropical Annual Report 1981." CIAT, Cali, Colombia.
Coleman, R. (1966). *Soil Sci.* **104**, 230–239.
Cunha, T. J. (1973). *Feedstuffs* **45**(20), 27–28.
Cunningham, I. J. (1950). In "Copper Metabolism: A Symposium on Animal, Plant and Soil Relationships" (W. D. McElroy and B. Glass, eds.), pp. 246–273. Johns Hopkins Press, Baltimore.
Dick, A. T. (1952). *Aust. Vet. J.* **28**, 30–33.
Ellis, W. G., Pfander, W. H., Muhrer, M. E., and Pickett, E. E. (1958). *J. Anim. Sci.* **17**, 180–188.
Fontenot, J. P., and Webb, K. E., Jr. (1974). *Fed. Proc., Fed. Am. Soc. Exp. Biol.* **33**, 1936–1937.
Hansard, S. L., and Mohammed, A. S. (1969). *J. Anim. Sci.* **28**, 283–287.
Hartmans, J. (1969). *Agric. Dig.* **18**, 42–48.
Harris, E. D., and O'Dell, B. L. (1974). In "Protein–Metal Interactions" (M. Friedman, ed.), pp. 267–284. Plenum, New York.
Hill, R., Thambyah, R., Wan, S. P., and Shanta, C. S. (1962). *J. Agric. Sci.* **59**, 409–413.
Idris, O. F., Tartour, G., and Babiker, S. A. (1976). *Trop. Anim. Health Prod.* **8**, 13 (abstr.).
International Fertilizer Development Center (IFDC) (1979). "Sulfur in the Tropics." The Sulfur Institute, Muscle Shoals, Alabama.
Jerez, M., McDowell, L. R., Martin, F. G., Hargus, W. A., and Conrad, J. H. (1984). *Trop. Anim. Prod.* **9**, 12–21.
Kandylis, K. (1984). *J. Dairy Sci.* **67**, 2179–2187.
Kennedy, L. G., Mitchell, G. E., and Little, C. O. (1968). *Sulphur Inst. J.* **4**(1), 8–9.
Kennedy, P. M., Williams, E. R., and Siebert, B. D. (1975). *Aust. J. Biol. Sci.* **28**, 31–42.
Lebdosoekojo, S., Ammerman, C. B., Raun, N. S., Gomez, J., and Litell, R. C. (1980). *J. Anim. Sci.* **15**, 1249–1260.
McDowell, L. R. (1976). In "Beef Cattle Production in Developing Countries" (T. Smith, ed.), pp. 216–241. Centre for Tropical Veterinary Medicine, Edinburgh, Scotland.

McDowell, L. R., Conrad, J. H., Thomas, J. E., Harris, L. E., and Fick, K. R. (1977). *Trop. Anim. Prod.* **2**, 273–279.

McDowell, L. R., Bauer, B., Galdo, E., Koger, M., Loosli, J. K., and Conrad, J. H. (1982). *J. Anim. Sci.* **55**, 964–970.

McDowell, L. R., Conrad, J. H., and Ellis, G. L. (1984). *In* "Symposium on Herbivore Nutrition in Sub-Tropics and Tropics—Problems and Prospects" (F. M. C. Gilchrist and R. I. Mackie, eds.), pp. 67–88. Craighall, South Africa.

Matrone, G. (1970). *Trace Elem. Metab. Anim. Proc. WAAP/IBP Int. Symp.* 1969, pp. 354–362.

Miles, W. H., and McDowell, L. R. (1983). *World Anim. Rev.* **46**, 2–10.

Miller, W. J. (1979). "Dairy Cattle Feeding and Nutrition." Academic Press, New York.

Mills, C. F. (1966). *World Rev. Anim. Prod.* **1**, 51–57.

Mills, C. F., Dalgarno, A. C., and Wenham, G. (1976). *Br. J. Nutr.* **35**, 309–311.

Miltimore, J. E., and Mason, J. L. (1971). *Can. J. Anim. Sci.* **51**, 193–200.

Moir, R. J. (1979). *In* "The Second Annual International Minerals Conference," pp. 93–109. International Minerals and Chemical Corporation, Mundelein, Illinois.

Mtimuni, J. P. (1982). Ph.D. thesis, Univ. of Florida, Gainesville.

Munro, I. B. (1957). *Vet. Rec.* **69**, 125–129.

National Research Council (NRC) (1978). "Nutrient Requirements of Dairy Cattle," 5th rev. ed. Nutrient Requirements of Domestic Animals, No. 3. Natl. Acad. Sci., Washington, D.C.

National Research Council (NRC) (1980). "Mineral Tolerance of Domestic Animals." Natl. Acad. Sci., Washington, D.C.

National Research Council (NRC) (1984). "Nutrient Requirements of Beef Cattle," 6th rev. ed. Nutrient Requirements of Domestic Animals, No. 4. Natl. Acad. Sci., Washington, D.C.

Neal, W. M., Becker, R. B., and Shealy, A. L. (1931). *Science* **74**, 418–419.

Netherlands Committee on Mineral Nutrition (NCMN) (1973). "Tracing and Treating Mineral Disorders in Dairy Cattle." Centre for Agricultural Publishing and Documentation, Wageningen, The Netherlands.

Peducassé, C. A., McDowell, L. R., Parra, L. A., Wilkins, J. V., Martin, F. G., Loosli, J. K., and Conrad, J. H. (1983). *Trop. Anim. Prod.* **8**, 118–130.

Phillippo, M., Humphries, W. R., Bremner, I., Atkinson, T. G., and Henderson, G. (1984). *Trace Elem. Metab. Anim. Proc. Int. Symp. 5th,* p. 17 (abstr.).

Rakes, A. H., and Clark, A. K. (1984). *In* "Proceedings, Florida Nutrition Conference," pp. 153–160. Univ. of Florida, Gainesville.

Rees, M. C., Minson, D. J., and Smith, F. W. (1974). *J. Agric. Sci.* **82**, 419–422.

Russell, F. C., and Duncan, D. L. (1956). "Minerals in Pasture: Deficiencies and Excesses in Relation to Animal Health," Technical Communication No. 15. Commonwealth Bureau of Animal Nutrition, Rowett Institute, Aberdeen, Scotland.

Sousa, J. C. de, Conrad, J. H., McDowell, L. R., Ammerman, C. B., and Blue, W. G. (1980). *Pesqui. Agropecu. Bras.* **15**, 335–341.

Stobbs, T. H., and Minson, D. J. (1980). *In* "Digestive Physiology and Nutrition of Ruminants" (D. C. Church, ed.), pp. 257–277. O & B Books, Corvallis, Oregon.

Suttle, N. F., and Field, A. C. (1974). *Vet. Rec.* **95**, 165–168.

Suttle, N. F., and Peter, W. (1984). *Trace Elem. Metab. Anim. Proc. Int. Symp. 5th,* p. 68 (abstr.).

Tejada, R. (1984). Ph.D. thesis, Univ. of Florida, Gainesville.

Thornton, I., Kershaw, G. F., and Davies, M. K. (1972). *J. Agric. Sci.* **78**, 165–171.

Todd, J. R. (1969). *Proc. Nutr. Soc.* **28**, 189–198.

Todd, J. R. (1970). *Trace Elem. Metab. Anim. Proc. WAAP/IBP Int. Symp. 1969*, pp. 448–451.
Underwood, E. J. (1977). "Trace Elements in Human and Animal Nutrition," 4th ed. Academic Press, New York.
Underwood, E. J. (1981). "The Mineral Nutrition of Livestock." Commonwealth Agricultural Bureaux, London.
Vargas, R., McDowell, L. R., Conrad, J. H., Martin, F. G., Buergelt, C., and Ellis, G. L. (1984). *Trop. Anim. Prod.* **9**, 103–113.
Ward, G. M. (1977). *J. Anim. Sci.* **46**, 1078–1085.

12

Cobalt, Iodine, and Selenium

L. R. McDOWELL
Department of Animal Science
University of Florida
Gainesville, Florida

I.	General	259
II.	Cobalt	260
	A. Introduction	260
	B. Metabolism	260
	C. Requirement	262
	D. Deficiency	263
	E. Prevention and Control	267
	F. Toxicity	268
III.	Iodine	268
	A. Introduction	268
	B. Metabolism	268
	C. Requirement	270
	D. Deficiency	271
	E. Prevention and Control	273
	F. Toxicity	275
IV.	Selenium	275
	A. Introduction	275
	B. Metabolism	276
	C. Requirement	277
	D. Deficiency	277
	E. Prevention and Control	281
	F. Toxicity	283
	References	287

I. GENERAL

Research from both tropical and temperate regions of the world has illustrated the devastating effects of trace mineral deficiencies to grazing livestock. A review of research from tropical regions concludes that copper (Cu) is the trace mineral most likely deficient, followed by cobalt (Co, iodine (I), and selenium (Se) (McDowell et al., 1984). The need for ruminant livestock for I supplementation was established in the nineteenth century and for Co in the 1930s, but not until the late 1950s was the importance for Se considered apart from toxicity considerations. Unless specific research has indicated the contrary, all free-choice mineral mixtures for grazing ruminants should contain adequate quantities of Co, I, and Se.

II. COBALT

A. Introduction

Prior to discovery of the need for Co in the 1930s, grazing ruminants could not be produced in many regions due to deficient concentrations of this element in forages. In 1935, Australian research indicated that Co was effective in preventing two debilitating diseases of sheep known as "coast disease" and "wasting disease" (Underwood, 1977). For cattle, the first report of a Co deficiency was from Florida in 1937 (Fig. 12.1), which showed this element as responsible in part for a condition known as "salt sick" (Becker et al., 1965). Iron (Fe) was first suggested as the mineral responsible for preventing the debilitating disease since large doses of supplemental Fe gave positive responses. Cobalt was eventually isolated as the contaminant in Fe responsible for preventing the wasting diseases in Australia. However, from the mineral-deficient, sandy soils of Florida, the "salt sick" condition could be prevented only if all three of the deficient elements, Co, Cu, and Fe, were supplied in adequate quantities. Although Co was recognized as an essential microelement for ruminants, the action of Co in the body and the reason for its necessity were not discovered until the simultaneous discovery of vitamin B_{12} by Smith (1948) in England and Rickes et al. (1948) in the United States. Both studies reported the presence of Co in their compounds, which were effective against pernicious anemia in humans.

B. Metabolism

The Co requirement of ruminants is actually a Co requirement of rumen microorganisms. The microbes incorporate Co into vitamin B_{12}, which is utilized by both microorganisms and animal tissues. The main source of

12. Cobalt, Iodine, and Selenium

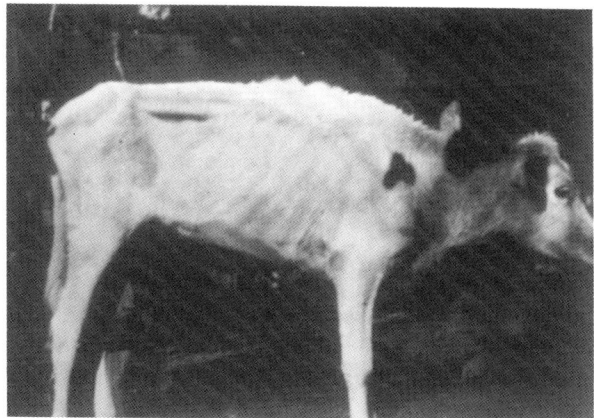

Fig. 12.1. A Co-deficient heifer (top) in Florida that had access to an Fe–Cu salt supplement. Note the severe emaciation. Her blood contained 6.6 g of hemoglobin per 100 ml on February 25, 1937. The same heifer (bottom) fully recovered with an Fe–Cu–Co salt supplement while on the same pastures. (Courtesy of R. B. Becker, Univ. of Florida, Gainesville.)

energy to ruminants is not glucose but primarily acetic and propionic acids. Vitamin B_{12} is of key importance in the utilization of propionic acid. The main physiological manifestation of Co or B_{12} deficiency is impaired propionate metabolism, since it is needed for activity of methylmalonyl–CoA isomerase, an enzyme that catalyzes the conversion of methylmalonyl–CoA to succinyl–CoA (Marston *et al.*, 1961). Vitamin B_{12} is also needed for an enzyme that catalyzes the recycling of methionine from homocysteine and for normal liver folacin metabolism (NRC, 1984).

About 3% of ingested Co is converted to vitamin B_{12} in the rumen

(Smith and Marston, 1970). Of the total vitamin B_{12} produced, only 1–3% is absorbed in the lower portion of the small intestine [National Research Council (NRC), 1984]. Both Co and vitamin B_{12} are mainly excreted in the feces, although variable amounts are secreted in urine (Smith and Marston, 1970).

C. Requirement

Although vitamin B_{12} injections will prevent Co deficiency in ruminants, it is more convenient and cheaper in Co-deficient areas to supplement the diet with this element, allowing the microorganisms to synthesize the vitamin for subsequent absorption by the host.

Dietary Co requirements for ruminants have been established at 0.1 ppm for sheep (NRC, 1975), dairy cattle (NRC, 1978) and beef cattle (NRC, 1984). The Agricultural Research Council (ARC) (1980) estimated the requirement for cattle and sheep is 0.11 ppm. Under grazing conditions, lambs are the most sensitive to Co deficiency, followed by mature sheep, calves, and mature cattle, in that order (Andrews, 1956). Field experience suggests that only small species differences exist among ruminants in Co requirements.

It has been shown that the ruminant animal has a higher vitamin B_{12} requirement than the nonruminant, presumably because of its involvement in the metabolism of propionic acid. The evidence of quantitative needs for Co rests on studies of its content in forage where deficiency does and does not occur. It has been reported that the dry matter of grass in healthy areas contains around 0.1 ppm of Co or more, on the average, as compared with 0.004–0.07 ppm for deficient areas. As little as 0.1 ppm has restored sick animals to health. Further evidence of 0.10 ppm as the dietary requirement for sheep was provided by Mohammed (1983), who fed various levels of Co and found that vitamin B_{12} and propionic acid in rumen concentrations were maximal at 0.10 ppm.

A staggers syndrome in sheep and cattle that have grazed pastures in which *Phalaris tuberosa* predominated have developed a disease known as "Phalaris staggers" (Underwood, 1977). A number of world regions have reported this condition, with affected animals suffering irreversible and commonly fatal nervous degeneration. Cobalt supplementation has been shown to alleviate the "Phalaris staggers" condition and would, therefore, influence the Co requirement (Lee *et al.*, 1957). Cobalt, but not vitamin B_{12}, is required to neutralize those neurotoxins in the rumen; borderline Co deficiency would result in "Phalaris staggers" in ruminants because there is not enough Co to neutralize the neurotoxin and to meet the daily requirement of the animal.

D. Deficiency

Prior to the recognition of Co deficiency in livestock in many parts of the world, cattle could be maintained on deficient pastures only if they periodically moved to "healthy" ground. Cobalt deficiency can be prevented by taking animals every year for a few months to a "healthy" region, preferably during the rainy season. An example of the necessity of periodically moving animals was illustrated in a disease condition known as "togue" in Espirito Santo, Brazil (Tokarnia et al., 1971). The disease was observed when animals stayed for a period longer than 60 to 180 days on certain pastures. Sick animals isolated themselves from the rest of the herd, were apathetic, showed loss of appetite, rough haircoat, dry feces, and lost body condition. If the animals were not moved from the pasture, they died, but if they were taken to a pasture where the disease did not occur, the animals recovered quickly.

Cobalt deficiency occurs in large areas of many countries and is largely, but not exclusively, restricted to grazing ruminants that have little or no access to concentrates. With the exception of phosphorus (P) and Cu, Co deficiency is the most severe mineral limitation to grazing livestock in tropical countries (McDowell et al., 1984). A total of 24 developing tropical countries or regions in Latin America, Africa, and Asia have noted Co deficiencies or low forage concentrations of this element (see Chapter 16 of this volume).

Deficiency signs for Co are not specific, and it is often difficult to distinguish between an animal having a deficiency and malnutrition due to low intake of energy and protein, and an animal that is diseased or parasitized. Acute clinical signs of Co deficiency include lack of appetite, rough hair coat, thickening of the skin, anemia (normocytic and normochromic), wasting away (Fig. 12.2), and eventually death, if the animals are not moved to "healthy" pastures or if Co supplements are not made available. These clinical signs are identical to those of simple starvation and may indicate that the effect of a lack of Co may be simply the effect on appetite, rather than a direct effect of the mineral on the body itself. Cobalt-deficient livestock respond quickly to Co treatment, recovering appetite, vigor, and weight, and this serves as an easy, practical test to determine whether a Co deficiency exists.

Cobalt subclinical deficiencies or borderline states are extremely common and are characterized by low production rates unaccompanied by clinical manifestations or visible signs. Subclinical deficiencies of Co often go unnoticed, thereby resulting in great economic losses to the livestock industry (Latteur, 1962). No estimate can be made of the effect of Co subdeficiencies on animal performance in general, but, in many areas of the world, it is one of the major causes of poor production.

Fig. 12.2. A Co-deficient lamb (top) fed a ration containing 0.05 ppm of Co and weighing 22 kg at the end of the experimental period. A positive control lamb (bottom) that daily received, in addition, 0.1 mg of Co as the sulfate and weighs 42 kg. (Courtesy of S. E. Smith, Cornell Agricultural Experiment Station, Ithaca, New York.)

Incidence of Co deficiency can vary greatly from year to year, from an undetectable mild deficiency to an acute stage. Lee (1963) illustrates this variation in a 14-year experiment with sheep in southern Australia. Half the ewes, replacements, and progeny were dosed with Co and remained healthy. The undosed half had the following performance for the 14 years: in 2 years, lambs were unthrifty, but there were no deaths; in 3 years, growth rate of the lambs was slightly retarded; in 4 years, 30–100% of the lamb crop was lost; in 5 years, the performance of the remaining stock was as good as that of dosed members of the flock.

Cobalt deficiency is found on soils of diverse origin, including coarse, volcanic, sandy loams and leached sands. Soil containing less than 2 ppm of Co is generally considered deficient for ruminants (Corrêa, 1957). Raising the pH by liming reduces the Co uptake by the plant and may increase the severity of the deficiency. Plants grown on a 15 ppm soil that is neutral or slightly acid may contain more Co than those grown on a 40-ppm alkaline soil (Latteur, 1962). High rainfall tends to leach Co from the topsoil. This problem is often aggravated further by rapid growth of forage during the rainy season, which dilutes the Co content.

Plants have varying degrees of affinity for Co, some being able to concentrate the element much more than others. Legumes, for example, generally have greater ability to concentrate Co than do grasses (Underwood, 1977). Of 140 forage averages in the 1974 "Latin American Tables of Feed Composition" with Co values, 43.1% had concentrations less than the requirement of 0.1 ppm (McDowell et al., 1977). In a review of mineral deficiencies in Latin America, Fick et al. (1978) reported deficient concentrations of forage Co in Argentina, Colombia, Guyana, Haiti, Surinam, and Peru.

Peducassé et al. (1983) evaluted the mineral status of two cattle-producing regions of Bolivia. From one region, all forages were above the Co requirement, while, for the second region, 47.6% of forages contained less than 0.1 ppm of Co. Vargas et al. (1984) analyzed forages for mineral content from three regions in the llanos of Colombia and found 72 and 31% of all forages below 0.1 ppm for the wet and dry seasons, respectively. On the contrary, of forage samples collected from northern Mato Grosso, Brazil, mean Co concentrations were deficient (0.08 ppm) in the dry season but adequate during the wet season (0.17 ppm) (Sousa et al., 1981).

In Malaysia, Co deficiency was reported in cattle that were consuming a grass–legume pasture containing 0.01 ppm of Co ('t mannetje et al., 1976). There was a dramatic increase in liveweight gain as a result of either vitamin B_{12} or Co administration.

Fig. 12.3. Cobalt-deficient cattle in northern Mato Grosso, Brazil. (Courtesy of Jürgen Döbereiner and Carlos H. Tokarnia, EMBRAPA, Rio de Janeiro, Brazil.)

Cobalt deficiency in ruminants, in its milder forms, is impossible to diagnose with certainty on the basis of clinical and pathological observations alone. Forage, soil, and animal tissue concentrations have all been used to determine the status of Co (see Chapter 15 of this volume). Reduced ruminant liver stores of Co and vitamin B_{12} are indicative of a dietary Co deficiency, and concentrations are frequently used to determine the Co status of ruminants. From the state of São Paulo, Brazil, Corrêa (1957) studied a disease condition in cattle with clinical signs suggesting a Co deficiency. Soil, plant, and animal tissue analysis each clearly indicated a Co deficiency. Average Co concentrations in soil, forage, and liver samples were 0.797, 0.046, and 0.058 ppm for Co-deficient areas versus 20.37, 0.334, and 0.201, respectively, for Co-adequate regions. Plasma or urine concentrations of methylmalonic acid are successfully used as a diagnostic tool for determining Co deficiency (McMurray et al., 1984).

Tokarnia and Döbereiner (1978) have carried out extensive research concerning mineral deficiencies in Brazil and also recently reviewed the literature concerning mineral deficiency diseases. For Brazil, Co deficiencies were widespread and were noted in São Paulo, Ceara, Amapa, Espirito Santo, Maranhão, and Rio Grande do Sul. Figs. 12.3 and 12.4 illustrate Co-deficient cattle in Brazil. From the research of Tokarnia and Döbereiner (1978), Co deficiencies were confirmed from cattle liver Co concentrations, with below 0.05 ppm indicating deficiency, from 0.05 to 0.12 ppm indicating marginal deficiency, and above 0.12 ppm as adequate.

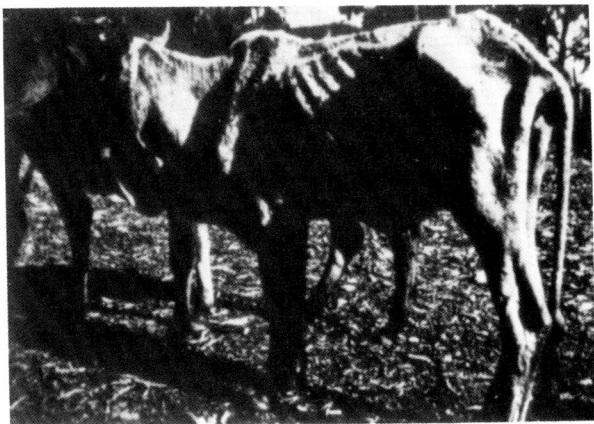

Fig. 12.4. An extreme condition of Co-deficiency of an Indu–Brazil steer in São Paulo, Brazil. (Courtesy of Nelson dos Santos Fernández, Instituto Biológico de São Paulo, São Paulo, Brazil.)

E. Prevention and Control

Cobalt deficiency in ruminants can be cured or prevented through treatment of soils or pastures with Co-containing fertilizers or by direct oral administration of Co to the animal. In deficient areas where the pastures require regular fertilizer applications to increase their yields, adequate Co intakes by grazing ruminants can usually be ensured by including Co salts or oxide ores in the fertilizers used. In this way, Co in the treated herbage can be raised to satisfactory concentrations and maintained there for long periods of time (Underwood, 1981). Underwood (1977) states that on most deficient soils, as little as 100–150 g of cobalt sulfate per acre applied annually or biennially provides adequate levels of Co in the forage. Often, however, use of fertilizer Co is impractical and uneconomical for extensive range conditions.

Cobalt deficiency in grazing ruminants can best be prevented by direct oral administration of Co through free-choice mineral supplements. Large and frequent injections of vitamin B_{12} can effectively prevent or cure Co deficiency, but are much more expensive. Oral dosing or drenching with dilute Co solutions are likewise satisfactory if the doses are regular and frequent.

An additional method of providing Co developed in Australia is through the use of an orally administered, heavy pellet ("bullet") made of cobalt oxide plus finely divided Fe that remains in the reticulorumen for an extended period of time. Pellets have the limitations of being lost through regurgitation or of becoming ineffective because of formation of a surface coating of calcium phosphate. The addition of a steel grinder, which provides an abrasive action, reduces the surface coating and extends the usefulness of the pellet. In a review of Co nutrition, Ammerman (1981) pointed out that if Co is the only mineral that is deficient in an area or if it is severely deficient, the use of pellets may serve as an effective means of providing supplemental Co to grazing livestock. If, however, there are also fairly critical deficiencies of sodium (Na), P, and Cu, for example, which require the use of a mixed supplement for the grazing area, there appears to be little advantage in the use of Co pellets.

Although only limited research has been done concerning the biological availability of inorganic Co compounds, the carbonate, chloride, sulfate, and oxide forms have been proposed as satisfactory dietary sources of the mineral (see Chapter 17 of this volume).

F. Toxicity

The maximum tolerable dietary level for ruminants is estimated at 5 ppm (NRC, 1984). Cobalt toxicity for cattle is characterized by a mild polycythemia; excessive urination, defecation, and salivation; shortness of breath; and increased hemoglobin, red cell count, and packed cell volume (NRC, 1984). Although most researchers consider the margin between safe and toxic doses to be wide enough to render a toxicity under natural conditions to be unlikely, several reports have appeared in the literature concerning Co toxicity. These reports almost invariably involve management mistakes in formulating mineral mixtures.

III. IODINE

A. Introduction

Iodine deficiency in man and farm animals as endemic goiter (enlarged thyroid) is one of the most prevalent deficiency diseases, and it occurs in almost every country in the world (Fig. 12.5). The incidence of deficiency has declined as a result of the widespread use of iodized salt. However, in many tropical countries of the world, endemic goiter remains an exceedingly serious human and livestock problem.

B. Metabolism

The physiological requirement for I is for synthesis of hormones by the thyroid gland that regulate energy metabolism. The only known role of I is

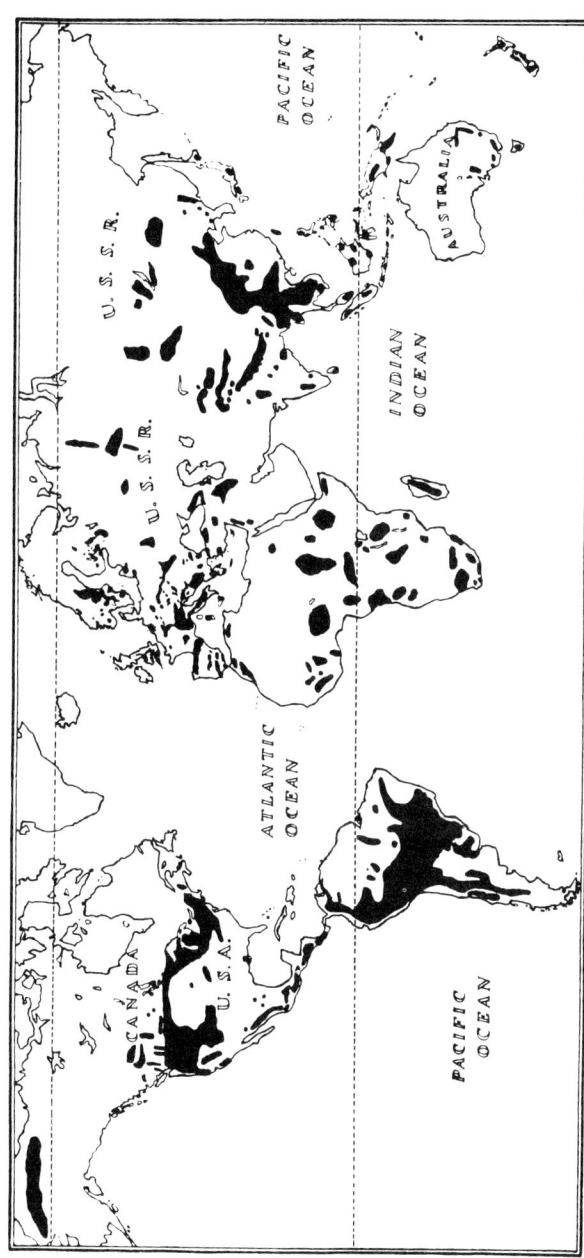

Fig. 12.5. World map showing occurrence of endemic goiter. Black indicates areas where endemic goiter has been found.

in the synthesis of the thyroid hormones, thyroxine, and triiodothyronine. Thyroid hormones have an active role in thermoregulation, intermediary metabolism, reproduction, growth and development, circulation, and muscle function.

Iodine is readily absorbed from the gastrointestinal tract and is excreted mainly in the urine. The rumen is the major site of absorption of I, and the abomasum, the major site of endogenous secretion; i.e., for the re-entry of circulating I into the digestive tract (Underwood, 1977). Over 90% of the administered I can be accounted for by thyroid uptake and urinary secretions. Iodine occurs throughout the body, but a very high percentage of the total amount is stored in the thyroid gland. In animal tissue, two forms of I exist—inorganic iodide and organically bound I (Herrick, 1982).

C. Requirement

Iodine requirements of ruminant species have not been well established and have varied from 0.05 to 0.8 ppm (see Chapter 2 of this volume). Iodine requirements for growth are not necessarily identical with those for reproduction and lactation, or for maintenance of the integrity of thyroid structure and function (Underwood, 1981). The thyroid gland has a great capacity to adapt to low dietary I, with this ability influencing I requirements.

Perhaps the presence of goitrogenic substances is of equal or greater importance than is low dietary I as a contributing factor toward I deficiencies. Goitrogenic substances are much more prevalent in feeds than is generally recognized and include most cruciferous plants, soybeans, flaxseed, cassava, peas, and peanuts. In addition, brassicas (i.e., kale and rape) are known to produce active goitrogens, with many species occurring as minor components of pastures (McDonald, 1968). The alkaloid mimosine, found in the tropical legume *Leucaena leucocephala,* is also a goitrogenic factor. The feeding of other feedstuffs, such as a high intake of corn silage, has been shown to result in calves with goiter (Hemken *et al.,* 1971). Goitrogens have been demonstrated in the milk of cows consuming cruciferous plants, thus increasing the likelihood of I deficiency in both calves and humans consuming the milk (Clements, 1957).

The net effect of goitrogens in most instances is to increase the I requirement. If the diet contains as much as 25% strongly goitrogenic feeds, supplemental I should be at least doubled (Miller, 1979). Thiocyanate-type goitrogens are easily counteracted by additional I, but thioracil types can only partially be suppressed by additional I. In addition to goitrogens, Underwood (1977) reported interrelationships that interfere with I metabolism and thus affect requirements such as (1) high dietary arsenic (As), fluorine (F), or calcium (Ca) levels; (2) deficient or high Co levels; and (3) low manganese (Mn) intakes.

D. Deficiency

Kelly and Snedden (1960) estimated the number of goitrous people in the world to be approximately 200 million. Follis (1966) reported that endemic goiter exists in all of the continental Latin American countries. In Ceylon and Nigeria, the incidence of goiter among females in selected villages ranged up to 56 and 72%, respectively (Wilson, 1953). Goiter or other I deficiency signs are reported in animals in most areas where human goiter is endemic. Since animals usually receive feeds produced locally, they are often more susceptible to goiter than are human beings. In fact, I deficiency has been stated to be the most widespread of all mineral deficiencies in grazing stock (Allman and Hamilton, 1949).

Iodine deficiency has declined in more recent years for both humans and animals due to the use of iodized salt and other preventive measures. However, there are no reliable estimates on incidence of goiter for ruminants in tropical countries or indications of how subclinical I deficiency may be affecting grazing livestock production.

The principal clinical sign of I deficiency is goiter, which is illustrated in young ruminants in the tropical regions of Minas Gerais, Brazil (Fig. 12.6); Mato Grosso, Brazil (Fig. 12.7); the Philippines (Fig. 12.8); and Indonesia (Fig. 12.9). Iodine deficiency in young ruminants is also manifested by general weakness, and animals may be born blind, hairless, or dead, with the exact clinical signs depending on the severity of the deficiency. It would appear that goiter may be a clinical sign of a less severe deficiency than the lack of hair or wool (Underwood, 1981).

Iodine deficiency in the young lamb has been reported to permanently impair the quality of the adult fleece, since the normal development of the wool-producing secondary follicles requires thyroid activity in excess of that needed for general body growth (Ferguson et al., 1956). Iodine deficiency in breeding animals severely reduced productivity, resulting in suppression of estrus periods in the female and lack of libido in the male (Church, 1971). Hemken (1970) reports that I-deficient dairy cattle are less able to resist stress and may even have a higher incidence of ketosis.

Factors influencing the availability of I include the level of I in soils, the distance of regions from the sea, which influences the level of I in rain, effect of light soils that facilitate leaching, variation of plant species in their ability to absorb I, and interrelationships between other nutrients and goitrogens. Increased forage yields have resulted in I deficiencies in

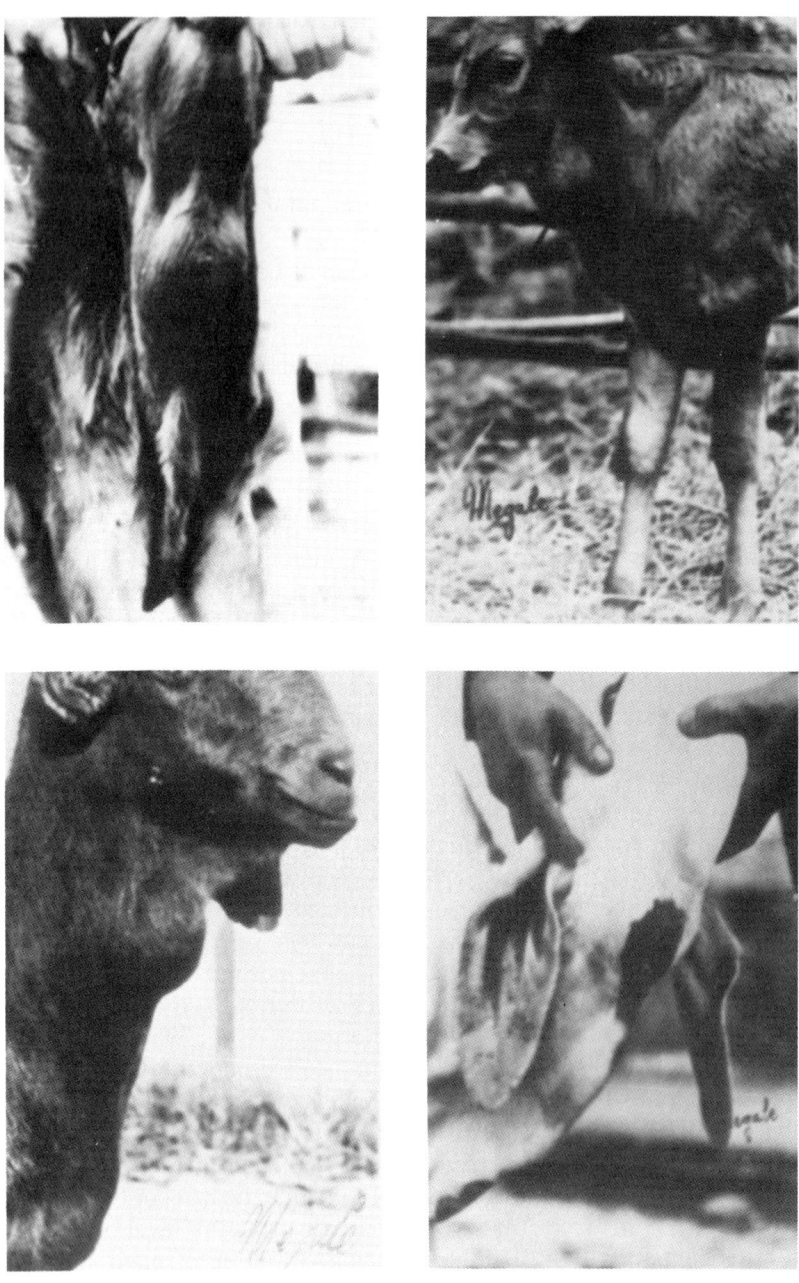

Fig. 12.6. Goiter caused by iodine deficiency in calves and goats in Minas Gerais, Brazil. (Courtesy of Francisco Megale, Federal Univ. of Minas Gerais, Belo Horizonte, Brazil.)

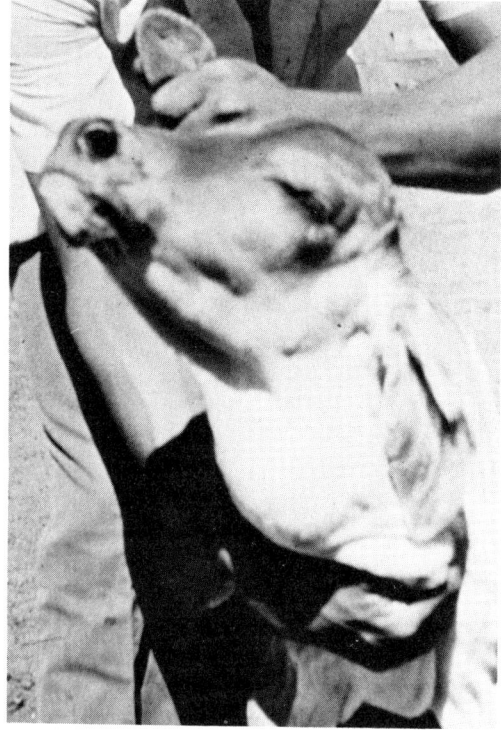

Fig. 12.7. Calf with severe goiter in Rondonapolis, Mato Grosso, Brazil.

Lebanon (Cowan *et al.*, 1967), and higher nitrogen (N) applications greatly increased plant concentrations of goitrogens in Rhodesia (Rodel, 1972).

Severe I deficiency can be diagnosed on the clinical evidence of goiter alone. Less severe forms of goiter or I deficiency are more difficult to diagnose, and thus weight and histological structure of the thyroid gland as well as serum I (largely thyroxine) are used to establish status of this element. Milk I concentration is extremely responsive to changes in dietary intakes and is used to establish the status of animals (see Chapter 15 of this volume).

E. Prevention and Control

For grazing livestock, the use of iodized salt has eliminated I deficiency in many parts of the world. Availability studies on various I compounds indicate that several sources are relatively equal in availability (see Chapter 17 of this volume). Potassium iodide, sodium iodide, and calcium

Fig. 12.8. Goiter in calf found in Mindanao, Philippines.

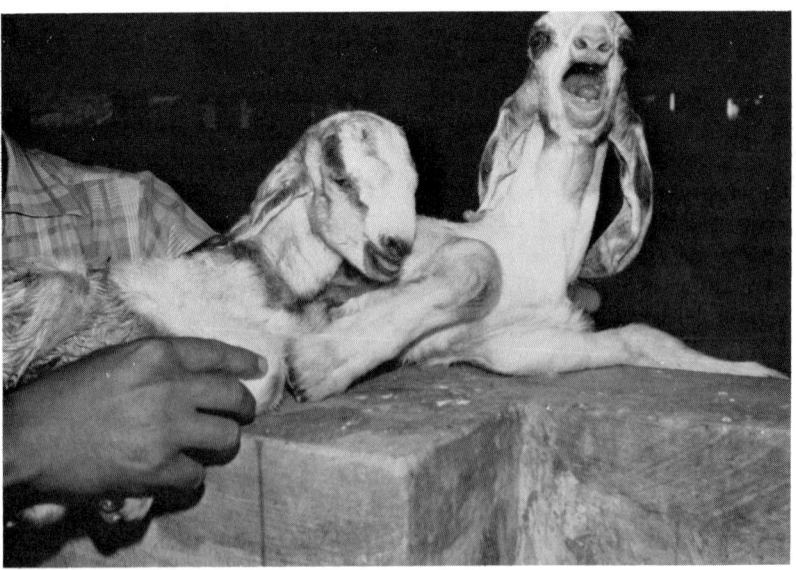

Fig. 12.9. Goiter in goats as a result of iodine deficiency in Yogyakarta, Indonesia.

iodate are readily available to ruminants but will leach or evaporate from salt blocks under hot, humid, tropical conditions. Johnson and Frederick (1940) found that a salt containing one part of I in 5000 parts of salt lost 40% or more of the I in 18 months. The loss is due to oxidation, with subsequent volatilization of I. Potassium iodate, stabilized potassium iodide, or pentacalcium orthoperiodate are equally available to livestock but are much more stable forms of I and not as rapidly lost from free-choice mineral mixtures. Ethylenediamine dihydriodide (EDDI) is an organic iodide compound that is used at relatively high levels for treatment of footrot, lumpy jaw, and other conditions. The I is available but is not as stable as in other I compounds, and should be kept dry and cool.

Injectable I compounds have proven effective for preventing goiter in both human and livestock studies. S. Lebdosoekojo (personal communication) studied an injectable iodized poppyseed oil administered to breeding goats in Yogyakarta, Indonesia. One injection was found to last for approximatley 2 years or until after the third kidding. These goats had received coconut oil meal and rice bran that was thought to contain goitrogenic substances.

F. Toxicity

Iodine toxicosis may result when high levels of I are used over long periods of time to correct or prevent diseases such as footrot and lumpy jaw. Toxicosis signs include depressed appetite, dull, listless appearance, scaliness and sloughing of the skin, excessive lacrimation, difficulty in swallowing, and hacking cough. Feeding excessive I for an extended period to cattle resulted in impaired function of the humoral and cell-mediated immune systems, with reduced ability to form antibodies in response to disease organisms (Haggard *et al.*, 1980). When the diet consistently contains over 50 ppm of I, toxic signs may appear (NRC, 1980). Recovery from I toxicity is rapid after the excess I is eliminated from the diet.

IV. SELENIUM

A. Introduction

Consumption of feedstuffs containing both toxic (>5 ppm) and deficient (<0.1 ppm) concentrations of Se presents a worldwide problem for grazing livestock. Prior to 1957, the only consideration given to Se was its toxicity. Effects of Se toxicity were observed in Chinese grazing livestock as early as 1295 by Marco Polo; however, Se was not known to be the causative agent until 1934. Franke and Potter (1934) showed Se as the toxic agent responsible for "alkali disease" or "blind staggers" in livestock of the United States. Selenium is among the few elements known to be absorbed by food and forage plants in sufficient amounts to create toxicity hazards to animals.

In 1957, an essential role for Se began to emerge when Schwarz and Foltz (1957) demonstrated that it was the effective component of "factor 3" in preventing liver degeneration in rats. Soon thereafter, Se was shown to prevent nutritional muscular dystrophy in calves (Muth *et al.*, 1958) and lambs (Hogue, 1958; McLean *et al.*, 1959). Disorders resulting from Se inadequacy have been recognized in practically all major livestock-producing countries of the world. Recent research has now established that the total areas of the world affected by Se deficiency are far greater and the consequences are more economically important than those afflicted with Se excess (Underwood, 1977).

B. Metabolism

Selenium is essential for such body functions as growth, reproduction, prevention of various diseases, and protection of the integrity of tissues. The metabolic function of Se is closely linked to vitamin E. Both Se and vitamin E function to protect biological membranes from oxidative degeneration. Lack of these nutrients results in tissue breakdown and degeneration.

Selenium-responsive diseases are accompanied by decreased tissue levels of glutathione peroxidase, an Se-containing enzyme (Rotruck *et al.*, 1973). Among other functions, this enzyme aids in protecting cellular and subcellular membranes from oxidative damage. The relationship of this function to that of vitamin E, which also acts as an antioxidant, apparently is that vitamin E functions as a specific lipid-soluble antioxidant in the membrane and that Se functions as a component of glutathione peroxidase that destroys peroxides before they can attack cellular membranes.

Additional biochemical roles for Se, unrelated to glutathione peroxidase, have been suggested (NRC, 1983) as follows: (1) it is a specific selenoprotein of spermatozoa that serves as a mitochondrial strucutral protein or as an enzyme; (2) it plays a role in RNA, since Se can be incorporated into purine or pyrimidine bases; (3) both Se and vitamin E are needed for adequate immune responses by livestock; and (4) it may have a specific role in prostaglandin synthesis and essential fatty acid metabolism. Selenium also has a strong tendency to complex with heavy metals and exerts a protective effect against the heavy metals, including cadmium (Cd), mercury (Hg), and silver (Ag) (Underwood, 1981; McDowell *et al.*, 1978).

Essentially, no absorption of Se occurs in the rumen or abomasum, the greatest absorption of Se occurring in the small intestine (duodenum) and cecum of sheep (Wright and Bell, 1966). These workers found Se absorbed less in ruminants than in monogastrics, with oral selenite Se retained 77% in swine but only 29% in sheep. Lower absorption for sheep may be due to selenite being reduced to insoluble compounds in the rumen. Absorbed Se is carried in the plasma (associated with protein) until it enters tissues and is excreted in feces, urine, and exhalation. Fecal excretion is generally greater than urinary excretion in ruminants, with exhalation a major route of Se excretion only when toxic concentrations are consumed (NRC, 1983).

Selenium retention is dependent on tissue demands, with Se more efficiently retained from Se-deficient than from Se-adequate animals. There was no significant difference in Se retention associated with different sources of Se when fed at levels of less than 0.10 ppm of Se (Cary et al., 1973). Beyond this level, organic forms of Se result in higher tissue levels than do inorganic forms. Selenium from grains (0.5 ppm) resulted in four to five times as much Se in muscles than the same amount of Se from the selenite form.

C. Requirement

It is probable that Se requirements for grazing ruminants fall in the range of 0.05–0.3 ppm. The minimum requirements of ruminants vary with the form of Se ingested, the criteria of adequacy employed, and the nature of the rest of the diet, particularly its content of vitamin E. The commonly accepted dietary Se requirement for lambs and calves of about 0.1–0.2 ppm does not apply with all types of feeding. Selenium requirements are higher when legumes are fed, sulfur (S) intake is high, dietary vitamin E is low, and diets contain heavy elements such as Cd, Hg, As, and Ag (McDowell et al., 1978; Hansard, 1983).

The presence of unsaturated fatty acids, which may become rancid, can increase the need for Se. Rancidity results in the formation of peroxides, which increase the need for Se and vitamin E since both act as antioxidants. Feed rancidity would be more frequent in tropical and subtropical countries of the world (Cunha, 1982).

D. Deficiency

White muscle disease (WMD), also known as nutritional muscular dystrophy, is the major clinical sign of Se deficiency in newborn ruminants. The disease is a degeneration of the striated muscles that occurs, without

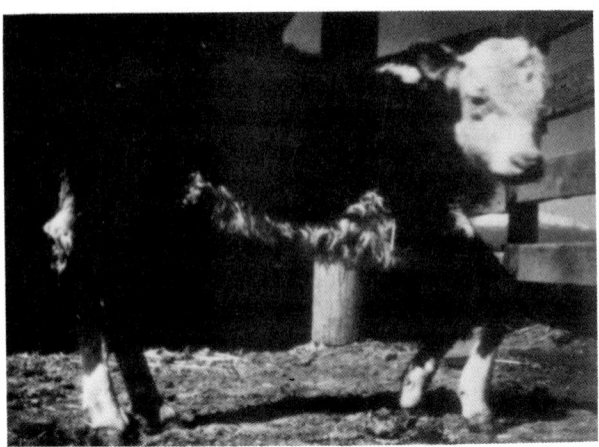

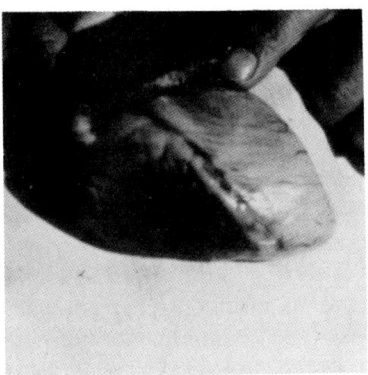

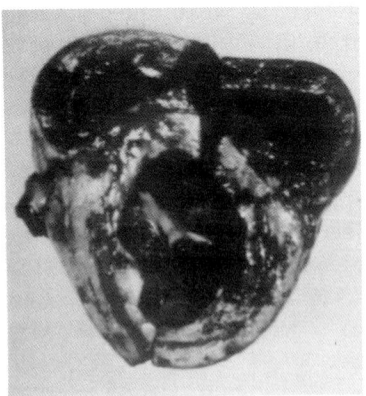

Fig. 12.10. The calf in the top photograph is about 3 months old and has white muscle disease. Lameness and generalized weakness of muscles can be seen. The bottom photographs show abnormal white areas in the heart muscles. (Courtesy of O. H. Muth, School of Veterinary Medicine, Oregon State Univ., Corvallis, Oregon.)

neural involvement, in a wide range of animal species. Unthriftiness, poor growth rates in calves and lambs, and poor reproductive performance in cows and sheep have been observed. Some calves or lambs with WMD have difficulty standing and others have difficulty in getting up due to stiffness of the legs. Figure 12.10 illustrates a calf with WMD, which was produced by a cow grazing low-Se pastures. Calves with WMD have chalky white striations, degeneration, and necrosis in the skeletal muscles and heart (Fig. 12.10). The same condition in lambs is known as "stiff lamb disease" (Fig. 12.11). Godwin (1975) reported that electrocardiograms of WMD showed a progressive development of a characteristic abnormality and a fall in blood pressure, therefore suggesting that the fundamental change occurring in Se deficiency is circulatory failure. During some years, incidence of WMD in certain regions is sporadic, with less than 1% of the herds affected. In other areas such as Turkey and New Zealand, a 20–30% incidence of WMD may occur.

White muscle disease is relatively rare in mature animals but has been observed in sheep in New Zealand and Scotland and in cattle in Canada and the Scandinavian countries (Judson and Obst, 1975). Other Se-responsive conditions are not restricted to young animals and relate to

Fig. 12.11. Selenium–vitamin E deficiency in sheep, known as stiff lamb disease or white muscle disease. The lamb is unable to stand as a result of tissue degeneration. (Courtesy of O. H. Muth, School of Veterinary Medicine, Oregon State Univ., Corvallis, Oregon.)

unthriftiness ("illthrift"), occurring in lambs and hoggets at pasture and can occur in beef and dairy cattle of all ages (Underwood, 1981). Cattle grazing on peaty muck soils in the Florida Everglades developed anemia associated with the presence of Heinz bodies and suboptimal blood Se (Morris et al., 1984). Selenium supplementation of these cattle corrected anemia, prevented Heinz body formation, and increased body weight of both cows and calves.

In subclinical Se deficiency, performance may be reduced, with slower gains and lowered reproduction involving an increased number of services needed per conception. Poor reproductive performance has been shown to include retained placenta, with high incidence of retained placentas in cattle greatly reduced by the administration of adequate dietary levels of Se, as shown by research in the United States, Scotland, and Brazil. From Ohio, incidence of retained placenta was reduced from a mean of 51.2% in untreated cows to 8.8% in cows injected with a combination of Se–vitamin E (Julien et al., 1976). More recent studies from Ohio (Harrison et al., 1984) reported 17.5% retained placenta for control dairy cows, with no incidence for cows receiving both Se and vitamin E (neither vitamin E nor Se was effective alone). From the same study, control versus Se administration reduced cystic ovaries (47 versus 19%) and incidence of metritis (84 versus 60%).

High incidences of retained placentas have been reported when blood serum levels were below 0.04 ppm of Se. Selenium levels below 0.04 ppm were reported in 75% of serum samples from 974 dairy cattle in 12 regions of the state of São Paulo, Brazil (Lucci et al., 1983b). Lucci and Moxon (1982) have analyzed serum Se from 43 dairy herds in São Paulo, Brazil, with only 5 herds having an average concentration greater than 0.04 ppm; likewise, confirmed WMD in calves was also encountered in this region.

White muscle disease can be diagnosed by tissue Se levels, gross and histological examination of the affected muscles, and elevation of serum glutamic-oxalacetic transaminase (SGOT), an indirect indicator of tissue degeneration. Detection of Se status of grazing livestock is relatively easy since animal tissues reflect the dietary Se level over a wide range from deficient to toxic intakes (see Chapter 15 of this volume). The kidney and the liver are the most sensitive indicators of the Se status of the animal, with blood Se and glutathione peroxidase, an Se-containing enzyme, often used for diagnostic purposes.

Although it is now known that the total areas of the world affected by Se deficiency far exceed those afflicted with excess, prior to the late 1970s, there were few reported Se deficiencies in developing tropical countries. A possible explanation for lack of reported Se deficiencies in livestock of developing regions is that in the majority of these countries, the veterinary staffs were preoccupied with the prevention of major endemic diseases, so that disorders arising from Se-responsive diseases were easily overlooked. Also, techniques for analyzing small quantities of Se are relatively new, and an extremely small number of tissue and feed samples had been analyzed in these countries. Illustrative of this fact is that, of 3362 feeds included in the 1974 "Latin American Tables of Feed Composition," only 0.06% contained Se analyses (McDowell et al., 1977). With recent improvements in analytical methods for Se and an increased awareness by researchers and veterinarians of the likelihood of a deficiency, it is expected that reports in the future will continue to confirm Se deficiency in grazing ruminants in more areas of the world.

Selenium deficiency is confirmed or suspected in 20 tropical countries of Latin America, Africa, and Asia (McDowell et al., 1984; see Chapter 16 of this volume). A number of Latin American countries, including Brazil, Ecuador, Mexico, and Uruguay, have reported clinical cases of calves exhibiting WMD. From Peru, Terry (1964) reports a disease condition in ruminants that resembles Se–vitamin E deficiency. When treated with Se–vitamin E, 70% of the animals recovered from the condition. In Paraguay, Se administration increased weight gains and increased pregnancy rate in heifers over controls (Boggino et al., 1973). Low Se levels in feeds from Mexico (Gutierrez et al., 1974; Barradas et al., 1984), the Bahamas (A. Dorsett, personal communication), Brazil (Moxon, 1971), and Honduras (A. L. Moxon, personal communication) have been reported. Moxon (1971) analyzed corn samples for Se from 19 locations in Brazil and found average values of 0.060 and 0.036 ppm for 1969 and 1971, respectively. Corn from locations within Minas Gerais, Rio Grande do Sul, and São Paulo had particularly low Se concentrations. Forage analyses from 80 farms in São Paulo, Brazil, revealed low Se concentrations, with a mean of 0.072 for the rainy season and 0.054 ppm for the dry season, with corn silage averaging 0.034 ppm of Se (Lucci et al., 1983a). Table 12.1 summarized forage Se concentrations from seven studies in Latin America since 1972; the majority of samples were low in Se.

E. Prevention and Control

The principal methods of increasing Se intake by livestock include (1) a free-choice Se mineral supplement, (2) Se fertilization, (3) injections of Se, and (4) Se ruminal pellets. The use of Se-fortified salt mixtures appears to be the most promising procedure for prevention of deficiency of this element (McDowell et al., 1984).

Sources of supplemental Se currently in use are sodium selenite and sodium selenate. Physical characteristics of the two salts are similar to

TABLE 12.1

Forage Selenium Concentrations (ppm) for Selected Warm Climate Regions

Location	Season	Sample no	Mean	Percentage of samples below 0.1 ppm
Malawi[a]	Dry	21	0.05	95.8
	Wet	48	0.10	55.4
Bolivia[b]	Dry	8	0.11	33.3
	Wet	16	0.14	22.2
Bolivia[c]	Dry	17	0.07	88.0
	Dry	42	0.10	47.0
Dominican Republic[d]	Dry	33	0.14	48.0
Colombia[e]	Dry	34	0.11	74.0
	Wet	35	0.12	38.0
Florida[f]	Dry	10	0.05	90.0
	Wet	19	0.07	84.2
Guatemala[g]	Dry	84	0.14	63.0
	Wet	84	0.40	49.0

[a] Mtimuni (1982).
[b] McDowell et al. (1982a).
[c] Peducassé et al. (1983).
[d] Jerez et al. (1984).
[e] Vargas et al. (1984).
[f] McDowell et al. (1982b).
[g] Tejada et al. (1984).

those of sodium chloride, and they mix uniformly with various other ingredients (Ammerman, 1981). The selenate form, which is less commonly used, has been considered preferable by some, because selenite is more readily reduced to less available elemental Se that can form insoluble compounds with other metals. Selenium from animal sources has an availability of about 35%, but plant Se sources show an availability of 80–140% when compared to sodium selenite as 100% available (NRC, 1983).

The use of heavy ruminal pellets (similar to those earlier developed for Co), consisting of 95% finely divided Fe and 5% elemental Se, has been shown by Australian workers to release sufficient Se to maintain adequate blood values for several months and to prevent the occurrence of WMD in sheep grazing deficient pastures (Underwood, 1981). Evidence is accumulating that these pellets cannot always be relied on to provide adequate protection to sheep on a deficient diet for periods beyond 12 months. Larger pellets of similar composition will raise the blood Se levels in cattle where concentrations have been found to be low (Dodson and Judson, 1973).

F. Toxicity

In tropical regions, Se deficiency is much more of a problem for grazing livestock than is toxicity of this element. Nevertheless, there are specific regions in selected countries where Se toxicity is a serious detriment to grazing livestock. The minimum toxic level for cattle suggested by the NRC (1980, 1983, 1984) is 2 ppm. Unfortunately, this estimate was based on data from monogastric animals and failed to sufficiently evaluate ruminant Se toxicity studies from natural diets or to take into account that monogastric animals may absorb over 2.5 times as much Se as ruminants from their diets (Wright and Bell, 1966). Therefore, until more specific ruminant Se toxicity studies are made, a maximum tolerable level of 4–5 ppm of Se for ruminants seems more reasonable.

Selenium toxicity conditions termed "alkali disease" or "blind staggers" have resulted in extreme losses of livestock. "Alkali disease" has resulted from livestock grazing forages with excess Se in the range of 5–40 ppm. In the acute form of toxicosis known as "blind staggers," much larger quantities of Se are consumed. Acute Se poisoning is most often associated with consumption of Se accumulator plants (e.g., *Austragalus racemous*), which may contain from 100 to over 9000 ppm of Se. Accumulator plants rarely occur outside seleniferous areas and may therefore be used as indicators of Se-toxic regions (Underwood, 1981).

Clinical signs of selenosis in cattle include loss of appetite, lack of thriftiness, atrophy and cirrhosis of the liver, chronic nephritis, myocardium necrosis, loss of vitality, loss of hair, lameness, and elongated hooves; in prolonged toxicity, the hooves may be sloughed off (Underwood, 1977). Selenium toxicity as evidenced by abnormal hooves is illustrated for cattle in Colombia (Fig. 12.12), Ecuador (Fig. 12.13), and

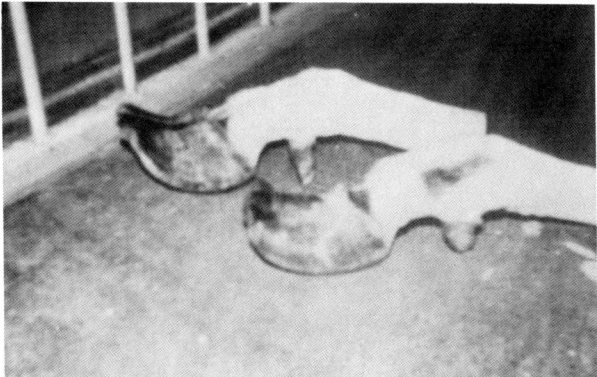

Fig. 12.12. Selenium toxicity in the Se-toxic region of Puerto Boyacá, Colombia. Misshapen hoof is due to selenium injury.

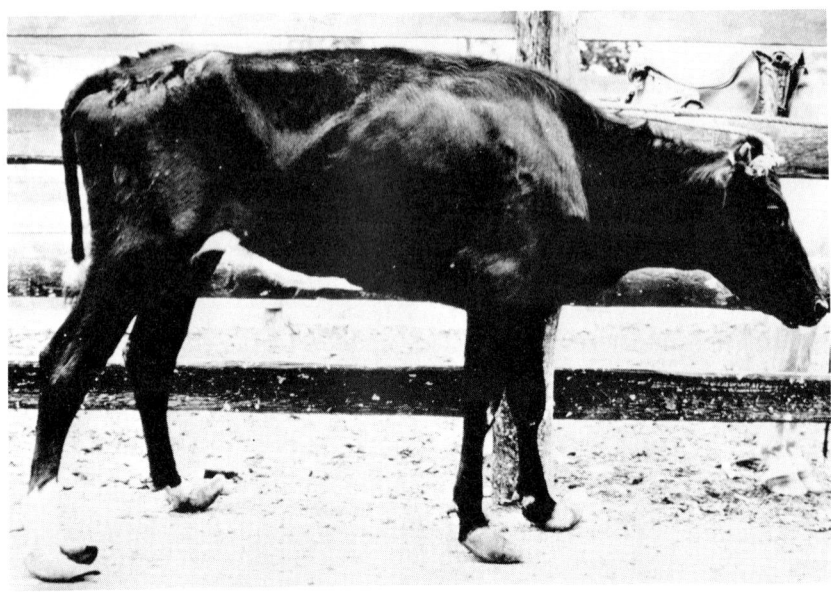

Fig. 12.13. An animal showing signs of Se toxicity in the Pichilingue region of Ecuador. The main clinical sign of toxicity is misshapen hooves.

Fig. 12.14. Selenium toxicity in holstein cattle in the region of Saltillo, Mexico. Misshapen hooves were found frequently, particularly in older animals.

Fig. 12.15. Malformed lambs have been observed in several areas in Wyoming, resulting from high Se intake by mature sheep. (Courtesy of O. A. Beath, Univ. of Wyoming, Laramie, Wyoming.)

Mexico (Fig. 12.14) and for sheep in Wyoming (Fig. 12.15). In acute Se poisoning, cattle exhibit blindness, abdominal pain, excessive salivation, teeth grating, paralysis, and respiratory failure. Respiration is ultimately disturbed, and death is the result of respiratory failure along with starvation and thirst.

Selenium toxicity has been listed in 21 geographical locations in developing tropical countries of Latin America, Africa, and Asia (see Chapter 16 of this volume). Latin American countries in which selenosis is encountered include Argentina, Chile, Colombia, Ecuador, Honduras, Mexico, Peru, and Venezuela (Lemley, 1943; Russell and Duncan, 1956; Benavides and Mojica, 1965; Kerdel-Vegas, 1966; Dickson, 1969; Jaffe et al., 1969). Lemley reports rather heavy livestock losses in Argentina due to acute Se poisoning caused by ingestion of *Astragalus bergi*. Throughout Argentina, descriptions of both acute and chronic Se poisoning are common. In the Irapuato area of Mexico, "saliman disease," which is similar to alkali disease, occurs on vegetation containing 1–120 ppm of Se. From Irapuato, it is reported that a drover lost an entire herd of 126 cattle that rested for one night in an alfalfa field (Russell and Duncan,

1956). The sites for Se toxic regions and the economic importance in Colombia have been described by Benavides and Mojica (1965). The loss to the livestock industry lies not so much in livestock with acute Se poisoning as in the subclinical, chronic, low-grade poisonings, which probably are of greater economic importance.

Animals with clinical signs of Se toxicity (e.g., abnormal hooves) likewise sometimes may exhibit clinical signs of molybdenum (Mo) toxicity (e.g., faded hair coat). In the Saltillo region of Mexico (Gartenberg, 1985) and a region north of Bogotá, Colombia (Gil et al., 1975), ruminants graze pasture deficient in Cu and toxic in both Mo and Se. These conditions would be expected more from alkaline soils that favor plant uptake of both Se and Mo while depressing forage Cu concentrations.

Preventing or minimizing Se toxicity for grazing livestock is obviously a much more difficult challenge than providing supplementary Se for deficient livestock. Underwood (1981) suggested three possible methods for prevention of Se toxicity: (1) treatment of the soil to reduce Se uptake by plants, (2) treatment of the animal so that Se absorption is reduced or excretion increased to limit accumulation in the tissues, and (3) modification of the diet by inclusion of substances that inhibit or antagonize the toxic effects of Se within the body tissues.

At present, practical measures for controlling Se poisoning for livestock rely on pasture rotation and the use of supplemental feeds from nonseleniferous areas. However, relying on feeds from non-Se-toxic regions is an impractical solution for grazing livestock. Locating seleniferous soils and mapping them in sufficient detail is essential. The Se content of grasses is highest in young plants and declines rapidly at later stages of maturity (Olson et al., 1942). Brown and De Wet (1967) report that wheat on a 30-ppm Se soil contained 1120 ppm in the young stages and 220 ppm at maturity. Therefore, use of pastures of low Se content during the growing season and high-Se forages toward the end of the growing season is a very effective control measure (Olson, 1969).

Management procedures to minimize Se toxicity in grazing cattle can be summarized as follows (Olson, 1969): (1) Map portions of the ranch and determine which fields are most seleniferous. (2) Delay using high-Se pastures until late summer versus spring, when the Se content of the plants has dropped considerably (plants decrease in Se as they mature). (3) Breed to calf in early spring, to minimize the effects of the Se grasses. (4) Avoid overgrazing (allow selection on animal's part; seleniferous indicator plants are rather unpalatable to livestock). (5) Remove fences, which results in larger pastures; this will also allow animals greater feed selection.

REFERENCES

Allman, R. T., and Hamilton, T. S. (1949). "Nutritional deficiencies in livestock." FAO Agricultural Studies No. 5, Washington, D.C.
Agricultural Research Council (ARC) (1980). "The Nutrient Requirements of Ruminant Livestock." Commonwealth Agricultural Bureaux, Slough. England.
Ammerman, C. B. (1981). *Anim. Nutr. Health* **37,** 26–28.
Andrews, E. D. (1956). *N. Z. J. Agric.* **92,** 239–244.
Barradas, H. V., Cook, R. M., and Shimada, A. S. (1985). *J. Dairy Sci.* (in press).
Becker, R. B., Henderson, J. R., and Leighty, R. B. (1965). *Bull—Fl. Agric. Exp. Stn.* **699.**
Benavides, S. T., and Mojica, F. S. (1965). "Selenosis," 2nd Bd. Instituto Geográfico "Augustin Codazzi," Bogotá, Colombia.
Boggino, E. V., Romero, J., and Cano, M. A. (1973). "Informe Anual 1973." Programa Nacional de Investigación y Extensión (Ganadera), Asunción, Paraguay.
Brown, J. M. M., and De Wet, P. J. (1967). *Onderstepoort J. Vet. Res.* **34,** 161–217.
Cary, E. E., Allaway, W. H., and Miller, M. (1973). *J. Anim. Sci.* **36,** 285–292.
Church, D. C. (1971). "Digestive Physiology and Nutrition of Ruminants," Vol. 2. Nutrition. D. C. Church and Oregon State Univ. Book Stores, Inc., Corvallis, Oregon.
Clements, F. W. (1957). *Med. J. Aust.* **2,** 645–646.
Corrêa, R. (1957). *Arq. Inst. Biol. São Paulo* **24,** 199–227.
Cowan, J. W., Silabian, A., and Djibelian, W. (1967). *In* "Proceedings Third Mideast Nutrition Conference." Berlin.
Cunha, T. J. (1982). *Anim. Nutr. Health* **38,** 12–15.
Dickson, J. D. (1969). *Econ. Botany* **23,** 133–134.
Dodson, M. E., and Judson, G. J. (1973). *Aust. Vet. J.* **49,** 320.
Ferguson, K. A., Schinckel, P. G., Carter, H. B., and Clarke, W. H. (1956). *Aust. J. Biol. Sci.* **9,** 575–585.
Fick, K. R., McDowell, L. R., and Houser, R. H. (1978). *In* "Proceedings Latin American Symposium on Mineral Nutrition Research with Grazing Ruminants," pp. 149–162. Univ. of Florida, Gainesville.
Follis, R. J., Jr. (1966). *Bol. Of. Sanit. Panam.* **60,** 28–38.
Franke, K. W., and Potter, V. R. (1934). *J. Nutr.* **10,** 213–221.
Gartenberg, P. (1985). M. S. thesis. Univ. of Florida, Gainesville (in preparation).
Gil, A., McDowell, L., and Tritschler, J. (1975). *In* "Informe Anual del Programa Nacional de Nutricion Animal," pp. 110–115. ICA-Tibaitata, Bogotá, Colombia.
Godwin, K. O. (1975). *In* "Trace Elements in Soil-Plant-Animal Systems" (D. J. D. Nicholas and A. R. Egan, eds.), pp. 259–270. Academic Press, New York.
Gutierrez, J. L., Smith, G. S., Wallace, J. D., and Nelson, A. B. (1974). *J. Anim. Sci.* **38,** 1331 (Abstr.).
Haggard, D., Stowe, H. D., Conner, G. H., and Johnson, D. W. (1980). *Am. J. Vet. Res.* **4,** 539–543.
Hansard, S. L. (1983). *Nutr. Abstr. Rev., Ser. B,* 1–25.
Harrison, J. H., Hancock, D. D., and Conrad, H. R. (1984). *J. Dairy Sci.* **67,** 123–132.
Hemken, R. W. (1970). *J. Dairy Sci.* **53,** 1138–1143.
Hemken, R. W., Vandersall, J. H., Sass, B. A., and Hibbs, J. W. (1971). *J. Dairy Sci.* **54,** 85–88.
Herrick, J. (1982). *Anim. Nutr. Health* **38** (July–Aug.), 13–16.
Hogue, D. E. (1958). *Proc.—Cornell Nutr. Conf. Feed Manuf.* pp. 32–39.

Jaffe, W. G., Chavez, J. F., and Mondragon, M. C. (1969). *Arch. Latinoam. Nutr.* **19**(3), 299–307.
Jerez, M., McDowell, L. R., Martin, F. G., Hargus, W. A., and Conrad, J. H. (1984). *Trop. Anim. Prod.* **9**, 12–21.
Johnson, F. F., and Frederick, E. R. (1940). *Science* **92**, 315–316.
Judson, G. J., and Obst, J. M. (1975). *In* "Trace Elements in Soil-Plant-Animal Systems" (D. J. D. Nicholas and A. R. Egan, eds.), pp. 385–406. Academic Press, New York.
Julien, W. E., Conrad, H. R., Jones, J. E., and Moxon, A. L. (1976). *J. Dairy Sci.* **59**, 1954–1959.
Kelly, F. C. and Snedden, W. W. (1960). "Endemic goiter," World Health Organization Monograph Series, No. 44. Geneva. Switzerland.
Kerdel-Vegas, F. (1966). *Econ. Botany* **20**, 187–195.
Latteur, J. P. (1962). "Cobalt Deficiencies and Sub-deficiencies in Ruminants." Centre D'Information du Cobalt, Brussels. Belgium.
Lee, H. J. (1963). *In* "Animal Health, Production and Pasture" (A. H. Worden, K. O. Sellers, and D. E. Tribe, eds.), p. 662. Longmans, Green, New York.
Lee, H. J., Kuchel, R. E., Good, B. F., and Trowbridge, R. F. (1957). *Aust. J. Agric. Res.* **8**, 502–510.
Lemley, R. E. (1943). *J. Lancet* **63**, 257–259.
Lucci, C. S., and Moxon, A. L. (1982). *In* "Anais da Sociedade Brasileira de Zootecnia," p. 104 (abstr.). Sociedade Brasileira de Zootecnia. Piracicaba, São Paulo, Brazil.
Lucci, C. S., Moxon, A. L., Zanetti, M. A., Schalch, E., Pettinati, R. L., Fukushima, R. S., Franzolin Neto, R., and Marcomini, D. G. (1983a). *In* "Anais da Sociedade Brasileira de Zootecnia," p. 192 (abstr.).Sociedade Brasileira de Zootecnia. Pelotas, Brazil.
Lucci, C. S., Moxon, A. L., Zanetti, M. A., Schalch, E., Pettinati, R. L., Fukushima, R. S., Franzolin Neto, R., and Marcomini, D. G. (1983b). *In* "Anais da Sociedade Brasileira de Zootecnia," p. 193 (abstr.).Sociedade Brasileira de Zootecnia. Pelotas, Brazil.
McDonald, I. W. (1968). *Nutr. Abstr. Rev.* **38**,(2), 381–400.
McDowell, L. R., Conrad, J. H., Thomas, J. E., and Harris, L. E. (1974). "Latin American Tables of Feed Composition." Univ. of Florida, Gainesville.
McDowell, L. R., Conrad, J. H., Thomas, J. E., Harris, L. E., and Fick, K. R. (1977). *Trop. Anim. Prod.* **2**, 273–279.
McDowell, L. R., Froseth, J. A., and Piper, R. C. (1978). *Nutr. Rep. Int.* **17**, 19–33.
McDowell, L. R., Bauer, B., Galdo, E., Koger, M., Loosli, J. K., and Conrad, J. H. (1982a). *J. Anim. Sci.* **55**, 964–970.
McDowell, L. R., Kiatoko, M., Bertrand, J. E., Chapman, H. L., Pate, F. M., Martin, F. G., and Conrad, J. H. (1982b). *J. Anim. Sci.* **55**, 38–47.
McDowell, L. R., Conrad, J. H., and Ellis, G. L. (1984). *In* "Symposium on Herbivore Nutrition in Sub-Tropics and Tropics—Problems and Prospects" (F. M. C. Gilchrist and R. I. Mackie, eds.), pp. 67–88. Craighall, South Africa.
McLean, J. W., Thomson, G. G., and Claxton, J. H. (1959). *N. Z. Vet. J.* **7**, 47–52.
McMurray, C. H., Rice, D. A., McLoughlin, M., and Blanchflower, W. J. (1984). *Trace Elem. Man Anim. Proc. Int. Symp. 5th*, p. 37 (Abstr.).
Marston, H. R., Allen, S. H., and Smith, R. M. (1961). *Nature (London)* **190**, 1085–1091.
Miller, W. J. (1979). *In* "Dairy Cattle Feeding and Nutrition," pp. 74–180. Academic press, New York.
Mohammed, R. (1983). Ph.D. dissertation. Univ. of Clermont, Clermont, France.
Morris, J. G., Cripe, W. S., Chapman, H. L., Jr., Walker, D. F., Armstrong, J. B., Alexan-

der, J. D., Jr., Miranda, R., Sanchez, A., Jr., Sanchez, B., Blair-West, J. R., and Denton, D. A. (1984). *Science* **223**, 491–493.
Moxon, A. L. (1971). "Deficiencia Nutricional de Selenio," p. 10 (abstr.). Suplemento Agricola 855, Estado de São Paulo. São Paulo, Brazil.
Mtimuni, J. P. (1982). Ph.D. dissertation. Univ. of Florida, Gainesville.
Muth, O. H., Oldfield, J. E., Remmert, L. F., and Schubert, J. R. (1958). *Science* **128**, 1090.
National Research Council (NRC) (1975). "Nutrient Requirements of Sheep," 5th rev. ed. Nutrient Requirements of Domestic Animals, No. 5. Natl. Acad. Sci., Washington, D.C.
National Research Council (NRC) (1976). "Nutrient Requirements of Beef Cattle," 5th rev. ed. Nutrient Requirements of Domestic Animals, No. 4. Acad. Sci., Washington, D.C.
National Research Council (NRC) (1978). "Nutrient Requirements of Dairy Cattle," 5th rev. ed. "Nutrient Requirements of Domestic Animals, No. 3. Natl. Acad. Sci., Washington, D.C.
National Research Council (NRC) (1980). "Mineral Tolerance of Domestic Animals." Natl. Acad. Sci., Washington, D.C.
National Research Council (NRC) (1983). "Selenium in Nutrition," rev. ed. Natl. Acad. Sci., Washington, D.C.
National Research Council (NRC) (1984). "Nutrient Requirements of Beef Cattle," 6th rev. ed. Nutrient Requirements of Domestic Animals, No. 4. Natl. Acad. Sci., Washington, D.C.
Olson, O. E. (1969). *In* "Proceedings Georgia Nutrition Conference," pp. 68–78. Univ. of Georgia, Athens.
Olson, O. E., Jornlin, D. F., and Moxon, A. L. (1942). *J. Am. Soc. Agron.* **34**, 607–615.
Peducassé, C. A., McDowell, L. R., Parra, L. A., Wilkins, J. V., Martin, F. G., Loosli, J. K., and Conrad, J. H. (1983). *Trop. Anim. Prod.* **8**, 118–130.
Rickes, E. L., Brink, N. G., Koniuszky, F. R., Wood, T. R., and Folkers, K. (1948). *Science* **108**, 134.
Rodel, M. G. W. (1972). *Rhod. Agric. J.* **69**(3), 59–60.
Rotruck, J. T., Pope, A. L., Ganther, H. E., Swanson, A. B., Hafeman, D. G., and Hockstra, W. G. (1973). *Science* **179**, 588–590.
Russell, F. C., and Duncan, D. L. (1956). "Minerals in Pasture: Deficiencies and Excesses in Relation to Animal Health." Commonwealth Bureau of Animal Nutrition Technical Communication No. 15, Rowett Institute, Aberdeen, Scotland.
Schwarz, K., and Foltz, C. M. (1957). *J. Am. Chem. Soc.* **79**, 3292–3293.
Smith, E. L. (1948). *Nature (London)* **162**, 144–145.
Smith, R. M., and Marston, H. R. (1970). *Br. J. Nutr.* **24**, 857–877.
Sousa, J. C. de, Conrad, J. H., Blue, W. G., Ammerman, C. B., and McDowell, L. R. (1981). *Pesqui. Agropecu. Bras.* **16**(5), 739–746.
Tejada, R., McDowell, L. R., Martin, F. G., Conrad, J. H., and Ellis, G. L. (1985). *Trop. Agric.* (submitted).
Terry, T. (1964). *In* "Annales del II Congreso Nacional de Medicina Veterinaria y Zootecnia," p. 163 (abstr.). San Marcos Univ., Lima, Peru.
't mannetje, L., Singh Sidhu, A., and Murugaiah, M. (1976). *MARDI Res. Bull.* **4**(1), 90–98.
Tokarnia, C. H., and Döbereiner, J. (1978). *In* "Proceedings Latin American Symposium on Mineral Nutrition Research with Grazing Ruminants" (J. H. Conrad and L. R. McDowell, eds.), pp. 163–169. Univ. of Florida, Gainesville.
Tokarnia, C. H., Guimaraes, J. A., Canella, C. F. C., and Döbereiner, J. (1971). *Pesqui. Agropecu. Bras.* **6**, 61–77.

Underwood, E. J. (1977). "Trace Elements in Human and Animal Nutrition." Academic Press, New York.
Underwood, E. J. (1981). "The Mineral Nutrition of Livestock." Commonwealth Agricultural Bureaux, London.
Vargas D., R., McDowell, L. R., Conrad, J. H., Martin, F. G., Buergelt, C., and Ellis, G. L. (1984). *Trop. Anim. Prod.* **9,** 103–113.
Wilson, D. C. (1953). *Br. J. Nutr.* **8,** 90–97.
Wright, P. L., and Bell, M. C. (1966). *Am. J. Physiol.* **211,** 6–10.

13

Iron, Manganese, and Zinc

L. R. McDOWELL
Department of Animal Science
University of Florida
Gainesville, Florida

I.	General.	291
II.	Iron.	292
	A. Introduction.	292
	B. Metabolism.	292
	C. Requirement.	294
	D. Deficiency.	294
	E. Prevention and Control.	297
	F. Toxicity.	297
III.	Manganese.	297
	A. Introduction.	297
	B. Metabolism.	298
	C. Requirement.	299
	D. Deficiency.	300
	E. Prevention and Control.	303
	F. Toxicity.	303
IV.	Zinc.	304
	A. Introduction.	304
	B. Metabolism.	304
	C. Requirement.	306
	D. Deficiency.	307
	E. Prevention and Control.	310
	F. Toxicity.	311
	References.	311

I. GENERAL

Research from tropical regions suggests that trace mineral mixtures for grazing ruminants should contain cobalt (Co), copper (Cu), iodine (I), and sometimes selenium (Se). These minerals are known with certainty to be required in large areas and, under some conditions, in amounts greater than those available from forage. The advisability of supplementing iron (Fe), manganese (Mn), and zinc (Zn) has been less certain. However, grazing ruminant research has indicated the definite need for supplemental Zn, but most studies have indicated that the majority of tropical forages contain Fe and Mn in excess of requirements.

The majority of tropical soils are acid, favoring high forage uptake of Fe and Mn. In addition, soil consumption will provide substantial quantities of these minerals to grazing livestock diets, particularly Fe. As a consequence, mineral imbalances typified by excesses of Fe and Mn, often associated with tropical forages, may interfere with metabolism of other minerals (Lebdosoekojo *et al.*, 1980). Table 13.1 illustrates forage concentrations of Fe, Mn, and Zn and the percentage considered deficient in five tropical countries. Very few forages had low concentrations of Fe and Mn, suggesting no deficiency. In contrast, the vast majority of forages contained deficient Zn concentrations.

II. IRON

A. Introduction

Iron deficiency is one of the most commonly occurring deficiency diseases of swine and humans. Nevertheless, it is rarely of practical concern in grazing cattle and sheep, except in circumstances involving blood loss or disturbance in Fe metabolism as a consequence of parasitic infestation or disease. Even during the suckling period, Fe deficiency is unlikely to occur in ruminant species, but if they are maintained on milk alone for an abnormally long period, anemia may occur.

B. Metabolism

Iron plays a vital role in animal metabolism, mainly confined to the process of cellular respiration, as a component of hemoglobin, myoglobin, and cytochrome, and in certain enzymes. The role of Fe in the body as components of myoglobin and hemoglobin serve to store and transport molecular oxygen by coordination with Fe atoms in these compounds (Underwood, 1977).

Iron absorption occurs throughout the gastrointestinal tract including the stomach and colon, but the major sites of active absorption are located in the duodenum and jejunum. Thomas (1970) summarizes Fe ab-

TABLE 13.1
Mean Forage Iron, Manganese, and Zinc Concentrations (ppm) by Season and Percentage of Samples below Critical Concentrations (cc) for Five Tropical Countries[a]

Country	Season	Iron			Manganese			Zinc		
		Mean	cc	Percentage below cc	Mean	cc	Percentage below cc	Mean	cc	Percentage below cc
Colombia[b]	Wet	139	30	0	209	40	0	24	30	89
	Dry	171	30	0	264	40	0	24	30	74
Dominican Republic[c]	Dry	154	30	0	151	30	10	22	30	86
Bolivia[d]	Wet	178	30	0	365	20	0	24	30	89
	Dry	122	30	0	339	20	0	26	30	81
Guatemala[e]	Wet	378	30	0	92	40	32	32	30	61
	Dry	661	30	0	93	40	24	33	30	49
Malawi[f]	Wet	207	50	6	98	20	6	44	40	96
	Dry	195	50	3	245	20	3	19	40	94

[a] Critical concentrations (cc) are based on ruminant needs (McDowell and Conrad, 1977).
[b] Based on 35 and 36 samples for the wet and dry seasons, respectively (Vargas et al., 1984).
[c] Based on 69 samples (Jerez et al., 1984).
[d] Based on 16 samples for the wet season (McDowell et al., 1982a) and 84 for the dry season (Peducassé et al., 1983).
[e] Based on 84 samples for both the wet and dry seasons (Tejada, 1984).
[f] Based on 48 and 21 samples for the wet and dry seasons, respectively (Mtimuni, 1982).

sorption concepts as follows: (1) absorption is more efficient when body stores are low, (2) the amount absorbed is usually a small portion of that ingested, (3) many dietary factors influence the amount absorbed, (4) absorption occurs directly into blood, with limited amounts into the lymphatic system, (5) the quantity absorbed from the duodenal area exceeds that from other portions of the tract, and (6) orally administered ferrous (Fe^{2+}) salts are more effective for hemoglobin regeneration than ferric (Fe^{3+}) salts. High dietary levels of phosphorus (P) reduce Fe absorption, presumably by the formation of insoluble ferric phosphate and phytate, and high dietary levels of several divalent metals, including Cu, Mn, lead (Pb), and cadmium (Cd), increase Fe requirements by competing for absorption sites in the intestinal mucosa (Underwood, 1981).

Iron is excreted in the urine and feces, in addition to losses through sweat, hair, and nails. However, absorbed Fe is retained with great tenacity and is not readily lost from the body except through hemorrhage. Iron is released from hemoglobin during red blood cell breakdown and is carried to the liver and secreted in the bile. Most of the Fe in the bile is reabsorbed and can be used again to form hemoglobin.

C. Requirement

The Fe requirements of ruminants are not well established (Underwood, 1981). The suggested Fe requirement for ruminants is between 30 and 100 ppm [National Research Council (NRC) 1975, 1978, and 1984]. It is known that young animals have higher requirements than adults. Calves fed on exclusive whole milk diets (milk is low in Fe) can develop Fe-deficiency anemia within 2–3 months.

D. Deficiency

Iron deficiency seldom occurs in adult livestock unless there is a considerable blood loss from parasites or disease. Signs of a lack of Fe, in addition to anemia and related blood changes, include lower weight gains, listlessness, inability to withstand circulatory strain, labored breathing after mild exercise, reduced appetite, and decreased resistance to infection (Miller, 1979a).

Hibbs et al. (1961) reported that calves may be deficient in Fe at birth since 30% of those born in a 12-year U.S. study had low hemoglobin values (<9 g/100 ml). Newborn calves receiving milk have increased hemoglobin formation as a result of supplemental Fe (Thomas et al., 1954). It is well known that anemia develops in calves restricted to a milk diet for long periods and that the condition can be alleviated by administration of Fe (Blaxter et al., 1957; Roy et al., 1964).

Heavy infestation with intestinal parasites results in an Fe-deficiency type of anemia in lambs and calves (Campbell and Gardiner, 1960). Anemia can result from the direct loss of blood via blood-sucking parasites or through a rise in the rate of degradation of blood cells and also through a depression of hematopoiesis from the action of toxic substances produced by the parasites.

Departure from normal levels of serum Fe, total Fe-binding capacity, percentage transferrin saturation value, and hemoglobin and hematocrit values can be used to diagnose Fe deficiency in livestock. Low hemoglobin and hematocrit values are not sensitive indicators of early Fe-deficiency stages because they only occur when storage Fe is severely depleted. Their use is often limited to diagnosis and confirmation of Fe deficiency (Miller and Stake, 1974). Percentage of saturation of transferrin is the best criterion of Fe-deficient erythropoiesis (Underwood, 1977). Under practical conditions, an assessment of Fe deficiency can be made with reduced transferrin saturation (<13–15%), serum Fe (<1.1. mg/liter), and hemoglobin levels (<10 g/100 ml) (McDowell, 1976).

Iron deficiency is considered rare for grazing cattle due to generally adequate pasture concentrations together with contamination of plants by soil (see Chapter 8 of this volume). Mitchell (1963) indicated that soil contains 20–100 times the Fe content found in pastures grown on a particular soil. Under New Zealand grazing conditions, annual ingestation of soil can reach 75 kg for sheep and 600 kg for dairy cows (Healy, 1974).

Acid soil conditions favor availability and plant uptake of Fe. Even plants grown on neutral or slightly alkaline soils often contain quite high levels of Fe. Since most of the soils are not alkaline in tropical regions, Fe deficiency would not be expected to be widespread (Cox, 1973).

Table 13.1 illustrates high Fe concentrations in forages from Bolivia, Colombia, the Dominican Republic, Guatemala, and Malawi. The majority of all forages contained Fe concentrations considerably in excess of requirements. Of 256 forage averages in the 1974 "Latin American Tables of Feed Composition," only 3.5% contained less than 30 ppm of Fe (McDowell et al., 1974). Forage analyses from São Paulo (Andreasi et al., 1968), Minas Gerais (Gomide et al., 1969), and Rio Grande do Sul (Gavillon and Quadros, 1973), Brazil, Guatemala (Scaillet, 1969), and Panama (Blue et al., 1969) all indicated Fe concentrations in excess of ruminant requirements. Of 192 and 120 forage samples collected during the dry and wet seasons in northern Mato Grosso, Brazil, mean Fe concentrations were 212 and 263 ppm, respectively (Sousa et al., 1981). However, other reports from Panama note that 7 out of 28 locations averaged less than 30 ppm Fe for samples of *Hyparrhenia rufa* (Chicco, 1972).

Iron deficiency may be a problem where ruminants are fed on low-quality forages, such as straw, for extended periods (Sen and Ray, 1964).

Fig. 13.1. This weak (Fe-deficient) 12-year-old cow, which was grazing on a Blanton fine sand (yellow) soil in Florida, had to be helped up. Her hemoglobin was only 4.8 mg/100 ml of whole blood. After being given supplemental Fe (ferric ammonium citrate), the hemoglobin value increased to 12.6 g/100 ml, and she regained body condition and strength. (Courtesy of R. B. Becker, Univ. of Florida, Gainesville.)

These investigators noted that Fe supplementation reduced weight losses in lactating cattle and produced more rapid gains in suckling calves. Iron deficiency has been reported in Florida when cattle grazed forages grown on white and gray, sandy loam and fine, sand soils (Becker et al., 1965). Fig. 13.1 illustrates an Fe-deficient cow from Florida whose hemoglobin was increased from 4.8 to 12.6 g/100 ml after ferric ammonium citrate administration. When Florida cattle are pastured on light, sandy soils, and often when they have been subjected to heavy insect or parasite infestations, the addition of Fe has served a useful purpose, and the hemoglobin levels, as well as the condition of the animals, have improved under such treatment (Davis, 1951). Other investigators who fed supplemental Fe have revealed no increase in weight gain or productivity and little or no change in hematological characterisitics when older animals had been consuming typical rations. However, Fe deficiency exists and, in some cases, supplemental Fe has produced marked improvement (Thomas, 1970).

E. Prevention and Control

Supplementation with Fe is much less important than for other trace minerals. Iron supplementation is most warranted for grazing livestock when forages contain less than 100 ppm of Fe and/or if insects or parasites are causing substantial blood loss.

The utilization of different Fe compounds by ruminants has only been studied to a limited extent. Dietary Fe in the ferrous form is considered to be absorbed more efficiently than that in the ferric form (Thomas, 1970). Ammerman and Miller (1972) reported that for ruminants, orally administered ferrous sulfate and ferric citrate were equal in value. Using these sources as a standard of 100, ferrous carbonate and ferric oxide would have relative values of 60 and 10%, respectively.

F. Toxicity

The maximum tolerable level for cattle is suggested as 1000 ppm and for sheep, 500 ppm (NRC, 1980). Standish et al. (1969, 1971) indicated that high dietary levels of Fe as ferrous sulfate resulted in reduced performance and changes in mineral composition of steer tissues. Average daily feed intake and average daily gain were depressed by as little as 400 ppm. This indicates that under certain situations levels below NRC suggested maximum tolerable levels should be considered. Standish et al. (1971) reported a decrease in plasma P by high levels of dietary Fe. Iron toxicity is characterized by reduced feed intake, lowered daily gain, diarrhea, hypothermia, and metabolic acidosis.

III. MANGANESE

A. Introduction

The degree to which Mn deficiencies, excesses, or imbalances are practical problems for grazing livestock is not well-defined. With the exception of Fe, Mn requires less attention than the commonly supplemented trace minerals Co, Cu, I, Se, and Zn for grazing livestock. Although Mn deficiency has been produced experimentally in calves and cattle, with effects on skeletal development and reproductive performance, until recently, doubt had been expressed whether this deficiency arises under field conditions. Manganese deficiency for ruminants under grazing conditions has been suggested in the United States (Bentley and Phillips, 1951; Dyer et al., 1964; Shupe et al., 1967), Holland (Grashuis et al., 1953), the

United Kingdom (Munro, 1957; Wilson, 1965), Union of South Africa (Bisschop and Groenewald, 1963), and Germany (Anke *et al.*, 1973).

B. Metabolism

Manganese is needed in the body for normal bone structure, for reproduction, and for the normal functioning of the central nervous system. Everson (1970) summarized the functions of Mn as follows: (1) this element is essential for optimum growth; (2) a deficiency of Mn during gestation results in poor viability of the offspring; (3) the young of severely deficient animals have skeletal abnormalities; (4) many of the Mn-deficient offspring are ataxic at birth, showing incoordination and poor equilibrium; and (5) manganese deficiency causes sterility in the male.

Like the other trace elements, Mn is closely involved with several enzyme systems. Liver arginase contains Mn as an essential component, with numerous enzyme systems being activated by Mn *in vitro*. The fact that Mn is concentrated in the mitochondria has led to the suggestion that *in vivo*, Mn is involved in the partial regulation of oxidative phosphorylation. Manganese is needed in bone formation, being essential for the development of the organic matrix, which is composed largely of mucopolysaccharides. Many of the defects associaed with Mn deficiency may be accounted for by the need for Mn in glycosyltransferase activity (Leach, 1974).

The absorption of ingested Mn is generally considered to be very low at about 1% (Abrams *et al.*, 1977). However, Howes and Dyer (1971) and Carter *et al.* (1974) have shown that Mn absorption in young calves was considerably higher than generally reported in other species. Howes and Dyer (1971), working with calves, showed that Mn is preferentially absorbed under low dietary intakes. Their data also suggest that Mn in the tissues of nonsupplemental calves has a slower turnover rate versus that in calves with adequate Mn status.

Britton and Cotizas (1966) previously suggested that variable excretion rather than variable absorption regulates tissue Mn. Animals on a low Mn diet continue to excrete Mn. The Mn body pool is small and the body does not ordinarily accumulate Mn, so that total body excretion is continuously very nearly equal to intake. Twenty-five to fifty percent of the body pool is often ingested daily (Lassiter *et al.*, 1970).

Oral Mn is excreted mainly in the feces (95–98%), with only 0.1–3% usually excreted in the urine (Thomas, 1970). Even toxic levels of Mn administered orally to lambs were excreted almost exclusively by way of the feces (Watson *et al.*, 1973). It is concluded from cattle research data that both variable excretion and absorption are major homeostasis mechanisms by which animals maintain normal health and performance over the wide range of Mn intake found under field conditions.

C. Requirement

The minimum dietary requirements of Mn depend upon the species, the criteria of adequacy employed, the chemical forms in which the element is ingested, and the nature of the rest of the diet (Underwood, 1977). Necessary levels for growth of cattle are less than for reproduction. Bentley and Phillips (1951) concluded that concentrations of 10 ppm are adequate for growth but marginal for optimum reproductive performance. Anke et al. (1973) further confirmed the lower requirements for growth than for fertility in female goats. Goats fed diets containing 20 ppm Mn in the first year and 6 ppm in the second year grew as well as those receiving 100 ppm of additional Mn, but the former exhibited greatly impaired reproductive performance. For ruminants, the estimated Mn requirement ranges between 20 and 40 ppm (Underwood, 1981; NRC, 1984).

Dyer et al. (1964) and Rojas et al. (1965) concluded that the Mn requirements of cows for maximum fertility are in excess of 16 ppm. Cuthbertson (1969) reviewed the literature on the effects of Mn on growth, reproduction, milk yield, and blood hemoglobin levels and suggested that Mn should be present in the diet at a level of 40 ppm.

The level of calcium (Ca) and P in the diet affects Mn requirements. Hignett (1959) suggested that Ca and P affect the utilization of Mn required for enzyme systems essential for establishing and maintaining pregnancy. Excessively high intakes of Ca and P over a long period reduced weight gains of growing calves reared mainly on a milk diet, possible due to interference with Mn absorption, but additional Mn (50 ppm) tended to counteract the excess (Hawkins et al., 1955).

On the basis of field evidence with cattle, unidentified factors other than Ca and P exist that are capable of producing "conditioned" Mn deficiency and thereby increasing Mn requirements (Underwood, 1977). Shupe et al. (1967) reported that cows ingesting *Lupinus caudatus*, *Lupinus sericeus*, lupine extracts, and sparteine sulfate gave birth to deformed calves. The alkaloids in the lupinus plants were toxic to animals, with the teratogenic effects on calves similar to anomalies due to Mn deficiency (Dyer et al., 1964; Rojas et al., 1965). Howes et al. (1973) showed that lupines (sparteine sulfate) had a significant effect on mineral metabolism. The reduction and/or increase in ^{54}Mn and ^{65}Zn activity caused by the addition of sparteine sulfate suggests that this alkaloid causes a change in the tissue turnover rate of Mn and Zn.

D. Deficiency

Clinical signs and conditions observed when the diet contains insufficient Mn include suboptimal soft tissue and skeletal growth; decreased breaking strength of bones; abnormal bone shape; ataxia; muscular weakness; excess accumulation of body fat; reduced tissue storage of Mn in bone, liver, hair, and ovary; reduced milk production; reduced level of Mn in milk; delayed, irregular, or absence of estrus; resorption of fetus; fetal deformities; aplasia or hypoplasia of pancreatic cells; diabetic-like blood glucose disappearance; decrease in liver arginase; and decrease in bone and plasma alkaline phosphatase (Thomas, 1970).

The reproductive processes are particularly susceptible to lack of Mn. A report from Wisconsin notes that sterility exists in about 10% of the cattle of certain herds where low Mn rations (<20 ppm) are fed (Bentley and Phillips, 1951). The effect of Mn supplementation on cattle infertility in Great Britain was reported by Wilson (1965) in an experiment with 350 cows. Of those given Mn, 63% conceived after the first service, compared with 51% of those not given Mn. Also, in Great Britain, Munro (1957) previously observed that conception rates were raised from 48 to 72% following Mn supplementation. In South Australia, Mn supplementation increased the lambing percentage in flocks in which reproductive performance had declined (Egan, 1972). Anke et al. (1973) reported that cows on rations low in Mn weakly expressed estrus, and those pregnant from the first insemination were 31% compared with 64% in the control herd. There was a significant prolongation of gestation, from 287 to 290 days.

The nature and severity of the skeletal abnormalities that arise in Mn-deficient animals vary from a mild and generalized rarefaction of bone to gross and crippling deformities, particularly of the long bones (Underwood, 1981). Reports from western United States show a positive relationship between a low Mn intake of gestating cows and the incidence of neonatal deformities in their calves (Dyer et al., 1964). Shupe et al. (1967) report a "crooked calf" disease in nine western states that is similar to the clinical signs of Mn deficiency. A survey conducted over a 4-year period with a total of 4000 head of cattle indicated that 2.7% of the calves born over this period were deformed (Shupe et al., 1967).

Rojas et al. (1965) fed pregnant cows rations containing 15.8 or 25 ppm of Mn. Calves born to cows fed the low Mn ration were born with a general weakness and deformities characterized by enlarged joints, stiffness, and twisted legs (See Fig. 13.2). Deformed calves had shorter humeri, with greatly reduced breaking strength (Rojas et al., 1965). The changes in epiphyseal cartilage of growing bones and the developing fetus have led to studies showing a decrease in the mucopolysaccharide content

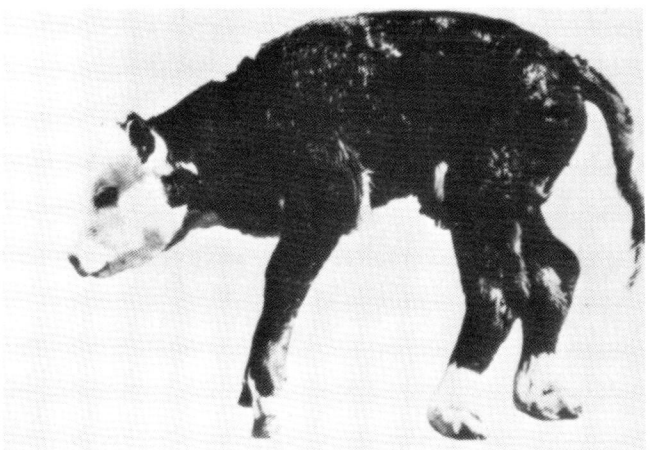

Fig. 13.2. Manganese deficiency in a newborn calf; legs are weak and deformed. (Courtesy of I. A. Dyer, Washington State Univ., Pullman, Washington.)

and radioactive sulfur (S) uptake of bone in Mn deficiency (Thomas, 1970). Although rarely reported for tropical regions, clinical signs suggesting a Mn deficiency have been observed in Costa Rica (C. Lang, personal communication) and Mato Grosso, Brazil (Mendes, 1977).

Concentration of Mn in forages is dependent on soil factors, plant species, stage of maturity, yield, pasture management, climate and soil pH. Gomide *et al.* (1969), studying mineral composition of tropical grasses in Minas Gerais, Brazil, found significant year × species × age interaction in Mn content. Poor drainage conditions increase forage Mn (Mitchell, 1963), but increasing soil pH decreases plant availability and uptake of Mn (Cox, 1973). From the Amazon Valley of Brazil, Sutmöller *et al.* (1966) report a very high intake of Mn by Brazilian cattle on soils that are deeply flooded during a long period of the year, while low intake occurs on nonflooded or slightly flooded soils. In Germany, Mn deficiency in cattle developed only after heavy liming (Anke *et al.*, 1973).

From Peru, Beeson and Gomez (1970) noted that Mn concentrations tended to be excessively high in forages grown on acid soils. Cox (1973) notes that in order to develop a Mn-deficient soil, soils must be leached of reduced forms of Mn, and soils frequently are limed, often in excess of crop requirements. In Brazil, Sutmöller *et al.* (1966) report a relationship between Mn content of soils and Mn intake by cattle. Soils of the semiarid region of the state of Pernambuco, Brazil, are characterized by adequate Mn, with the chances of a Mn deficiency remote (Horowitz and Dantas,

1966). However, in soils near the coastal Zona de Mata, where most of the cattle are located, available Mn is much less than 20 ppm, which is considered deficient (Horowitz and Dantas, 1966; Dantas, 1971).

Table 13.1 illustrates high forage Mn from the five tropical countries of Bolivia, Colombia, the Dominican Republic, Guatemala, and Malawi, Forage Mn in these countries far exceeded the requirement of this element by grazing livestock. Of 293 average forage values from the 1974 "Latin American Tables of Feed Compositon," 21% contained less than 40 ppm Mn, while only 5.2% contained less than 20 ppm (McDowell et al., 1974). Andreasi et al. (1968) reported adequate Mn concentrations of three common forages in the state of São Paulo, Brazil. Mean values taken from different locations ranged from a low of 89 to 326 ppm, while the distribution of all values was between 22 and 487 ppm. In Minas Gerais, Brazil, Gomide et al. (1969) reported the mineral composition of six tropical grasses and found the Mn species means to range from 64 to 248 ppm. Of 192 and 120 forage samples from Northern Mato Grosso, Brazil, collected during the dry and wet seasons, mean Mn concentrations were 204 and 87 ppm, respectively (Sousa et al., 1981). From Guatemala, only 1 of 28 species had mean Mn values of less than 20 ppm (Scaillet, 1969). Blue et al. (1969) reported adequate Mn forage concentrations in four regions of Panama, with values ranging from 28 to 213 ppm. However, Chicco (1972) reported low average Mn concentrations (<20 ppm) for 7 out of 28 locations in Panama.

There is no single, simple diagnostic test that will permit the early detection of Mn deficiency in animals. Other than forage and soil analysis, Egan (1975) reported that the sampling of stream bed sediments was used to reveal previously unsuspected Mn deficiency in cattle. Underwood (1981) believes that the level of Mn in the liver is a useful but not an entirely reliable indicator of Mn deficiency, unless the deficiency is severe. Van Koetsveld (1958) and Egan (1975) considered liver Mn values below 10 ppm and 6 ppm, respectively, as indicating a deficiency. Contrary to other tissues, hair Mn levels continue to rise as dietary Mn intake increases. Costa Rican cattle consuming high Mn forages (>100 ppm) had very elevated hair Mn levels compared with controls, 83.3 versus 18.6 ppm, respectively. McDowell (1976) concluded from the literature that Mn deficiency can best be detected by the combination of liver (<6 ppm) and diet (<20–40 ppm) analyses, while a toxicity is suspected when hair samples contain over 70 ppm Mn. The only certain way of detecting mild Mn deficiency is by measuring the response in reproductive performance, skeletal development, and growth that occurs as a result of feeding a Mn supplement.

E. Prevention and Control

Manganese-deficient soils have been discovered in many parts of the world on which plant growth response can be obtained from the application of Mn-containing fertilizer. In areas where Mn deficiency in crops has been demonstrated, attention should be given to the possible inadequacy of supplies of dietary Mn for ruminants.

In areas where Mn deficiency or "conditioned" Mn deficiency occurs in ruminants, remedial measures could normally include a careful control of liming, the use of crop sprays or fertilizers containing Mn, and the inclusion of Mn-containing supplements in the ration. Supplementation of the feed with Mn sulfate at the rate of 4 g for cows, 2 g for heifers, or 1 g per day for calves is sufficient for either the prevention or the cure of the deficiency (Underwood, 1981). Underwood (1981) also reports that treatment of pastures with Mn sulfate at the rate of 15 kg/heçtare has been found completely effective under the conditions that exist in the Netherlands.

On the basis of the literature, it appears that 40 ppm of forage Mn would be considered adequate, 20–40 ppm, borderline, and less than 20 ppm deficient for ruminants. Nevertheless, it should be remembered that soil and water likewise provide Mn toward meeting the animal's requirement. Dyer (1961) reported that water from a ranch with normal calves contained 1.7 ppm while that from a nearby ranch with deformed calves contained 0.02 ppm of Mn.

When roughages contain less than 40 ppm of Mn, it seems likely that most beef or dairy cattle rations would require supplemental Mn. It seems logical, as a preventive measure, to add 10–20 ppm of supplemental Mn to ruminant diets when forage analyses are routinely below 40 ppm. This level of Mn should be adequate to correct any likely borderline deficiency.

Manganese as sulfate, carbonate oxide, or chloride have been shown to be effective sources of Mn for ruminants. Black (1983) utilized tissue accumulation as a bioassay to compare three sources of Mn and found that Manganese sulfate was most available, followed by Manganese oxide, with Manganese Carbonate least available.

F. Toxicity

Experimental levels of 50 to over 4000 ppm of Mn in the diet of monogastric and ruminant animals have caused reduced hemoglobin, growth depression, abnormal bones, and other clinical signs (Thomas, 1970). It appears that 1000 ppm of dietary Mn is the maximum tolerable level for cattle and sheep (NRC, 1980). However, continuous grazing of forage containing 200 ppm or higher, produced on volcanic soils of Costa Rica, resulted in reproductive abnormalities in dairy cattle (Fonseca and Davis, 1968; Lang, 1971). These cattle that consumed high Mn forages had a lower calving percentage and required more services per pregnancy than did controls (Lang, 1971).

IV. ZINC

A. Introduction

Large areas of Zn-deficient soils exist in many countries in which significant responses have been demonstrated in the yield of pastures and crops to applications of Zn-containing fertilizers (Underwood, 1981). Zinc deficiency, under experimental conditions, has been produced by many workers. However, a deficiency under natural conditions was once thought unlikely. Since few reports of clinical signs of Zn deficiency in grazing animals appeared, it was generally assumed that the herbage of Zn-deficient soils carries enough Zn for the needs of animals. This assumption is no longer valid, with Legg and Sears (1960) first demonstrating a parakeratosis type skin disorder in grazing cattle of Guyana that responded to Zn therapy.

B. Metabolism

Zinc is essential in all animals, functioning in enzyme systems and being largely involved in nucleic acid metabolism, protein synthesis, and carbohydrate metabolism. Zinc has likewise been shown to be involved in vitamin A metabolism by maintaining normal concentrations of vitamin A in plasma (Smith *et al.*, 1973). These workers used animals deficient in both Zn and vitamin A and demonstrated that Zn is necessary for normal mobilization of vitamin A from the liver. In Australia, an extensive study with grazing cattle showed that a 12% annual cattle mortality rate was due, in part, to both a low forage carotene (vitamin A precursor) and a slow release of liver vitamin A (Geurin, 1981). Apparently, low forage Zn (25 ppm) was inadequate for synthesis of a protein (retinol binding protein) required for mobilizing hepatic vitamin A (retinol) to the blood and thus brought about the vitamin A deficiency.

The small intestine is the main site of Zn absorption. Absorption is affected by dietary level, amounts and proportions of several other elements and dietary components, and by the chemical form in which Zn is ingested (Underwood, 1977). A high percentage of all Zn is excreted through the feces by entrance via pancreatic secretions and very little by way of urine (Miller, 1970). Zinc excretion can also occur in sweat, especially in tropical countries.

The most important factor affecting absorption is the Zn content of the ration. Stake et al. (1975b) reported that Zn absorption for calves and cows ranged between 47.2 and 53.4% when fed low-Zn rations. Zinc absorption was directly reflective of the physiological demand for Zn in both calves and lactating cows (Miller, 1969; Stake et al., 1975a). Zinc-deficient animals absorb a higher percentage of administered Zn, with calves having a net absorption as high as 80% of an oral Zn dose (Miller, 1970). Neathery et al. (1973) reported that dairy cattle compensated for low-Zn diets within one week by a 50% increase in percentage absorption of Zn, a decrease in milk Zn, and a decline in excretion of Zn in feces.

Zinc is widely distributed throughout the body, with substantial amounts in all tissues containing considerable protein or calcified material (Miller, 1970). The prostate gland, choroid and the iris of the eye, and the pituitary gland accumulate particularly high concentrations of Zn. Zinc occurs in the animal and is transported almost entirely combined with organic compounds (Miller, 1971). In absorption and transport, Zn is apparently passed from one binding protein to another. Zinc may be passed as a metallo–ligand complex with such compounds as amino acids or EDTA as the nonprotein ligands. Metabolism after absorption can be affected by the ligands involved (Miller, 1975).

When young ruminants are fed a very Zn-deficient diet, the Zn content of some tissues decline, but in others, there is little or no change. With a severe deficiency, there is limited Zn reduction in hair, bone, liver, lung, kidney, spleen, pancreas, and blood plasma (Miller, 1970). Neathery et al. (1973) reported that Zn tissue contents of first lactation dairy cows fed 16.6 ppm or 39.5 ppm Zn for 6 weeks were very similar. Of 25 tissues analyzed, only rib cartilage and rumen wall from cows given the low-Zn diet had significantly less Zn.

Animals have a limited capacity for storing Zn in a form that can be mobilized rapidly to prevent deficiency signs. With a very low-Zn diet, there is an immediate sharp decline in plasma Zn, with reductions in feed intake and cessation of growth within less than 1 week (Mills et al., 1967). Plasma Zn concentrations fall below 0.04 mg/100 ml within 24 hours of feeding Zn-deficient diets. However, Spais and Papasteriadis (1974) reported plasma Zn between 0.04 and 0.06 mg/100 ml corresponding to severe Zn characteristic lesions, with 0.06 and 0.1 mg/100 ml related to slight Zn-deficiency signs.

Many natural occurring diets contain adequate amounts of Zn, but only small amounts of the element seem to be available to the animal. Therefore, factors must be present that reduce Zn absorption or impair its utilization in the animal. The metabolism of Zn may be influenced by interaction with other elements like cadmium, Ca, Cu, Fe, Se, and Mn (Strain and Pories, 1970; Ivan and Grieve, 1975). The presence of chelating materials, phytic acid, levels of certain minerals, particularly Ca, and many unknown factors affect Zn absorption in monogastric animals (Blackmon et al., 1967). However, in most instances, the factors that may affect the requirements of ruminants are unknown. In normally fed ruminants, there is at present lack of direct evidence that Ca, phytic acid, or other factors in plant proteins decrease Zn absorption (Mills et al., 1967; Miller, 1970).

C. Requirement

The minimum Zn requirement of ruminants varies with the chemical form, the combination in which the element occurs with other components of the diet, and with the criteria of adequacy employed. The suggested Zn requirement for cattle is 20–50 ppm [NRC, 1984, 1978; Agricultural Research Council (ARC) 1980].

For normal growth and Zn plasma levels in calves, 9–14 ppm of Zn has been found adequate (Miller et al., 1963; Miller et al., 1967). In contrast, Zn deficiency has been reported, under field conditions in various European locations, when forage contained from 19 to 83 ppm Zn (Miller, 1970). For dairy cattle, Haaranen (1963) suggests that the requirement for Zn is 45 ppm when the ration contains 0.3% Ca and is increased by 16 ppm for each 0.1% of additional Ca. In contrast, Kincaid (1979) reported that elevated amounts of dietary Ca, as ground limestone, reduced absorption of Zn in rats fed soyprotein but had no effect on absorption of Zn in lactating cows.

The Zn requirement would be dependent on the criterion of adequacy employed. For example, the minimum Zn requirements for spermatogenesis and testicular development in young male sheep are significantly higher than they are for body growth (Underwood and Somers, 1969). Testicular growth and spermatogenesis in ram lambs were markedly subnormal at 17 ppm dietary Zn and entirely normal at intakes of 32 ppm. If body growth is taken as the criterion of adequacy, Zn requirements would clearly be lower than for similar animals kept for breeding.

Zinc is known to provide protection against liver damage caused by toxins from the fungi, *Pithomyces chartarum* (Underwood, 1981) and *Phomopis leptostromiformis*. Intakes of Zn required to provide protection from these mycotoxins is some 50 times the normal Zn requirement and, thus, Zn toxicity is encountered.

D. Deficiency

Blackmon et al. (1967) summarized clinical signs of a Zn deficiency to include (1) inflammation of the nose and mouth with submucous hemorrhages, (2) unthrifty appearance, (3) rough coat, (4) stiffness of the joints, with soft edematous swelling of the feet in front of the fetlocks, (5) cracks in the skin of the coronary bands around the hooves, which later become deep fissures, (6) dry, scaly skin on the ears, (7) thickening and cracking of the skin around the nostrils, (8) development of horny overgrowths of the mucosa on the lips and dental pads, (9) gnashing of the teeth, (10) alopecia, (11) red, scabby and wrinkled scrotal skin, and (12) bowing of the hind legs.

Early effects of Zn deficiency include reduced feed intake, reduced growth rate and feed efficiency, followed by skin disorders. Parakeratosis of the skin is perhaps the most obvious clinical sign of severely Zn-deficient cattle and goats (Miller, 1979a). In calves, the scrotum, head, and the area around the nostrils, neck, and legs most often are parakeratitic. In lactating cows, teats may show considerable parakeratosis (Kirchgessner and Schwarz, 1975). The effects, including lesions, of a Zn deficiency in goats and sheep are similar to those in cattle (Miller, 1979a). Zinc-deficient calves grow more slowly and are lethargic, their wounds heal very slowly, if at all, and they are highly susceptible to nonspecific secondary infections (Miller, 1970). Testicular growth and development are often retarded in a Zn deficiency. Inhibited testicular growth has been observed in Venezuela in cattle consuming low-Zn rations (J. Perdomo, personal communication).

Legg and Sears (1960) demonstrated a parakeratosis skin disorder in cattle in Guyana that were consuming forage containing 18–42 ppm of Zn. In the more acute cases, parakeratosis spread rapidly over about 40% of the body surface. The skin and hair rapidly returned to normal after oral administration of Zn sulfate. Spais and Papasteriadis (1974) reported that 60% of 150 cattle herds in Greece had severe to mild symptoms of Zn deficiency, consisting mainly of unthrifty appearance, rough coat, loss of hair, and eczematous lesions with thickening and folding in various parts of the skin. From South Africa, clinical signs of Zn deficiency, characterized by hair loss, were reported in cattle, with animals responding to supplementation (Fig. 13.3). Likewise, hair loss and skin lesions are reported with sheep in the Sudan consuming forages containing between 16 and 27 ppm of Zn (Mahmoud et al., 1983) (Fig. 13.4). With the sheep,

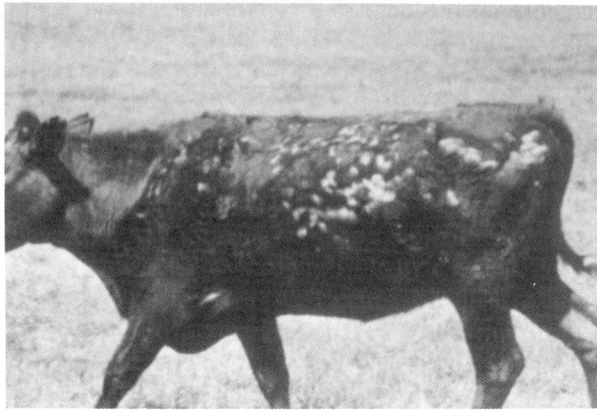

Fig. 13.3. Zinc-deficiency lesions in a grazing ruminant observed in Bethlehem, Orange Free State, South Africa. The major clinical sign is widespread alopecia. The most severe lesions were between the rear and front legs, which cannot be seen. Bleeding of the cracked skin in this area was sometimes observed. (Courtesy B. D. H. Van Niekerk, Voermol Products Ltd., Natal, South Africa.)

Fig. 13.4. Zinc-deficiency lesions in a sheep in the Sudan that responded to supplementation (Courtesy of O. M. Mahmoud, Univ. of Khartoum, Sudan.)

there was also a death loss of lambs. Animals given a Zn injection responded by elimination of skin lesions and reduced mortality compared with controls.

Masters and Fels (1980) suggested that pregnant sheep grazing pastures with less than 25 ppm of Zn may have suboptimum reproductive performance. Mayland *et al.* (1980) reported greater preweaning weight gain and higher blood Zn in calves from cows grazing western rangelands and supplemented with Zn than in calves from unsupplemented cows.

A number of factors, including soil, plant species, stage of maturity, yield, pasture management, and climate, may affect the likelihood of a Zn deficiency for ruminants. Cox (1973) reports low levels of Zn in soil and plants throughout much of Latin America. Poor drainage increases forage Zn, but increasing soil pH decreases plant availability and uptake of Zn. In most circumstances, the level of Zn declines as plants mature (Underwood, 1981; Mayland, 1975). Gomide *et al.* (1969) reported significant differences in forage-Zn concentrations between 2 consecutive years. Vast areas of the Campo Cerrado region of Brazil are severely deficient in Zn, as evidenced by substantial increases in crop yields when Zn was applied to soils (Britto *et al.*, 1971; Pereira *et al.*, 1973).

Table 13.1 illustrates forage-Zn concentrations from four Latin American and one African country. The majority of all forages was deficient, containing less than 30 ppm of Zn. The one exception was an average of 44 ppm for Malawi forages during the wet season but 19 ppm for the dry season (Mtimuni, 1982). Of 192 and 120 forage samples from northern Mato Grosso, Brazil, collected during the dry and wet seasons, mean concentrations were deficient, containing 28 and 25 ppm, respectively (Sousa *et al.*, 1982). Of 177 forage averages in the 1974 "Latin American Tables of Feed Composition," 61.5% contained less than 40 ppm of Zn (McDowell *et al.*, 1974). In 72 forage observations from Venezuela, the overall mean was 32 ppm (Perdomo, 1975). From Puerto Rico, forages averaged 34 ppm, with 82% of the grass species studied not meeting established Zn requirements for ruminants (Kayongo-Male *et al.*, 1974). Considering the Zn analyses of the different grasses of Panama, Chicco (1972) concluded that a marginal level of this element occurs particularly during the rainy season. Gomide *et al.* (1969) reported low Zn concentrations for six plant species in Minas Gerais, Brazil, ranging from 26.4 to 34.7 ppm.

Although severe clinical Zn deficiencies in ruminants have been described in different regions of the world, this likely is of much less economic importance than borderline subclinical deficiencies. The possibility of a widespread, mild or borderline deficiency of economic importance should be investigated. Marginal Zn deficiency in grazing sheep and cattle, characterized by subnormal growth, fertility, and serum Zn values but without other clinical signs, is more widespread than was earlier believed (Underwood, 1981). The first effects of a mild Zn deficiency would be expected to be decreased feed intake, growth, feed efficiency, and milk production, resistance to infection and stress, and lower reproductive efficiency (Miller, 1970). A borderline Zn deficiency, which is more likely to occur, would be difficult to diagnose clinically. As an illustration, Mayland (1975) reported increased weight gains for cattle supplemented with Zn, in the absence of Zn-deficiency signs. A total of 25 tropical countries have reported definite or probable Zn deficiencies on the basis of clinical signs and/or low serum and forage concentrations of this element (see Chapter 16 of this volume).

Biochemical changes with the most promising diagnostic value are decline in Zn concentration of plasma, hair, and bone, with a tissue decline of alkaline phosphatase (Blackmon et al., 1967). Kirchgebner and Roth (1981) report the Zn-binding capacity of serum and the activity of alkaline phosphatase as good indicators of Zn status. Under survey conditions for determining the likelihood of deficiencies for large numbers of ruminants, Zn concentrations in plasma (<0.6–0.8 µg/ml) and forage (<40 ppm) would be indicators (McDowell et al., 1984).

E. Prevention and Control

Mineral supplementation with Zn is advised if tissue levels are low, when clinical signs are suggestive of deficiencies, or if forage concentrations are less than 40 ppm. Supplementary Zn can be provided either by injections of Zn salts such as sulfate, by feeding Zn in mineral salts containing 1–2% Zn, or by treatment of the soils with Zn-containing fertilizers (Underwood, 1981). Zinc supplementation by means of intraruminal Zn pellets has also been effective in improving reproductive performance in sheep (Underwood, 1981). Australian research on Zn-deficient soils indicated that 2.2–3.2 kg of zinc sulfate per acre every 2 or 3 years would maintain increased Zn concentrations in forage. Zinc as the metal, sulfate, carbonate, oxide, and in several natural ores has been shown to be relatively available as supplements (Ammerman and Miller, 1972).

The Zn content of most feeds is variable, with data on most being unavailable and not readily obtainable. Of 2615 forages included in the 1974 "Latin American Tables of Feed Composition," only 6.8% contained Zn values (McDowell et al., 1974). Since 1974, considerable forage analyses have been made in various tropical countries, with the vast majority yielding values less than 40 ppm (McDowell et al., 1982b). Under tropical conditions, it seems logical to add 20–30 ppm of supplemental Zn to ruminant diets, unless forage analyses suggest high levels of this element. This level of Zn should be adequate to correct any likely borderline deficiency. Much higher levels of Zn, although not directly toxic to cattle, might aggravate borderline deficiency problems of other essential elements. Ivan and Grieve (1975) reported that 100 ppm of supplemental Zn significantly lowered liver Cu to 38 ppm compared with 148 ppm for control calves.

The majority of trace mineral mixtures in tropical countries contain Zn. Nevertheless, many commercial trace-mineralized salt mixtures that are widely distributed in Latin America contain an insignificant amount of Zn relative to the requirements of ruminants (McDowell et al., 1984). Free-choice mineral supplements need to be carefully evaluated to insure that they contain a minimum of 0.50% Zn, assuming an approximate consumption of 50 g/day (see Chapter 17 of this volume).

F. Toxicity

Cattle, sheep, and most mammals exhibit considerable tolerance to high intakes of Zn. The maximum tolerable level of dietary Zn has been set at 500 ppm for cattle and 300 ppm for sheep (NRC, 1980). Consumption by lambs of diets containing 1000 ppm of zinc oxide reduced weight gains and decreased feed efficiency; 1500 ppm depressed feed consumption; and 1700 ppm induced a depraved appetite and wood chewing (Ott et al., 1966).

The maximum safe Zn level probably is affected by the amount of Cu and Fe in the diet, since excessive Zn may aggravate borderline deficiencies of these elements (Miller, 1979b). An occurrence of Zn toxicosis would likely be accidental, such as from acid erosion of galvanized containers used to feed or water ruminants or from errors in adding Zn to ration mixtures.

REFERENCES

Abrams, E., Lassiter, J. W., Miller, W. J., Neathery, M. W., Gentry, R. P., and Blackmon, D. M. (1977). *J. Anim. Sci.* **45**, 1108–1113.

Ammerman, C. B., and Miller, S. M. (1972). *J. Anim. Sci.* **35**, 681–694.

Andreasi, F., Silva, M. V., Joao, P. F., and Xavier de Mendonca, C. (1968). *Rev. Fac. Med. Vet. Univ. Sao Paulo* **7**, 857–870.

Anke, M., Groppel, B., Reissig, W., Ludke, H., Grun, M., and Dittrich, G. (1973). *Arch. Tierernaehr.* **23**, 197–211.

Agricultural Research Council (ARC) (1980). "The Nutrient Requirements of Farm Livestock, No. 2 Ruminants." London.

Becker, R. B., Henderson, J. R., and Leighty, R. B. (1965). *Bull—Fla. Agric. Exp. Stn.* **699.**
Beeson, K. C., and Gomez, G. G. (1970). *Proc. Int. Grassl. Cong., 11th,* pp. 89–92.
Bentley, O. G., and Phillips, P. H. (1951). *J. Dairy Sci.* **34,** 396–403.
Bisschop, J. H. R., and Groenewald, J. W. (1963). *In* "Proceedings 1st World Conference on Animal Production," pp. 47–72. European Association of Animal Production, Rome.
Black, J. (1983). "Ph.D. dissertation. Univ. of Florida, Gainesville.
Blackmon, D. M., Miller, W. J., and Morton, J. D. (1967). *Vet. Med. Small Anim. Clin.* **62,** 265–270.
Blaxter, K. L., Sharman, G. A. M., and MacDonald, A. M. (1957). *Br. J. Nutr.* **11,** 234–246.
Blue, W. G., Ammerman, C. B., Loaiza, J. M., and Gamble, J. F. (1969). *BioScience* **19,** 616–618.
Britto, D. S., Castro, A. F., Mendes, W., Jacound, A., Ramos, D. P., and Costa, F. A. (1971). *Pesqui. Agropecu. Bras. Ser. Agron.* **6,** 81–89.
Britton, A. A., and Cotizas, A. C. (1966). *Am. J. Physiol.* **211,** 203–206.
Campbell, J. A., and Gardiner (1960). *Vet. Rec.* **72,** 1006–1010.
Carter, J. C., Jr., Miller, W. J., Neathery, M. W., Gentry, R. P., Stake, P. E., and Blackmon, D. M. (1974). *J. Anim. Sci.* **38,** 1284–1290.
Chicco, C. F. (1972). "Estudio de la Nutrición Mineral del Ganado de la Región Occidental de Panamá." Programa de Pastos y Forrajes, Project UNDP/SF No. 323, David, Panama.
Cox, F. R. (1973). *In* "Micronutrients. A Review of Soils Research in Tropical Latin America," pp. 182–189. North Carolina Agriculture Experiment Station, Raleigh.
Cuthbertson, D. (1969). *In* "Nutrition of Animals of Agricultural Importance," Part 2, p. 883–920. Pergamon, Oxford.
Dantas, H. da Silviera (1971). *Pesqui. Agropecu. Bras. Ser. Agron.* **6,** 27–30.
Davis, G. K. (1951). *J. Am. Vet. Med. Assoc.* **119,** 450–451.
Dyer, I. A. (1961). *Proc. Annu. Meet. Am. Soc. Anim. Sci. West. Sect.* **12,** 67.
Dyer, I. A., Cassatt, W. A., Jr., and Rao, R. R. (1964). *BioScience* **14**(3), 31–33.
Egan, A. R. (1972). *Aust. J. Exp. Agric. Anim. Husb.* **12,** 131–135.
Egan, A. R. (1975). *In* "Trace Elements in Soil-Plant-Animal Systems" (D. J. Nicholas and A. R. Egan, eds.), pp. 371–384. Academic Press, New York.
Everson, G. J. (1970). *Trace Elem. Metab. Anim. Proc. WAAP/IBP Int. Symp. 1969,* pp. 125–130.
Fonseca, H. A., and Davis, G. K. (1968). *In* "Proceedings 2nd World Conference on Animal Production," p. 371 (abstr.). Bruce Publishing Co., St. Paul.
Gavillon, O., and Quadros, A. T. (1973). *Pesqui. Agropecu. Bras. Ser. Zootec.* **8,** 47–54.
Geurin, H. B. (1981). *J. Anim. Sci.* **53,** 758–764.
Gomide, J. A., Noller, C. H., Mott, G. O., Conrad, J. H., and Hill, D. L. (1969). *Agron. J.* **61,** 120–123.
Grashuis, J., Lehr, J. J., Beuvery, L. L. E., and Beuvery, A. (1953). "De Schothorset." s40 Institut Moderne Veevolding, Netherlands.
Haaranen, S. (1963). *Feedstuffs* **35**(11), 17–18.
Hawkins, G. E., Wise, G. H., Matrone, G., Waugh, R. K., and Lott, W. L. (1955). *J. Dairy Sci.* **38,** 536–547.
Healy, W. B. (1974). *Trace Elem. Metab. Anim. Proc. Int. Symp. 2nd,* pp. 448–450.
Hibbs, J. W., Conrad, H. R., and Gale, C. (1961). *J. Dairy Sci.* **44,** (abstr.) 1184.
Hignett, S. L. (1959). *Vet. Rec.* **71,** 247–256.
Horowitz, A., and Dantas, H. da Silveira (1966). *Pesqui. Agropecu. Bras.* **1,** 383–390.
Howes, A. D., and Dyer, I. A. (1971). *J. Anim. Sci.* **32,** 141–145.

Howes, A. D., Dyer, I. A., and Haller, W. H. (1973). *J. Anim. Sci.* **37,** 455–458.
Ivan, M., and Grieve, C. M. (1975). *J. Dairy Sci.* **58,** 410–415.
Jerez, M. G., McDowell, L. R., Hargas, W., Conrad, J. H., and Ellis, G. L. (1984). *Trop. Anim. Prod.* **9,** 12–21.
Kayongo-Male, H., Thomas, J. W., and Ullrey, D. E. (1974). *Trace Elem. Metab. Anim. Proc. Int. Symp. 2nd,* pp. 455–457.
Kincaid, R. L. (1979). *J. Dairy Sci.* **62,** 1081–1085.
Kirchgebner, M., and Roth, H. P. (1981). *Trace Elem. Metab. Man Anim. Proc. Int. Symp. 4th,* pp. 327–330.
Kirchgessner, M., and Schwarz, W. A. (1975). *Zentralbl. Veterinaermed., Réihe A* **22,** 572–582.
Lang, C. E. (1971). B.S. thesis. Univ. de Costa Rica, San Jose.
Lassiter, J. W., Morton, J. D., and Miller, W. J. (1970). *Trace Elem. Metab. Anim. WAAP/ IBP Int. Symp. 1969.*
Leach, R. M., Jr. (1974). *Trace Elem. Metab. Anim. Proc. Int. Symp. 2nd,* pp. 51–60.
Lebdosoekojo, S., Ammerman, C. B., Raun, N. S., Gomez, J., and Litell, R. C. (1980). *J. Anim. Sci.* **51,** 1249–1260.
Legg, S. P., and Sears, L. (1960). *Nature (London)* **186,** 1061–1062.
McDowell, L. R. (1976). *In* "Conference on Beef Cattle Production in Developing Countries," p. 216–241. Centre for Tropical Veterinary Medicine, Univ. of Edinburgh, Edinburgh.
McDowell, L. R., and Conrad, J. H. (1977). *World Anim. Rev.* **24,** 24–33.
McDowell, L. R., Conrad, J. H., Thomas, J. E., and Harris, L. E. (1974). "Latin American Tables of Feed Composition." Univ. of Florida, Gainesville.
McDowell, L. R., Bauer, B., Galdo, E., Koger, M., Loosli, J. K., and Conrad, J. H. (1982a). *J. Anim. Sci.* **55,** 964–970.
McDowell, L. R., Conrad, J. H., and Ellis, G. L. (1982b). "Research in Mineral Deficiencies for Grazing Ruminants." Annual Report, AID Mineral Research Project. Univ. of Florida, Gainesville.
McDowell, L. R., Conrad, J. H., and Ellis, G. L. (1984). *In* "Symposium on Herbivore Nutrition in Sub-Tropics and Tropics—Problems and Prospects" (F. M. C. Gilchrist and R. I. Mackie, eds.), pp. 67–88. Craighall, South Africa.
Mahmoud, O. M., El Samani, F., Bakheit, A. O., and Hassan, M. A. (1983). *J. Comp. Pathol.* **93,** 591–595.
Mayland, H. F. (1975). *J. Anim. Sci.* **41,** 337 (abstr.).
Mayland, H. F., Rosenau, R. C., and Florence, A. R. (1980). *J. Anim. Sci.* **51,** 966–974.
Masters, O. G., and Fels, H. E. (1980). *Biol. Trace Elem. Res.* **2,** 281–290.
Mendes, M. O. (1977). Ph.D. thesis. Univ. of Florida, Gainesville.
Miller, W. J. (1969). *Am. J. Clin. Nutr.* **22,** 1323–1331.
Miller, W. J. (1970). *J. Dairy Sci.* **53,** 1123–1135.
Miller, W. J. (1971). *In* "Mineral Studies with Isotopes in Domestic Animals," pp. 23–41. International Atomic Energy Agency, Vienna.
Miller, W. J. (1975). *J. Dairy Sci.* **58,** 1549–1560.
Miller, W. J. (1979a). *In* "Copper and Zinc in Animal Nutrition" (B. O'Dell, E. R. Miller, and W. J. Miller, eds.), pp. 39–72. National Feed Ingredients Association, Des Moines, Iowa.
Miller, W. J. (1979b). "Dairy Cattle Feeding and Nutrition." Academic Press, New York.
Miller, W. J., and Stake, P. E. (1974). *In* "Proceedings Georgia Nutrition Conference," for Feed Industry, pp. 25–43. Univ. of Georgia, Atlanta.
Miller, W. J., Clifton, C. M., and Cameron, N. W. (1963). *J. Dairy Sci.* **46,** 715–719.

Miller, W. J., Blackmon, D. M., Gentry, R. P., Pitts, W. J., and Powell, G. W. (1967). *J. Nutr.* **92**, 71–78.
Mills, C. F., Dalgarno, A. C., Williams, R. B., and Quarterman, J. (1967). *Br. J. Nutr.* **21**, 751–768.
Mitchell, R. L. (1963). *J. R. Agric. Soc. Engl.* **124**, 75–86.
Mtimuni, J. P. (1982). Ph.D. thesis, Univ. of Florida, Gainesville.
Munro, I. B. (1957). *Vet. Rec.* **69**, 125–129.
National Research Council (NRC) (1975). "Nutrient Requirements of Sheep," 5th rev. ed. Nutrient Requirements of Domestic Animals, No. 5. Natl. Acad. Sci., Washington, D.C.
National Research Council (NRC) (1976). "Nutrient Requirements of Beef Cattle," 5th ed. Nutrient requirements of Domestic Animals, No. 4. Natl. Acad. Sci., Washington, D.C.
National Research Council (NRC) (1978). "Nutrient Requirements of Dairy Cattle," Nutrient Requirements of Domestic Animals, No. 3. Natl. Acad. Sci., Washington, D.C.
National Research Council (NRC) (1980). "Mineral Tolerance of Domestic Animals." Natl. Acad. Sci., Washington, D.C.
National Research Council (NRC) (1984). "Nutrient Requirements of Beef Cattle," 6th rev. ed. Nutrient Requirements of Domestic Animals, No. 4. Natl. Acad. Sci., Washington, D.C.
Neathery, M. W., Miller, W. J., and Blackmon, D. M. (1973). *J. Anim. Sci.* **37**, 848–852.
Ott, E. A., Smith, W. H., Harrington, R. B., and Beeson, W. M. (1966). *J. Anim. Sci.* **25**, 414–418.
Peducassé, C. A., McDowell, L. R., Parra, L. A., Wilkins, J. V., Martin, F. G., Loosli, J. K., and Conrad, J. H. (1983). *Trop. Anim. Prod.* **8**, 118–130.
Perdomo, J. T. (1975). Ph.D. thesis. Univ. of Florida, Gainesville.
Pereira, J., Viera, I. F., Araujo Moraes, E., and Rego, A. S. (1973). *Pesqui. Agropecu. Bras. Ser. Agron.* **8**, 187–191.
Rojas, M. A., Dyer, I. A., and Cassatt, W. A. (1965). *J. Anim. Sci.* **24**, 664–667.
Roy, J. H. B., Gaston, H. J., Shillam, K. W. G., Thompson, S. Y., Stobo, F. J. F., and Greatorex, J. C. (1964). *Br. J. Nutr.* **18**, 467–502.
Scaillet, M. (1969). "Programa Forrajero de Guatemala." Ministerio de Agricultura, FAO, Guatemala.
Sen, K. C., and Ray, S. N. (1964). *Bull. Indian Council Agric. Res.* **25**.
Shupe, J. L., James L. F., and Binns, W. (1967). *J. Am. Vet. Med. Assoc.* **151**, 191–197.
Smith, C. J., McDaniel, E. G., Fan, F. F., and Halsted, J. A. (1973). *Science* **181**, 954–955.
Sousa, J. C. de, Conrad, J. H., Blue, W. G., Ammerman, C. B., and McDowell, L. R. (1981). *Pesqui. Agropecu. Bras.* **16**(5), 733–740.
Sousa, J. C. de, Conrad, J. H., Mott, G. O., McDowell, L. R., Ammerman, C. B., and Blue, W. G. (1982). *Pesqui. Agropecu. Bras.* **17**(1), 11–20.
Spais, A. G., and Papasteriadis, A. A. (1974). *Trace Elem. Metab. Anim. Proc. Int. Symp. 2nd*, pp. 628–631.
Stake, P. E., Miller, W. J., Gentry, R. P., and Neathery, M. W. (1975a). *J. Anim. Sci.* **40**, 132–137.
Stake, P. E., Miller, W. J., Neathery, M. W., and Gentry, R. P. (1975b). *J. Dairy Sci.* **58**, 78–81.
Standish, J. F., Ammerman, C. B., Simpson, C. F., Neal, F. C., and Palmer, A. Z. (1969). *J. Anim. Sci.* **29**, 496–503.
Standish, J. F., Ammerman, C. B., Palmer, A. Z., and Simpson, C. F. (1971). *J. Anim. Sci.* **33**, 171–178.

Strain, W. H., and Pories, W. J. (1970). *Trace Elem. Metab. Anim. Proc. WAAP/IBP Int. Symp. 1969*, pp. 77–84.

Sutmöller, P., Vahia de Abreu, A., van der Grift, J., and Sombroek, W. G. (1966). "Mineral Imbalance in Cattle in the Amazon Valley." The Netherlands Communication No. 53. Department of Agricultural Research, Royal Tropical Institute, Amsterdam.

Tejada. R. (1984). Ph.D. thesis, Univ. of Florida, Gainesville.

Thomas, J. W. (1970). *J. Dairy Sci.* **53**, 1107–1123.

Thomas, J. W., Okamoto, M., Jacobson, W. C., and Moore, L. A. (1954). *J. Dairy Sci.* **37**, 805–812.

Underwood, E. J. (1977). "Trace Elements in Human and Animal Nutrition," 4th ed. Academic Press, New York.

Underwood, E. J. (1981). "The Mineral Nutrition of Livestock." Commonwealth Agricultural Bureaux, London.

Underwood, E. J., and Somers, M. (1969). *Aust. J. Agric. Res.* **20**, 889–897.

Van Koetsveld, E. E. (1958). *Tÿdschr Diergeneesk* **83**, 229.

Vargas D., R., McDowell, L. R., Conrad, J. H., Martin, F. G., Buergelt, C., and Ellis, G. L. (1984). *Trop. Anim. Prod.* **9**, 103–113.

Watson, L. I., Ammerman, C. B., Feaster, J. P., and Roessler, C. E. (1973). *J. Anim. Sci.* **36**, 131–136.

Wilson, J. G. (1965). *Vet. Rec.* **77**, 489–490.

14

Newly Discovered and Toxic Elements

SCOT N. WILLIAMS AND L. R. McDOWELL
Department of Animal Science
University of Florida
Gainesville, Florida

I. Introduction.	317
II. Toxic Elements.	318
A. General.	318
B. Arsenic.	318
C. Lead.	319
D. Cadmium.	323
E. Mercury.	325
F. Aluminum.	326
III. Newly Discovered Trace Elements.	327
A. General.	327
B. Boron and Silicon.	328
C. Chromium.	329
D. Lithium.	331
E. Nickel.	331
F. Tin.	332
G. Vanadium.	333
IV. Significance of Newly Discovered and Toxic Elements for Grazing Livestock.	334
References.	335

I. INTRODUCTION

At present, the exact number of essential trace elements is not known, however, there are 14 definitely known essential trace elements—iron (Fe), copper (Cu), manganese (Mn), zinc (Zn), cobalt (Co), iodine (I), molybdenum (Mo), selenium (Se), fluorine (F), chromium (Cr), tin (Sn), nickel (Ni), vanadium (V), and silicon (Si). Some elements, such as arsenic (As), lead (Pb), cadmium (Cd), mercury (Hg), and aluminum (Al) are frequently classified as toxic elements because their biological activity is largely confined to toxic reactions. However, all the trace elements can be toxic if consumed in large enough quantities or for long enough periods of time (Underwood, 1973). A further 20–30 trace elements occur regularly in feeds and animal tissue, and it is unknown whether they serve some useful purpose or are merely contaminants. Schwarz (1974a) suggests that even Pb is essential, and speculated that titanium might be found deficient if purified diets can be developed that contain lower levels of this element.

An attempt will be made in this chapter to briefly mention the most recently established essential trace elements [As, boron (B), Cr, lithium (Li), Ni, Si, Sn, V] and toxic elements (Al, As, Cd, Hg, Pb) that are detrimental to both man and livestock.

II. TOXIC ELEMENTS

A. General

Under typical conditions, mineral elements that are most devastating to grazing livestock in relation to their toxic properties are Se, Mo, and F. Toxicity aspects of F are discussed in Chapter 9, Mo and Se in Chapters 11 and 12 of this volume, respectively. Lead and As are generally considered to be the most frequent causes of poisoning in ruminants, whereas Al, Cd, and Hg toxicoses are not very common in most practical situations with grazing ruminants (Table 14.1). Some recent findings suggest that As may be essential in the ruminant, and a limited amount of data indicates that Cd and Pb may be essential to rats.

B. Arsenic

The most important uses of As compounds depend upon their poisonous character. They are used as weed killers, cattle and sheep dips, insecticides, and antihelminthics. The misuse of such compounds or the accidental exposure of animals to these materials has resulted frequently in toxicity. Most feeds contain less than 0.5 ppm and rarely exceed 1 ppm on a fresh basis.

In research with growing lambs, Bucy *et al.* (1955) reported that arsanilic acid and potassium arsenite were more toxic than was 3-nitro-4-hydroxyphenylarsonic acid to the ruminant. In these studies, severe pathological damage occurred at the higher levels of As without evidence of outward clinical signs of toxicosis. Frost (1967) has stated that "the most toxic arsenicals are well tolerated at levels which supply 10 to 20 ppm arsenic in the diet. The least toxic arsenicals can be fed without injury at levels which contribute up to at least 1000 ppm As in the diet." Arsenic in the forms in which it ordinarily occurs in feeds is well absorbed and rapidly eliminated, mainly in the urine. The As of organic compounds, such as arsanilic acid used as growth stimulants in the diets of pigs and poultry, is also well absorbed, disappears rapidly from tissues, and is excreted mostly in the feces (Underwood, 1973).

Arsenic has long had the reputation of the classical poison; however, new evidence has shown a possible essential role of this element. In a review by Nielsen (1982), it was suggested that As was an essential element for the rat, chick, pig, and goat, and may also be essential for humans. Frost (1983) has reported on the beneficial role of As in prevention of different cancers in humans. Uthus and Nielsen (1980) reported that As deprivation depressed kidney arginase and plasma urea in Zn-deficient chicks and that supplementation of arginine increased this effect.

Anke et al. (1980a) reported that female goats fed a diet containing 2 ppb of As had a significantly reduced number of viable kids compared with control animals fed 350 ppb of As, and mortality of the As-deficient group was 77%. Anke et al. (1984a) reported findings from an ongoing 10-year study with growing, gravid, and lactating goats fed either <10 or 350 ppb of As in dry matter (DM) in the diet. The deficient group showed increased abortion rate, reduced birth weight of kids, reduced life expectancy, sudden death in lactating goats (in the second lactation), reduced milk fat production, and reduced success of first insemination and conception rate. Anke et al. (1983a) theorized a minimum requirement for arsenic of <50 ppb of diet.

C. Lead

Lead poisoning (plumbism) in farm stocks is well documented (Ammerman et al., 1973; Neathery, 1984). Lead is one of the most frequently observed causes of poisoning in farm animals, especially in cattle, and it is believed that many more cases of Pb toxicosis occur than are actually diagnosed. Lead poisoning in domestic animals is mainly one of acute toxicity rather than an accumulative Pb problem (Ammerman et al., 1977; Miller, 1979). The natural curiosity and licking habits of cattle make any available Pb-containing material a potential source of poisoning. Livestock exposure to Pb is mainly derived from Pb in storage battery plates, water pipes, putty, linoleum, asphalt roofing, used gasoline, engine oil, feed containers, paints, insecticides, motor vehicle exhausts, and contaminated feeds in highly motorized urban communities [National Academy of Science (NAS), 1972].

TABLE 14.1
Relative Toxicity of Certain Mineral Elements for Animals[a]

Form of element	Estimated % absorption of oral intake	Prevalence of sources	Efficiency of excretion of absorbed element at low dosage	"Tolerable" level for human consumption[b]
Aluminum	Very low	Soil and water contamination, Al additives to diets	Good; readily cleared from body in feces	Relatively low toxicity
Arsenic Inorganic (As^{3+}, As^{5+}) Organic	±5–70%	Ambient air, industrial. Methylated arsenicals, natural or synthetic. Little or no As transferred from soil to plants	Good; readily cleared from blood and tissues. Animals have a "taste" for contaminated forage	Meat, 0.5 ppm As; other, 2 ppm As. Moderate tolerance
Lead[c] Inorganic (Pb^{2+}) Organic	Adults, 1–10%; young, 15–50%	Ambient air; dust, industrial, gasoline, inks, paints, batteries (alkyl lead of gasoline cleaved to Pb^{2+})	18–50% retained. Lead is largely sequestered in skeleton. May accumulate in liver and kidney	Controversial. Adults, 300 μg/day; Young, ?. Milk, up to 0.5 ppm
Cadmium[d] Organic Inorganic (Cd^{2+})	2–8%	Plant and animal tissues. Aquatic biota concentrate by 10^2, 10^3. Concentrates in aerial portions, some plants largely in roots of others. Occurrence is highly correlated with Zn. Source is mainly industrial	Negligible excretion until acute toxicity. Tissue retention longer than Pb or Hg	Little deposition in eggs, milk, or fetus

Mercury Inorganic (Hg^{2+}, Hg; may be a metabolite *in vivo* of organic Hg)	±15%	Vapor, salt, and dust from both natural and industrial sources. May be deposited on soil, plants, water. Phenyl and ethyl salts cleaved readily to Hg^{2+}	Negligible excretion until acute episode. Animal half time retention, 80–150 days. (Vapor exhaled)	Little deposition in eggs, milk, or fetus. 0.02–0.05 ppm Hg 100 μg/day
Organic Methylmercury[e] (CH_3Hg^+)	60–100%	Aquatic biota methylate Hg^{2+} and CH_3Hg^+ accumulates. Present in fish protein concentrates up to 90% of Hg. (Formerly in grain fungicide dressings)	Hair, feathers, eggs, and milk are routes of excretion. Animal half time retentions, about 20 days. C–Hg bond cleaved to Hg^{2+} *in vivo* about 5%.	0.5 ppm Hg (Se accumulates in fish at 1:1 molar ratio with CH_3Hg, moderates toxicity)

[a] Modified from McDowell *et al.* (1978).
[b] Underwood (1977).
[c] Toxicity: Minimal accumulation of Pb in plants or aquatic biota. Levels may be increased by external deposition. Toxicity insidious in animals.
[d] Toxicity: Accumulation in soft tissues, kidney, and liver. Turnover very slow. Major lesion is renal and intestinal epithelial destruction. Anemia and reproductive effects. Animals are Cu, Zn, and Fe deficient with time and low dosage.
[e] Toxicity: Mercuric ion, as for Cd. Methylmercury enters all tissues, crosses brain and placental "barriers," is lipid soluble. Biliary Hg is reabsorbed 46% for protein-bound and 100% as cysteine dipeptide. There is a delay period of several weeks to months between exposure and acute episodes for both Cd and Hg toxicity, permitting continued ingestion of the intoxicant.

In all species, young growing animals appear to be more susceptible to Pb toxicity than do older animals, and ruminants more so than nonruminants (Neathery and Miller, 1976a; Neathery, 1984). However, ruminants are commonly most affected with Pb poisoning, and this may be due in part to the tendency for particulate matter to settle in the reticulum and be converted to soluble Pb acetate by the action of the acid medium of the forestomach (Blood and Henderson, 1968). Morbidity is usually low (10–15%), whereas mortality is usually from 75 to 100% (Buck, 1970).

Studies from Northern Ireland showed Pb poisoning to be responsible for 1.7% of the deaths of adult animals (over 136 Kg) and 4.5% of the deaths in calves (Todd, 1962). In a survey of United States–Canadian veterinary laboratories, cattle had the highest incidence of Pb poisoning with 3 cases per 1000 animals tested for all causes (Priester and Hays, 1974). Contamination of pasture with Pb may occur in the vicinity of smelters. Lead content of plants in Pb-mining areas varies widely depending upon their proximity to the mines and mills, the smelter, and the ore-truck routes.

Allcroft (1951) states that approximately 200–400 mg of Pb per kilogram of body weight ingested in any one day, as the acetate, basic carbonate, or oxide, is sufficient to cause death in calves up to 4 months old. A daily intake of approximately 6–7 mg per kilogram body weight appears to be the minimum dose that will eventually give rise to toxicosis in cattle (Hammond and Aronson, 1964). Fick *et al.* (1976) increased dietary Pb levels in sheep to 1000 ppm over an 84-day period and found decreased liver Cu concentrations (532 to 282 ppm of DM), but reported no evidence in the animals of a Cu, Fe, Mn, or Zn imbalance.

The NAS (1972) reported that Pb may enter the body not only through absorption in the gastrointestinal tract, but also through the respiratory tract and the skin. Recent evidence suggests that Pb absorption in newborn and very young animals may be higher than 50% of intake. However, in research with sheep, limited intestinal absorption of Pb was demonstrated by low values recorded for apparent absorption and net retention of stable Pb (Fick, 1974). Lead has a marked affinity for bone, and levels between 5 and 20 ppm on the fresh basis have been reported (Underwood, 1973). Very limited Pb is deposited in milk or muscle (Miller, 1979; Neathery, 1984).

At present, Pb is not considered to be an essential mineral for animals [National Research Council (NRC), 1980]. Schwarz (1974a) added 1 ppm of Pb acetate and found an increased growth rate in rats of 12% over those of the controls (1.79 versus 2.08 g/day). More recent work (Reichlmayr-Lais and Kichgessner, 1984) in a Pb-depletion experiment with rats, in which a basal diet containing <15 ppb of Pb was fed to a group of 10 females and supplemented 800 ppb of Pb (as Pb acetate) to another group of 10 females, found decreased corpuscular volume of erythrocytes and hemoglobin in the F_1 generation offspring of unsupplemented animals. Also a decreased growth rate (10% less than that of supplemented controls) was noted.

D. Cadmium

A limited amount of data from rat experimentation suggests that Cd may be an essential element (NRC, 1980), but the major concern with Cd is its possible role as an environmental hazard. Most Cd comes from Zn smelters and from the sludge obtained from the electrolytic refining of Zn. Cadmium-containing products such as plastics, pigments, batteries, and alloys terminate in junk piles, creating a potential for biological injury through contamination of water supplies, air, and ingested matter. Relatively large quantities of Cd are found in grains, in commercial fertilizers containing phosphates, and in water in galvanized or black polyethylene pipes (Flick et al., 1971; Underwood, 1977). Vegetables, nuts, and fruits are poor sources of Cd, with concentrations mostly ranging between 0.04 and 0.08 ppm.

Cadmium toxicosis under most practical conditions in ruminants is unlikely, as it would be difficult for ruminants to ingest enough Cd based on most feeding, management, and environmental conditions (NRC, 1980). A borderline Cd toxicosis is a possibility when animals are fed certain types of recycled sewage sludge in which Cd becomes concentrated (Neathery and Miller, 1975). Animals grazing near Zn mines or smelters may also be susceptible to large intakes of Cd (Neathery and Miller, 1976b).

Decreased feed intake, reduced growth, retarded testicular development or degeneration, liver and kidney damage, scaly skin, anemia, enlarged joints, and increased mortality are some of the signs of a Cd toxicosis (Powell et al., 1964) (see Fig. 14.1). Stebbings and Lewis (1984) fed adult wethers 5 or 25 ppm of Cd (dry basis) for $4\frac{1}{2}$ years and found varying degrees of foci fibrosis and mineralized foci in the wethers fed 25 ppm of Cd. They suggested that, even after prolonged exposure to a dietary Cd intake of 25 ppm, sheep were unlikely to show any clinical effects.

Some of the toxic and environmental health hazards of exposure to Cd reside in its antagonistic activity with respect to the metabolism of several essential elements, namely Cu, Zn, and Fe. Signs of Zn deficiency can be accentuated by high intake of Cd, and the toxic effect of higher Cd intake can be ameliorated by increasing the Zn status of the diet. A resemblance between the symptoms of Cd toxicity and Zn deficiency was observed in

Fig. 14.1. Ten-week-old calf had been fed 2650 ppm cadmium for 14 days. (Courtesy of W. J. Miller, University of Georgia, Athens.)

several species, and it was found that supplemental Zn could overcome the toxicity (Ammerman *et al.*, 1973).

Among the more significant studies are implications that tolerance to Cd may be related to the Cu status of the animals. Low-level (3.5 ppm) exposure of newborn lambs to Cd resulted in highly significant declines in liver Cu and Zn (Mills, 1974). The Cd–Cu relationship suggests that in specific situations a high risk may arise, for example, in ruminant animals already exposed to high intakes of Mo.

One aspect of the biological effects of Cd (and also Pb) may be linked to their effect on the biosynthesis of porphyrins, hemes, and cytochromes, and on the mobilization and release of Fe (Petering, 1974). Jacobs *et al.* (1974) reported that the Fe concentration of the duodenum was decreased significantly in quail receiving 10 ppm of Cd. Previous investigations at this laboratory showed inhibition of Fe absorption by Cd at much lower levels of Cd.

Absorption of dietary Cd from the gastrointestinal tract is very limited (Miller, 1973). Absorption of inhaled Cd has been estimated to be as high as 40%, but only about 5% of ingested Cd is absorbed (Ammerman *et al.*, 1973). Once absorbed, Cd accumulates first in the liver and then moves to the kidney where its biological half-life has been estimated to be 20–30 years (Fox, 1983). Cattle appear to have no homeostatic control mechanism to limit Cd retention below a nontoxic level (Miller, 1973); therefore, the amount of Cd in the tissues varies directly with the amount ingested (Miller, 1979).

In a review, Ammerman *et al.* (1973) report that ascorbic acid, vitamin D, a Zn chelate, cysteine, glutathione, Zn, Fe, and Se have demonstrated some protection against Cd toxicity. As with Hg toxicity, Se compounds are highly effective in decreasing Cd toxicity, including full protection against all known specific effects of Cd related to reproduction (Parizek *et al.*, 1974).

E. Mercury

Mercury has long been known as a toxic element associated with both ingestion and inhalation. Current knowledge does not indicate that Hg is an essential element for living organisms (Underwood, 1977; Miller, 1979). Mercury is present in the air, the soil, rocks, natural water, and in other parts of the environment, either directly or indirectly, as a result of human activity. However, little is known of normal Hg intakes by farm animals. Underwood (1977) has summarized data indicating that soils contain 0.1–0.3 ppm and fruits, vegetables, and cereal grains 0.005–0.035 ppm of Hg. The levels of Hg in feathers, hair, and erythrocytes are useful indications of the tissue Hg status of individuals and of their environment (Underwood, 1977). Since excretion is minimal, urine and feces do not reflect tissue Hg buildup with time.

Available information indicates that Hg is readily absorbed by way of the respiratory tract and gastrointestinal tract and through unbroken skin. Methylmercury is 60–100% absorbed in ruminants, fowls, and humans. The mercuric ion is only about 15% absorbed. Mercury acts as a cummulative poison, since it is very slowly excreted by the animal body (Lagerwerff, 1972). Mercury has an influence on the absorption and transport of Cu, Zn, and Cd. Acute and chronic Hg poisoning can result in renal failure and death.

Particular interest has centered upon the levels of methylated forms of Hg, such as methylmercury (CH_3Hg^+). These compounds are more hazardous than inorganic forms of the element because they are nearly completely absorbed and enter a wider number of tissues with a more specific effect upon the central nervous system (Friberg, 1959). It has been shown that inorganic mercuric chloride can be converted to the more toxic methylmercuric chloride by bacterial action (Underwood, 1977). Research has been conducted on ingested Hg in ruminant metabolism, but studies with other species suggest that methylmercury is metabolized differently than are inorganic Hg compounds. Ansari *et al.* (1973) reported a far higher pecentage of methylmercury (methylmercury chloride) absorbed than of mercuric chloride.

Selenium compounds protect against the toxicity of Hg compounds. Pathological changes typical for intoxication by inorganic Hg salts did not appear in animals given Se compounds (Parizek *et al.*, 1974). Ganther *et al.* (1972) reported that increased dietary Se protected against methylmercury toxicity in rats and Japanese quail, as measured by growth rates.

Miller (1979) concluded that the main concern with Hg (or other toxic elements) is to be sure that animal products do not contain toxic concentrations. Cow's milk may range from 3 to 10 ppb of Hg (Roh *et al.*, 1975). Under practical feeding conditions, fish protein concentrates and contaminated seed grain, which may have been used accidentally, are the most common Hg sources. Accumulation of Hg in hair and feathers, which are major excretory routes of Hg (Hollins *et al.*, 1975), could cause contamination of processed hair and feather meals used as protein supplements for livestock (NRC, 1980).

F. Aluminum

Aluminum essentiality in animals is yet to be definitely confirmed (Schroeder and Nason, 1971). Aluminum ingestion in the ruminant can come from soil, water and feed contamination, and plant accumulation (Allen, 1984). Aluminum is present in highly variable concentrations in food of plant origin, and even lower concentrations in animal tissues. The variability from sample to sample for forages is probably a result of ease of contamination from dust (Underwood, 1973). In a review, Underwood (1977) presents data noting concentrations of Al as high as 3000–4000 ppm for certain trees and ferns, and 10–50 ppm of Al in grasses and clover. Aluminum intake of grazing animals on high-Al soils could be as high as 1.5% of the diet DM. Aluminum has been shown to form chelates with other minerals affecting their utilization prior to absorption and also to affect mineral metabolism post absorption (Valdivia *et al.*, 1982).

Under tropical conditions, grazing livestock are suspected to consume extremely large intakes of Al. Moomaw *et al.* (1959) reported that certain plants are classified as Al accumulators, as evidenced by 13 of 23 Hawaiian species that contained more than 1000 ppm. From Guyana, high Al contents (over 1000 ppm) were obtained in forage samples from some pasture areas (Holder, 1972). Grazing livestock ingested large Al concentrations from soil-contaminated forages as well as direct soil consumption, due to soils in the humid tropics containing characteristically high percentages of Al. Under New Zealand conditions, annual ingestion of soil can reach 75 kg for sheep and 600 kg for dairy animals (Healy, 1974) (see Chapter 8, this volume).

Dennis (1971) suggested that Al may be a causative agent of grass tetany, based on the Al content of winter-grazed cereals (500–1000 ppm of Al). Allen and Robinson (1980) reported that forage samples in tetany producing ryegrass (*Lolium multiflorum* var. Lam.) pastures in Louisiana commonly contained 2000–8000 ppm of Al, and ruminal samples of animals on those pastures contained 1000–3000 ppm of Al. Cherney *et al.* (1983) indicated that most of the Al resulted from soil contamination of forage samples and soil ingestion by animals. Ray and Robinson (1984) in an experiment testing various levels and forms of Al (1000 and 2000 ppm Al as citrate or sulfate) found no healthy plants contained high Al, and those that did contain high Al levels showed signs of Al toxicosis.

Phosphorus (P) deficiency is often a prominent sign of Al toxicosis (Foy and Brown, 1964). The detoxification of Al by the addition of P, lime, or a synthetic chelating agent is closely associated with an increase in the P concentration of plants (Rees and Sidrak, 1961). Latin American soils and many other tropical soils are characteristically acid with high percentages of exchangeable Al that complexes with P making it unavailable to plants (Woodruff and Kamprath, 1965). Aluminum is the predominant cation in acid-mineral soils with a pH of 5 or lower, while most soils above pH 5.6 have relatively small amounts of exchangeable Al.

For grazing cattle, the most prevalent mineral deficiency throughout tropical regions is P (see Chapter 16 of this volume). Phosphorus deficiency is more of a problem in tropical than in temperate regions because soils contain less P, and acid soils contain high concentrations of Al and Fe, which form insoluble phosphate complexes.

III. NEWLY DISCOVERED TRACE ELEMENTS

A. General

In addition to the possible essentiality of As, Cd, and Pb, seven additional elements, recently discovered, will be covered. Although markedly different in their chemistry, mode of action, and effective levels, the newer essential trace elements have in common the facts that they were first known for their toxic effects and that the induction of a dietary deficiency is very difficult. The rapid sequence of discoveries identifying the essentiality of these elements was based on the concepts of improved procedures for purification of diets, the use of the metal-free isolator system for raising animals, and advances in trace element analytical tech-

TABLE 14.2

Newer Elements in Mineral Nutrition

Element	Requirement discovery date	Requirement (ppm)	Tolerance[a] (ppm)
Arsenic[b]	1975	<0.05[c]	50
Boron[d]	1981	>0.3–0.4	150
Chromium[e]	1959	>0.1	1000
Fluorine[e]	1972	2.5 (F^-)	40
Lithium[f]	1976	—	—
Nickel[e]	1973	0.03–3.0	50
Silicon[e]	1972	5–100	—
Tin[e]	1970	1.5–2.0	—
Vanadium[e]	1971	0.1	10

[a] NRC (1980).
[b] Nielsen et al. (1975).
[c] Anke et al. (1983a).
[d] Hunt et al. (1983).
[e] Modified from McDowell et al. (1978).
[f] Anke et al. (1983d).

niques. Table 14.2 presents the discovery date and approximate requirement of the newer trace elements.

B. Boron and Silicon

Essentiality of B in higher plants has been known for over 50 years, but only recently has a possible role in animal nutrition been suggested. Hunt and Nielsen (1981) suggested that B supplementation in chicks tended to abate signs of a vitamin-D deficiency, such as depressed growth and increased plasma alkaline phosphatase. Hunt et al. (1983), in a series of factorial experiments with day-old chicks fed for 29–30 days, indicated a relationship among B, calcium (Ca), magnesium (Mg), and vitamin D_3. They also suggested a regulatory role of B in the metabolism of Mg. Nielsen (1984) reported that B deprivation depressed growth, hematocrit, hemoglobin, and kidney weight to body weight ratio in the rat (basal diet contained 0.3–0.4 ppm of B).

Boron intake of grazing animals varies depending on soil type and plant species ingested. Boron content of legumes is usually greater than that of grasses. European pasture grasses have been reported to contain 4–7 ppm of B (Underwood, 1977). At present, there is no evidence that B is essential in the ruminant.

The high content of Si in soils, plants, and atmospheric dust maintains a high, although greatly variable, intake by animals and man, especially grazing ruminants. Cereal grains carry much lower concentrations of Si than do leaves and stems of the same species. Whole grasses and cereals may contain 30–40% of their total ash, or 3–4% or more of the whole dry plants, as SiO_2 (Underwood, 1977). Even higher levels occur in some tropical grasses (Oyenuga, 1958). Contamination of feeds with soil, especially in hay and pasture herbage, elevates silica intake in the ruminant. Gallup et al. (1945) showed evidence though that Si was not absorbed in significant amounts in the ruminant. However, Van Soest and Jones (1968) showed that Si in forages depressed DM digestibility in vivo in ruminants, and Smith et al. (1971) showed a depressed organic matter digestibility in vitro.

The essentiality of Si is demonstrated in chicks and rats through marked effects on growth and skeletal development (Schwarz and Milne, 1972; Carlisle, 1974). Long-bone joints of deficient chicks were smaller and had reduced strength, and the bones had an altered chemical composition. In rats, Si deficiency resulted in growth retardation and changes in the structure of the skull. Recent work by Carlisle and Alpenels (1984) with chick embryos demonstrated a metabolic role for Si in cartilage. Although it has been known for some time that Si is a dietary essential in lab animals, a deficiency in grazing ruminants is unlikely due to the wide distribution of the element as either a component or contaminant in feedstuffs.

C. Chromium

No evidence has been shown of Cr essentiality in plants; however, animal requirements can be met by typical plant concentrations (Cary, 1982). Even though Cr is distributed widely in nature, plant uptake and absorption by animals is limited, leading to very low tissue levels in animals. Chromium appears to be an essential trace element due to its function as a cofactor involved in activation of insulin (Mertz et al., 1974; Schwartz, 1974b; Mertz, 1981). Chromium was identified as the active ingredient of a dietary agent, glucose tolerance factor (GTF), which is required for maintenance of normal glucose metabolism (Mertz, 1974b).

The demonstration of beneficial effects of Cr supplementation in malnourished children suggests that this element is also essential for man and that deficiences exist (Mertz, 1974a). Chromium supplementation almost immediately restored glucose tolerance to normal in these children. The fact that Cr supplementation can significantly enhance the rate of recovery from malnutrition points out the need to identify those areas of the world in which protein–calorie malnutrition is complicated by Cr deficiency (Mertz, 1974a).

Hexavalent Cr compounds are inactive; only those that are derived from trivalent Cr, or Cr of lesser oxidation states, show activity with trivalent Cr as the active ingredient of the GTF (Schwarz, 1974a). When Cr exists in an organically bound form (GTF), it is absorbed better, has a different tissue distribution, and is available to the fetus (Mertz and Roginski, 1971). Chromium concentrations of selected feeds and their relative biological values are presented in Table 14.3. Most animal products contain much of their total Cr in the form of GTF, whereas leafy plants do not contain their total Cr in this form (Toepfer *et al.*, 1973). These data indicate that determination of total Cr in any material gives little information about its biological value. Recent technological advances have brought greater sensitivity to instruments for measuring very small concentrations (i.e., as little as 1 ppb), but unfortunately, many reports of Cr values are unreliable since samples can be easily contaminated even with

TABLE 14.3

Occurrence of Newer Essential Trace Elements[a]

Element	Source
Arsenic[b]	Cereals (0.18 ppm), milk (0.05 ppm), meat (uncooked pork, beef, and lamb) (0.10 ppm), oysters (3–10 ppm), mussels (up to 120 ppm). All values on fresh basis
Boron[b]	Cow's milk (0.5–1.0 ppm), avocados (7–10 ppm, fresh basis), citrus fruits and berries (0.3–2.4 ppm, fresh basis), cereal grains (1–5 ppm, fresh basis)
Chromium[c]	Liver (1.77 ppm), brewer's yeast (1.17 ppm), corn meal (0.11 ppm), wheat germ (0.24 ppm), wheat bran (0.42 ppm). Relative biological value of these sources is 4.52, 44.88, 2.35, 4.05, and 2.21, respectively
Lithium[b]	Timothy grass (0.07–0.28 ppm), clovers (0.023–0.23 ppm), cabbage (0.093 ppm), onions (0.23 ppm), liver (0.007 ppm, wet basis), muscle (0.005 ppm, wet basis)
Nickel[c]	Wheat (0.08–0.35 ppm), tobacco leaves (0.05–1 ppm), pasture plants (0.5–3.5 ppm), fruits and tubers (0.15–0.35 ppm)
Silicon[c]	Blood (2 ppm), liver, muscle, lung, and brain (2–20 ppm), nails (56 ppm), hair (90 ppm), epidermis (106 ppm), prairie grass (*Fectuca scabrella*) (6.27%), alfalfa (0.39%)
Tin[c]	Widespread in foods, feeds, and tissues. Kidney (0.23–0.7 ppm), liver (0.35–1 ppm), lung (0.44–1.2 ppm), corn meal (0.11 ppm), oats (2.28 ppm, wet weight), pasture herbage, Scotland (0.3–0.4 ppm)
Vanadium[c]	Liver, spleen, pancreas, and prostate (0.02–0.03 ppm), lung (0.6 ppm), red clover (0.03–0.16 ppm), rye grass (0.03–0.11 ppm), barley (0.028 ppm), wheat (0.046 ppm)

[a] All values on a dry basis unless otherwise specified.
[b] Underwood (1977).
[c] Modified from McDowell *et al.* (1978).

stainless steel utensils, and organic Cr (GTF) is readily lost at high-ashing temperatures (Wolf et al., 1974).

D. Lithium

Anke et al. (1983b) fed goats Li-deficient diets from 1976 to 1983 [in 7 trials, control (20 ppm DM) versus deficient (1 ppm DM)] and reported that Li-deficient animals gained 26% less than did controls, needed more inseminations to become pregnant, had an increased number of abortions in the last third of pregnancy, had a lowered life expectancy, and had decreased milk production. Pickett (1983) reported that rats fed diets containing 3–5 ppb of Li (low-Li diets) had no difference in growth rate and behavior when compared to the control group, which were fed diets containing 500 ppb of Li; however, they reported significantly reduced reproductive performance. Lithium has also been shown to be an effective agent in the recovery of animals with Bovine Spastic Paresis (B. S. P.), a disease of the central nervous system including cerebral structures regulating specific muscle motricity (Arnault, 1983).

Underwood (1977) reported that soils generally contain 8–40 ppm of Li. The Li content in grassland plants was within the range of 0.5–2.0 ppm, and high rates of nitrogen fertilization increased Li uptake by grasses and clover (Lambert et al., 1983).

E. Nickel

Nickel occurs regularly in soils and plants in concentrations substantially higher than those normally present in animal tissues and fluids. The Ni content of feed and foodstuffs is variable and depends on the site and species in question (Anke et al., 1983c). Within the past few years, Ni deficiency has been reported in several animal species (Nielsen, 1982). Nickel deficiency reduces growth rate in rats, pigs, and goats (Anke et al., 1983c). The level of Ni associated with the growth depression was significant in the second generation of these animals (Anke et al., 1983d). Nickel deficient goats (<100 ppb of Ni in the ration) had higher abortion rates (9 versus 1% for controls), decreased conception rates (71 versus 83% for controls), and a decreased viability of female goats and their offspring (43 versus 24% and 38 versus 3%, respectively) (Anke et al., 1983c).

Anke et al. (1974) reported that male and female pigs and goats fed a low-Ni diet (0.1 ppm) had significantly decreased rate of gain, with both sexes developing slower than the control animals fed 10 ppm of Ni. About 20–30% of the baby pigs and young goats from the deficient group developed a scaly and crusty skin that was similar to parakeratosis. The change in Zn tissue levels in connection with parakeratotic changes on a low-Ni diet suggests that Ni deficiency affects Zn metabolism, and may also alter Ca incorporation in the skeleton (Anke et al., 1983c; Hoffmann et al., 1983). Effects of Ni on Fe metabolism have also been reported (Nielsen and Shuler, 1979; Nielsen, 1980; Nielsen, 1983), whereby Ni has a synergistic interaction when Fe is in the ferric form and an antagonistic interaction when Fe is in the ferrous form.

Work by Spears et al. (1977) suggested that rumen bacterial urease requires Ni for maximum activity. Ruminal urease activity was 5 to 9 times greater in lambs fed a purified basal diet low in Ni supplemented with 5 ppm of Ni than in those lambs receiving only the basal ration (Spears et al., 1977). Stearnes et al. (1984) indicated that the rumen bacterial urease activity is depressed by lasalocid and monensin (28 and 66%, respectively) when fed to steers.

Based on the present knowledge, Anke et al. (1980b) suggested that the Ni requirement of ruminants may be <500 ppb. Because Ni availability in nature exceeds the hypothesized requirement, a primary deficiency is not likely to be seen under practical conditions (Anke et al., 1983c).

Nickel toxicity occurs to varying degrees in both plants and animals. Common pasture plants normally contain 0.5–3.5 ppm of Ni (Underwood, 1977). However, Ni in plant tissues may be toxic at 40–60 ppm (Painter et al., 1953). High Ni in plants reduced Fe and Mn concentrations (Hewitt and Smith, 1975). Normally, insoluble Ni compounds become soluble in soil of low pH, which causes Ni to accumulate in plants, but adding lime to Ni-treated soil counteracts the toxic effect on plant growth. O'Dell et al. (1970) reported reduction of feed intake and growth rate in calves supplemented with 250 ppm of Ni. These workers reported that even though calves fed 1000 ppm of Ni had greatly reduced feed intake and lost weight during the 8-week treatment period, they were not emaciated and appeared to be younger than the others.

F. Tin

Schwarz (1971) produced a Sn deficiency in rats, with growth enhanced by nearly 60% if 1–2 ppm of Sn as stannic sulfate was added to the low Sn diet. Tin also has an effect on the pigmentation of teeth (Milne et al., 1972). The biological chemistry of Sn is largely unknown. Ingested inorganic Sn is very poorly absorbed and is excreted mostly in the feces. Perry and Perry (1959) suggested that metabolic alkalosis aided Sn absorption and increased urinary Sn, whereas metabolic acidosis tended to reduce Sn absorption.

Tin could well contribute to the tertiary structure of proteins or other biologically important macromolecules, such as nucleic acids. Tin also could function as the active site of metalloenzymes (Schwarz, 1974b). Organic Sn sources are presumably biologically more available than are the inorganic sources, since smaller dietary levels of the former are capable of producing more toxic responses than are the latter. Trace amounts of Sn occur widely distributed in tissues and feeds (Table 14.3). Schwarz (1971) notes that values reported in the literature for Sn are highly questionable if they are obtained with high-temperature ashing methods. A deficiency of Sn, under practical conditions with grazing ruminants, does not seem likely.

G. Vanadium

Vanadium is believed to be an essential trace element for the chick and rat. A deficiency results in reduced body and feather growth, impaired reproduction and survival of the young, altered red blood cell levels and Fe metabolism, impairment of bone tissue metabolism, and altered blood lipid levels (Hopkins and Mohr, 1974).

Hopkins and Mohr (1974) indicated that the V requirement may be between 0.05 and 0.5 ppm and this requirement may be higher when the V comes from natural feeds. Vanadium deficiencies have been induced in the rat with diets containing less than 0.1 ppm; for the chicken, a more severely deficient diet (less than 0.03 ppm) appears to be necessary (Mertz, 1974b). Analysis of a few feed samples indicates levels of less than 0.1 ppm of V, thus raising the question of nutritional adequacy of the diet.

In experiments with growing, gravid, and lactating goats fed less than 10 ppb of V in DM, Anke et al. (1983e; 1984b) found no effect of V deficiency on growth. Anke et al. (1984b) reported increased abortion rate, reduced milk and milk fat concentration, and reduced life expectancy, with approximately 50% of V-deficient kids dying during the lactation period.

In regard to animal feeding, V toxicosis may be of consequence, depending on its content in phosphate rock used in supplementation (Ammerman et al., 1977). Berg (1963) has reported that as much as 1400 ppm of V in phosphate adversely affects growth rate in chicks. Potentially toxic levels of V may be encountered by grazing animals through ingestion of soil (Thornton, 1974). Limited absorption of V from the intestine and rapid clearance into the urine may allow animals a partial protection mechanism against V toxicity (Ammerman et al., 1977).

IV. SIGNIFICANCE OF NEWLY DISCOVERED AND TOXIC ELEMENTS FOR GRAZING LIVESTOCK

Only limited data are available on the practical nutritional importance of the more recently discovered essential trace elements. Analysis of a limited number of feeds in relation to the estimated requirements suggests that under some feeding conditions deficiencies of Cr, Sn, V, and Ni might occur. It does, however, seem highly improbable that grazing animals would consume inadequate supplies of Si. Nevertheless, the chemical forms of the trace elements can greatly determine their ability to meet animal requirements. To illustrate, total Cr in any material may give little information about its biological value.

Modern agricultural and industrial practices have increased the possibility of deleterious effects from Cd, Hg, Pb, and As. These toxic elements are frequently encountered in products such as insecticides, fungicides, batteries, paints, gasoline additives, phosphate fertilizers, etc. Toxic minerals are widely distributed in air, feed, water, and soil. Cattle may ingest up to 10 times the amount of Pb and As in the form of soil to that of herbage. In mining and smelting regions, agricultural soils may contain up to 2500 ppm of As and 1700 ppm of Pb (Thornton, 1974). Even mineral ores that are used in preparing mineral supplements can provide considerable and highly variable quantities of toxic elements.

Toxic concentrations of As, Pb, Hg, and Cd are known to affect the metabolism of required elements. Mercury and Cd are particularly antagonistic to Ca, Cu, Zn, Se, and Fe. Recently, there have been a large number of world reports of Cu and Se deficiencies. In addition to the accepted reasons, could an additional explanation be found in environmental contamination of heavy metals, including Cd and Hg, which increases the requirements for Se and Cu? In areas of geothermal activity (i.e., volcanic regions of Central America), natural effluxes of Hg and possibly other volatile heavy metals may provide environmental levels that exceed tolerance levels. In Hawaii, which exhibits high volcanic activity, fish were shown to accumulate methylmercury from such sources of Hg vapor (Siegal and Siegal, 1975).

Although much less toxic on a molar basis, Al likely has a greater influence on adequate mineral nutrition for tropical regions than that of As, Pb, Hg, or Cd. In tropical countries, P deficiency is the most severe mineral limitation to grazing cattle. Over 46 tropical countries have reported P deficiencies (see Chapter 16, this volume). Aluminum complexes with P, making it unavailable for plants, thereby increasing the incidence and severity of P deficiencies in grazing cattle.

REFERENCES

Allcroft, R. (1951). *Vet. Rec.* **63**, 583-590.
Allen, V. G. (1984). *J. Anim. Sci.* **59**, 836-844.
Allen, V. G., and Robinson, D. L. (1980). *Agron. J.* **72**, 957-960.
Ammerman, C. B., Fick, K. R., Hansard, S. L., II, and Miller, S. M. (1973). "Toxicity of Certain Minerals to Domestic Animals: A Review," Anim. Sci. Res. Rep. AL73-6. Institute of Food and Agricultural Sciences, University of Florida, Gainesville.
Ammerman, C. B., Miller, S. M., Fick, K. R., and Hansard, S. L. (1977). *J. Anim. Sci.* **44**, 485-508.
Anke, M., Grün, M., Dittrich, G., Groppel, B., and Hennig, A. (1974). *Trace Elem. Metab. Anim. Proc. Int. Symp. 2nd*, pp. 715-718.
Anke, M., Groppel, B., Grün, M., Hennig, A., and Meissner, D. (1980a). In "3. Spurenelement-Symposium Arsen" (M. Anke, H.-J. Schneider, and Chr. Brückner, eds.), pp. 25-32. Abteilung Wissenschaftliche Publikationen, Friedrich Schiller Universität, Jena.
Anke, M., Schneider, H.-J., and Brückner, Chr. (1980b). In "3. Spurenelement Symposium Nickel" (M. Anke, H.-J. Schneider, and Chr. Brückner, eds.), pp. 375-376. Abteilung Wissenschaftliche Publikationen, Friedrich Schiller Universität, Jena.
Anke, M., Schmidt, A., Groppel, B., and Kronemann, H. (1983a). In "4. Spurenelement-Symposium" (M. Anke, W. Baumann, H. Braünlich, and Chr. Brückner, eds.), pp. 97-104. Abteilung Wissenschaftliche Publikationen, Friedrich Schiller Universität, Jena.
Anke, M., Groppel, B., Kronemann, H., and Grün, M. (1983b). In "4. Spurenelement-Symposium Lithium" (M. Anke, W. Baumann, H. Braünlich, and Chr. Brückner, eds.), pp. 58-65. Abteilung Wissenschaftliche Publikationen, Friedrich Schiller Universität, Jena.
Anke, M., Groppel, B., Nordmann, S., and Kronemann, H. (1983c). In "4. Spurenelement-Symposium" (M. Anke, W. Baumann, H. Braünlich, and Chr. Brückner, eds.), pp. 19-28. Abteilung Wissenschaftliche Publikationen, Friedrich Schiller Universität, Jena.
Anke, M., Grün, M., Groppel, B., and Kronemann, H. (1983d). In "Biological Aspects of Metals and Metal-Related Diseases" (B. Sarkar, ed.), pp. 89-105. Raven, New York.
Anke, M., Groppel, B., Kronemann, H., and Fuhrer, E. (1983e). In "4. Spurenelement-Symposium" (M. Anke, W. Baumann, H. Braünlich, and Chr. Brückner, eds.), pp. 135-141. Abteilung Wissenschaftliche Publikationen, Friedrich Schiller Universität, Jena.
Anke, M., Schmidt, A., and Kronemann, H. (1984a). *Trace Elem. Metab. Man Anim. Proc. Int. Symp. 5th*, p. 132 (abstr.).
Anke, M., Groppel, B., and Kronemann, H. (1984b). *Trace Elem. Metab. Man Anim. Proc. Int. Symp. 5th*, p. 46 (abstr.).
Ansari, M. S., Miller, W. J., Gentry, R. P., Neathery, M. W., and Stake, P. E. (1973). *J. Anim. Sci.* **36**, 415-419.
Arnault, G. A. (1983). In "4. Spurenelement-Symposium" (M. Anke, W. Baumann, H. Braünlich, and Chr. Brückner, eds.), pp. 306-311. Abteilung Wissenschaftliche Publikationen, Friedrich Schiller Universität, Jena.
Berg, L. R. (1963). *Poult. Sci.* **42**, 766-769.
Blood, D. C., and Henderson, J. A. (1968). "Veterinary Medicine," 3rd ed. Williams and Wilkins, Baltimore, Maryland.

Buck, L. L. (1970). *J. Am. Vet. Med. Assoc.* **156**, 1468–1472.
Bucy, L. L., Garrigus, V. S., Forbes, R. M., Norton, H. W., and Moore, W. W. (1955). *J. Anim. Sci.* **14**, 435–445.
Carlisle, E. M. (1974). *Fed. Proc., Fed. Am. Soc. Exp. Biol.* **33**, 1758–1766.
Carlisle, E. M., and Alpenels, W. F. (1984). *Trace Elem. Metab. Man Anim. Proc. Int. Symp. 5th*, p. 27 (abstr.).
Cary, E. E. (1982). *In* "Biological and Environmental Aspects of Chromium" (S. Langard, ed.), p. 49. Elsevier, Amsterdam.
Cherney, J. H., Robinson, D. L., Kappel, L. C., Hembry, F. G., and Ingraham, R. H. (1983). *Agron. J.* **75**, 447–451.
Dennis, E. J. (1971). *Fert. Solutions* **15**, 44–45.
Fick, K. R. (1974). Ph.D. thesis, Univ. of Florida, Gainesville.
Fick, K. R., Ammerman, C. B., Miller, S. M., Simpson, C. F., and Loggins, P. E. (1976). *J. Anim. Sci.* **42**, 515–523.
Flick, D. F., Kraybill, H. F., and Dimitroff, J. M. (1971). *Environ. Res.* **4**, 71–85.
Fox, M. R. S. (1983). *Fed. Proc., Fed. Am. Soc. Exp. Biol.* **42**, 1726–1729.
Foy, C. D., and Brown, J. C. (1964). *Soil Sci. Soc. Am. J.* **28**, 27–32.
Friberg, L. (1959). *Arch. Ind. Health.* **20**, 42.
Frost, D. V. (1967). *Fed. Proc., Fed. Am. Soc. Exp. Biol.* **26**, 194–208.
Frost, D. V. (1983). *In* "4. Spurenelement-Symposium" (M. Anke, W. Baumann, H. Braünlich, and Chr. Brückner, eds.), pp. 89–96. Abteilung Wissenschaftliche Publikationen, Friedrich Schiller Universität, Jena.
Gallup, W. D., Hobbs, C. S., and Briggs, H. M. (1945). *J. Anim. Sci.* **4**, 68–71.
Ganther, H. E., Govdie, C., Sunde, M. L., Kopecky, M. J., Wagner, P., Oh, Sang-Hwan, and Hoekstra, W. G. (1972). *Science* **175**, 1122–1124.
Hammond, P. B., and Aronson, A. L. (1964). *Ann. N. Y. Acad. Sci.* **111**, 595.
Healy, W. B. (1974). *Trace Elem. Metab. Anim. Proc. Int. Symp. 2nd*, pp. 448–450.
Hewitt, E. J., and Smith, T. A. (1975). "Plant Mineral Nutrition." Wiley, New York.
Hoffman, G., Anke, M., Groppel, B., Gruhn, K., and Faust, H. (1983). *In* "4. Spurenelement Symposium" (M. Anke, W. Baumann, H. Braünlich, and Chr. Brückner, eds.), pp. 29–34. Abteilung Wissenschaftliche Publikationen, Friedrich Schiller Universität, Jena.
Holder, N. L. (1972). M.S. thesis, Univ. of Florida, Gainesville.
Hollins, J. G., Willes, R. F., Bryce, F. R., Charbonneau, S. M., and Munro, I. C. (1975). *Appl. Pharmacol.* **33**, 438.
Hopkins, L., Jr., and Mohr, H. E. (1974). *Fed. Proc., Fed. Am. Soc. Exp. Biol.* **33**, 1773–1775.
Hunt, C. D., and Nielsen, F. H. (1981). *Trace Elem. Metab. Man Anim. Proc. Int. Symp. 4th*, pp. 597–600.
Hunt, C. D., Shuler, T. R., and Nielsen, F. H. (1983). *In* "4. Spurenelement-Symposium" (M. Anke, W. Baumann, H. Braünlich, and Chr. Brückner, eds.), pp. 149–155. Abteilung Wissenschaftliche Publikationen, Friedrich Schiller Universität, Jena.
Jacobs, R. M., Fox, M. R., Fry, B. E., Jr., and Harland, B. F. (1974). *Trace Elem. Metab. Anim. Proc. Int. Symp. 2nd*, pp. 684–686.
Lagerwerff, J. V. (1972). *In* "Micronutrients in Agriculture" (J. J. Mortradi, P. M. Giordano, and W. L. Lindsay, eds.), pp. 593–636. Publication Soil Science Society of America, Madison, Wisconsin.
Lambert, J., Sapek, A., and Sapek, B. (1983). *In* "4. Spurenelement-Symposium" (M. Anke, W. Baumann, H. Braünlich, and Chr. Brückner, eds.), pp. 32–38. Abteilung Wissenschaftliche Publikationen, Friedrich Schiller Universität, Jena.

McDowell, L. R., Fick, K. R., Ammerman, C. B., Miller, S. M., and Houser, R. H. (1978). *In* "Proceedings Latin American Symposium on Mineral Nutrition Research with Grazing Ruminants" (J. H. Conrad and L. R. McDowell, eds.), pp. 117–127. Univ. of Florida, Gainesville.
Mertz, W. (1974a). *Trace Elem. Metab. Anim. Proc. Int. Symp. 2nd*, pp. 185–198.
Mertz, W. (1974b). *Proc. Nutr. Soc.* **33**, 307.
Mertz, W. (1981). *Science* **213**, 1332–1338.
Mertz, W., and Roginski, E. E. (1971). *In* "Newer Trace Elements in Nutrition" (M. Mertz and W. E. Cornatzer, eds.), pp. 123–153. Dekker, New York.
Mertz, W., Toepfer, E. W., Roginski, E. E., and Polansky, M. M. (1974). *Fed. Proc., Fed. Am. Soc. Exp. Biol.* **33**, 2275–2280.
Miller, W. J. (1973). *Fed. Proc., Fed. Am. Soc. Exp. Biol.* **32**, 1915–1920.
Miller, W. J. (1979). "Dairy Cattle Feeding and Nutrition." Academic Press, New York.
Mills, C. F. (1974). *Trace Elem. Metab. Anim. Proc. Int. Symp. 2nd*, pp. 79–90.
Milne, D. B., Schwarz, K., and Sogannaes, R. (1972). *Fed. Proc., Fed. Am. Soc. Exp. Biol.* **31**, 700 (abstr.).
Moomaw, J. C., Nakamura, M. T., and Sherman, G. D. (1959). *Pac. Sci.* **13**, 335.
National Academy of Sciences (NAS) (1972). "Lead. Airborne Lead in Perspective." Natl. Acad. Sci., Washington, D.C.
National Research Council (NRC) (1978). "Nutrient Requirements of Domestic Animals, No. 3. Nutrient Requirements of Dairy Cattle," 5th rev. ed. Natl. Acad. Sci., Washington, D.C.
National Research Council (NRC) (1980). "Mineral Tolerance of Domestic Animals." Natl. Acad. Sci., Washington, D.C.
Neathery, M. W. (1984). *Proc. Ga. Nutr. Conf. Feed Ind.*, pp. 70–76.
Neathery, M. W., and Miller, W. J. (1975). *J. Dairy Sci.* **58**, 1766–1781.
Neathery, M. W., and Miller, W. J. (1976a). *Feedstuffs* **48**(7), 36–41.
Neathery, M. W., and Miller, W. J. (1976b). *Feedstuffs* **48**(3), 30–32.
Nielsen, F. H. (1980). *J. Nutr.* **110**, 965–973.
Nielsen, F. H. (1982). *In* "Clinical, Biochemical and Nutritional Aspects of Trace Elements" (A. S. Prasad, ed.), pp. 379–404. Liss, New York.
Nielsen, F. H. (1983). *In* "4. Spurenelement-Symposium" (M. Anke, W. Baumann, H. Braünlich, and Chr. Brückner, eds.), pp. 11–18. Abteilung Wissenschaftliche Publikationen, Friedrich Schiller Universität, Jena.
Nielsen, F. H. (1984). *Trace Elem. Metab. Man Anim. 5th*, p. 26 (abstr.).
Nielsen, F. H., and Shuler, T. R. (1979). *Biol. Trace Elem. Res.* **1**, 337–346.
Nielsen, F. H., Givand, S. H., and Myron, D. R. (1975). *Fed. Proc., Fed. Am. Soc. Exp. Biol.* **34**, 923 (abstr.).
O'Dell, G. D., Miller, W. J., Kings, W. A., Moore, S. L., and Blackmond, D. M. (1970). *J. Nutr.* **100**, 1447–1453.
Oyenuga, V. A. (1958). *Nutr. Abstr. Rev.* **28**, 985.
Painter, L. I., Toth, S. J., and Bear, F. E. (1953). *Soil Sci.* **76**, 421–429.
Parizek, J., Kalovskova, J., Babicky, A., Benes, J., and Pevlik, L. (1974). *Trace Elem. Metab. Anim. Proc. Int. Symp. 2nd*, pp. 119–131.
Perry, H. M., and Perry, E. F. (1959). *J. Clin. Invest.* **38**, 1452.
Petering, H. (1974). *Trace Elem. Metab. Anim. Proc. Int. Symp. 2nd*, pp. 311–325.
Pickett, E. E. (1983). *In* "4. Spurenelement-Symposium Lithium" (M. Anke, W. Baumann, H. Braünlich, and Chr. Brückner, eds.), pp. 66–70. Abteilung Wissenschaftliche Publikationen, Friedrich Schiller Universität, Jena.
Powell, G. W., Miller, W. J., Morton, J. D., and Clifton, C. M. (1964). *J. Nutr.* **84**, 205–214.

Priester, W. A., and Hays, H. M. (1974). *Am. J. Vet. Res.* **35,** 567.
Ray, D. J., and Robinson, D. L. (1984). *J. Plant Nutr.* **7**(11), 1545–1554.
Rees, W. J., and Sidrak, G. H. (1961). *Plant Soil* **14,** 101.
Reichlmayr-Lais, A. M., and Kichgessner, M. (1984). *Trace Elem. Metab. Man Anim. 5th,* p. 25 (abstr.).
Roh, J. K., Bradley, R. L., Jr., Richardson, T., and Weckel, K. G. (1975). *J. Dairy Sci.* **58,** 1782–1788.
Schwarz, K. (1971). *In* "Newer Trace Elements in Nutrition" (W. Mertz and W. E. Cornatzer, eds.), pp. 313–326. Dekker, New York.
Schwarz, K. (1974a). *Trace Elem. Metab. Anim. Proc. Int. Symp. 2nd,* pp. 355–380.
Schwarz, K. (1974b). *Fed. Proc., Fed. Am. Soc. Exp. Biol.* **33,** 1748–1757.
Schwarz, K., and Milne, D. B. (1972). *Nature (London)* **239,** 333–334.
Schroeder, H. A., and Nason, A. P. (1971). *Clin. Chem.* **17,** 461–474.
Siegel, S. M., and Siegel, B. Z. (1975). *Environ. Sci. Technol.* **9,** 473.
Smith, G. S., Nelson, A. B., and Boggiao, E. J. A. (1971). *J. Anim. Sci.* **33,** 466–471.
Spears, J. W., Smith, C. J., and Hatfield, E. E. (1977). *J. Dairy Sci.* **60,** 1073–1076.
Stearnes, S. R., Spears, J. W., Froctschel, M. A., and Croom, W. J., Jr. (1984). *J. Nutr.* **114,** 518–525.
Stebbings, R. S. J., and Lewis, G. (1984). *Trace Elem. Metab. Man Anim. Proc. Int. Symp. 5th,* p. 280 (Abstr.).
Thornton, I. (1974). *Trace Elem. Metab. Anim. Proc. Int. Symp. 2nd,* pp. 451–454.
Todd, J. R. (1962). *Vet. Rec.* **74,** 116.
Toepfer, E. W., Mertz, W., Roginski, E. E., and Polansky, M. (1973). *J. Agric. Food Chem.* **21,** 69–72.
Underwood, E. J. (1973). "Trace Elements: Toxicants Occurring Naturally in Foods." Natl. Acad. Sci., Washington, D.C.
Underwood, E. J. (1977). "Trace Elements in Human and Animal Nutrition," 4th ed. Academic Press, New York.
Uthus, E. O., and Nielsen, F. H. (1980). *In* "3. Spurenelement-Symposium Arsen" (M. Anke, H.-J. Schneider, and Chr. Brückner, eds.), pp. 33–39. Abteilung Wissenschaftliche Publikationen, Friedrich Schiller Universität, Jena.
Valdivia, R. C., Ammerman, C. B., Henry, P. R., Feaster, J. P., and Wilcox, C. J. (1982). *J. Anim. Sci.* **55,** 402–410.
Van Soest, P. J., and Jones, L. H. P. (1968). *J. Dairy Sci.* **51,** 1644–1648.
Wolf, W., Mertz, W., and Masironi, R. (1974). *J. Agric. Food Chem.* **22,** 1037–1042.
Woodruff, J. R., and Kamprath, E. J. (1965). *Soil Sci. Soc. Am. Proc.* **29,** 148–150.

15

Detection of Mineral Status of Grazing Ruminants

L. R. McDOWELL
Department of Animal Science
University of Florida
Gainesville, Florida

I.	Introduction.	339
II.	Clinical and Pathological Evaluation.	340
III.	Analysis of Water, Soil, and Forage.	341
IV.	Examination of Tissues and Fluids.	343
V.	Response to Supplementation.	346
VI.	Analyses Most Indicative of Mineral Status.	347
	A. Calcium and Phosphorus.	347
	B. Magnesium.	350
	C. Potassium.	350
	D. Sodium.	350
	E. Sulfur.	351
	F. Cobalt.	351
	G. Copper and Molybdenum.	351
	H. Fluorine.	352
	I. Iodine.	353
	J. Iron.	353
	K. Manganese.	353
	L. Selenium.	354
	M. Zinc.	354
VII.	A Mapping Technique for Determining Mineral Deficiencies and Toxicities.	354
	References.	355

I. INTRODUCTION

The detection of mineral element deficiencies or excesses involves clinical, pathological, and analytical criteria as well as response from specific element supplementation. Clinical signs of mineral deficiencies along with soil, water, plant, and animal tissue analyses have all been used with varying degrees of success to establish mineral deficiencies and toxicities. The most reliable method to confirm mineral deficiencies is response derived from specific mineral supplementation. However, supplementation studies are costly in time and resources if conducted with adequate control and assessment.

For several decades, a major goal in mineral research has been to discover and/or develop simple and accurate biochemical measurements of the status of animals for the minerals in which there are important practical problems (Miller and Stake, 1974). Like soils and plants, animal tissue-mineral concentrations are influenced by many factors. Nevertheless, when appropriate interpretation is made, animal tissue concentrations are often better indicators of the mineral status of livestock than either plant or soil concentrations (McDowell, 1976). When the evidence obtained from clinical, pathological, and biochemical examinations of the animal and from chemical analysis of the diet and its components is combined and assessed, it is usually possible to detect and define any nutritional abnormality of mineral origin, even when it is mild (Underwood, 1981).

II. CLINICAL AND PATHOLOGICAL EVALUATION

Changes in animal appearance or level of production can often be an early indication of diet inadequacy. Where the nutritional abnormalities are acute, or severe, well-marked clinical and pathological stigmata appear making detection and correction relatively easy. As examples, severe or acute deficiencies of iodine (I), magnesium (Mg), and copper (Cu) and toxicities of selenium (Se) and fluorine (F) are often characterized by specific clinical signs, but nutritional disorders are often mild or marginal and expressed only as a vague unthriftiness (i.e., hair losing its sheen) or suboptimal growth, fertility, or productivity. Unfortunately, these changes are often nonspecific and indistinguishable from those resulting from inadequate energy–protein or vitamins, or from parasitism or toxic plants. Therefore, it often becomes necessary to resort to chemical analyses in order to adequately determine mineral insufficiencies.

Clinical and pathological observations of animals have become essential diagnostic tools in the investigations of all mineral element deficiencies and toxicities. Since sudden death in the absence of clinical signs may also be a consequence of acute infectious disease, it is important to conduct a postmortem examination for purposes of differential diagnosis.

When evidence of gross pathology is supported by histological findings, it is often possible to present a tentative diagnosis of mineral element deficiency (Ullrey, 1983).

It is important to recognize the limitations of clinical and pathological findings. For instance, anemia can be caused by deficiencies of iron (Fe), Cu, and cobalt (Co) as well as protein and certain vitamins and by toxicities of zinc (Zn), cadmium (Cd), Se, and molybdenum (Mo). Similarly, abnormalities in the size, shape, strength, and chemical composition of the bones can occur in calcium (Ca), phosphorus (P), Cu, manganese (Mn), Zn, silicon (Si), and vitamin A and D deficiencies as well as in toxicities of F and Mo. Since early diagnosis is the key to preventive treatment, and preventive treatment is clearly superior to curative treatment given after productivity losses or mortality have occurred, it is important to develop detection techniques capable of giving an early and secure diagnosis (Underwood, 1979).

III. ANALYSIS OF WATER, SOIL, AND FORAGE

Mineral deficiencies and excesses have been established by soil, water, and plant analyses. Although highly variable, all mineral elements essential as dietary nutrients occur to some extent in water (see Chapter 3 of this volume). Nevertheless, grazing livestock obtain the majority of their mineral requirements from forages that, under some conditions, are contaminated with soil.

Plants withdraw essential elements from the soil solution in quantities to satisfy their own requirements as well as satisfying many of the requirements of grazing livestock. Besides essential plant elements, plants also withdraw Se, Co, and I, which are essential for the grazing ruminant. The soil–plant relationship is direct in that the plant must obtain all mineral nutrients from the specific soil with which it has contact.

In some instances, a soil survey can provide clues to potential livestock deficiencies. Soil concentrations of Co, Mo, and I reflect the plant's concentrations of these elements to a certain degree. However, numerous factors affect forage mineral uptake from soils, including yield of plant, stage of maturity, species and strain differences, climatic and seasonal conditions, chemical forms of minerals, and factors of the soil, including pH and degree of aeration or waterlogging (see Chapter 8 of this volume).

The concentration of a mineral in a soil is an uncertain guide to its concentration in the forage. Soil analysis, though useful for pasture fertilization, has been eliminated in some investigations because of lack of direct relationship to mineral content of herbage growing on the soil (Van der Veen, 1973; Gitter et al., 1975). For instance, plants growing on Co-deficient soil may not necessarily be deficient in Co nor would a soil rich in Co necessarily yield plants with high levels of Co (Latteur, 1962). However, in the Netherlands, soil analysis is preferred to that of forage analysis to establish a Co deficiency [Netherlands Committee on Mineral Nutrition (NCMN), 1973].

A more satisfactory soil analysis to relate mineral concentrations for livestock is the use of soil extracts (i.e., 0.1 N HCl or 2.5% acetic acid), which contain the more available forms of soil minerals. Analyses to determine the available forms of soil minerals can sometimes provide evidence of livestock mineral deficiencies, but, unfortunately, more often they are unreliable and difficult to interpret. Hartmans (1970) reports the available Cu of the soil does not show any positive relationship with the Cu status of the animal. Data from Brazil (Conrad et al., 1980), Bolivia (McDowell et al., 1982), Guatemala (Tejada et al., 1985), Malawi (Mtimuni, 1982), and Florida (Kiatoko et al., 1982) have indicated that mineral correlations among soil, plant, and animal tissue concentrations were highly variable among locations, and are often low or nonexistent. Typical forage–soil correlations reported in Brazil were Fe ($r = 0.12$), Mn ($r = -0.12$), and Zn ($r = 0.30$) (Conrad et al., 1980). In Malawi, few correlation coefficients in soil, forage, and animal tissues were greater or equal to ±0.5 (Mtimuni, 1982).

Mineral analysis of the forage consumed by the grazing animal is basic to mineral status diagnosis. If mineral concentrations are below minimum requirements or above the minimum tolerance level, there is an immediate suggestion of a nutritional problem. However, relying on a forage mineral analysis to establish mineral status assumes that the sample is representative of what animals consume. Fick et al. (1979) has summarized procedures to obtain a representative sample as follows: (1) Carefully observe livestock grazing patterns and hand pluck a sample to represent the animals' diet. Samples should not be taken from areas where excessive excretion has accumulated or from large clumps of grass that have obviously been rejected by animals. (2) Separate samples from each of the major species should be taken, with estimates made as to the percentage each particular species represents of the total consumption. (3) The aerial parts should be clipped using stainless steel scissors, with the cutting representing the height being grazed.

Additional disadvantages of forage element analyses to assess mineral adequacy is the difficulty of estimating forage intake and digestibility. The majority of mineral requirements is given in percentage or ppm (mg/kg), which assumes the expected consumption as estimated by dietary standards [i.e., National Research Council (NRC) or Agricultural Research

Council (ARC)]. Unfortunately, commonly used dietary standards are based on temperate forage consumption data and, therefore, would overestimate the intake of minerals. It is generally accepted that tropical forages are less digestible than are temperate species, and, therefore, daily consumption by grazing ruminants is lower. For more accuracy, total grams of specific minerals consumed per day, and not forage concentrations, determine the true adequacy of a mineral. Likewise, relative adequacy based on forage mineral concentrations is dependent on interactions with other nutrient fractions, such as proteins, lipids, or other elements, that can greatly affect the availability of the respective elements for digestion, absorption, and retention (Egan, 1975).

Forage analysis for certain trace minerals will be erroneously high due to the inherent problem of sampling forages free from contaminating soil. Mineral elements such as Ca, potassium (K), P, and Mo would not be greatly affected by soil contamination, since soil levels would be approximately equal to or less than plant material concentrations. In contrast, soil mineral levels of Co, Fe, I, sodium (Na), Mn, and Se and, to a lesser extent, Zn and Cu, would be higher than forages, and even slight contamination caused by splashing rain could give an erroneously high impression of concentration of these elements (Healy, 1973). Mitchell (1963) indicated that soil contains 20–1000 times the Co and Fe content found in pastures grown on a particular soil.

IV. EXAMINATION OF TISSUES AND FLUIDS

Without question, forage analysis is a much better indicator of mineral status for ruminants than is soil analysis. Likewise, animal tissue-mineral concentrations are better indicators of the availability of minerals than are forage mineral analyses. Grazing livestock obtain part of their mineral supply from the consumption of water, soil, leaves, tree bark, etc. rather than entirely from forages. Livestock tissue-mineral concentrations, therefore, more accurately portray the contribution of the total environment in meeting the mineral requirements of grazing animals. As an illustration, neither the available Cu content of the soil nor the Cu content in the herbage show any positive relationship with the Cu status of the animal (Sutmöller *et al.*, 1966; Hartmans, 1970). However, liver Cu concentrations of less than 25 ppm coincide with Cu deficiency signs in cattle (Sutmöller *et al.*, 1966; Tokarnia *et al.*, 1961).

Animal tissue and fluid levels of minerals, in addition to concentrations of particular enzymes, metabolites, or organic compounds with which the mineral in question is associated functionally, are important indicators of mineral status. The diagnostic significance of tissue and fluid analysis is based upon evidence that mineral element deficiencies are ultimately reflected in subnormal concentrations of the element, in altered concentrations of related metabolites or in changed activities of affected enzymes. The concentrations of minerals in the tissues, or of their functional forms, such as thyroxine (I) and vitamin B_{12} (Co), must be maintained within narrow limits if the growth, health, and productivity of the animal are to be sustained. Departures from these normal limits, which are now well defined for most elements, therefore, constitute useful diagnostic indicators. A further valuable aspect of such tissue composition changes is that they frequently arise prior to the appearance of adverse clinical signs (Underwood, 1979).

Ideally, animal scientists would like to determine the mineral status of an animal by measuring the mineral content of one tissue that is readily available from a live animal (Conrad, 1978). Unfortunately, no mineral concentration of any one tissue or fluid will portray the status of all minerals. Blood, urine, saliva, milk, feces, and hair may be easily sampled, and even liver and bone may be routinely biopsied with a minimum of time and danger to the animal. Surgical techniques for the acquisition

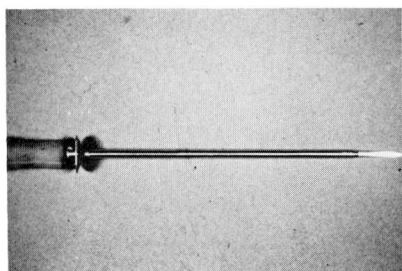
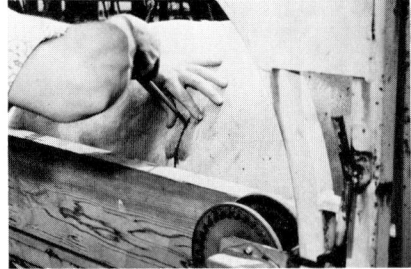
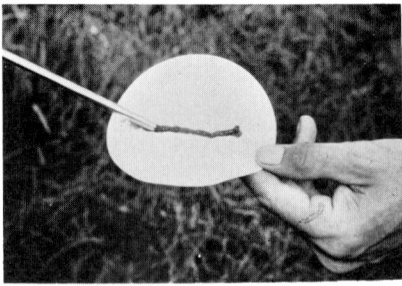

Fig. 15.1. Illustration of a liver biopsy sample taken for mineral analyses. Sample can be taken with a trocar and cannula, requiring only 5 min.

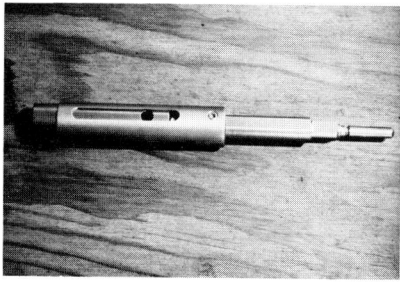
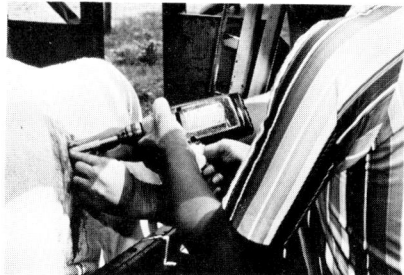

Fig. 15.2. Illustration of a bone biopsy procedure. The instrument used is a trephine (top left). The trephine is powered by an electric or battery drill (top right), with the rib sample (bottom) easily removed for analyses.

of biopsy samples of liver (Chapman *et al.*, 1963) and rib bone (Little, 1972) from cattle and sheep have been described. Fig. 15.1 illustrates the liver biopsy, and Fig. 15.2 illustrates rib bone biopsy in cattle. Liver taken either by biopsy or from sacrificed animals is an excellent indicator of the status of certain trace elements, but bone is the preferred tissue for evaluating bone-forming minerals, particularly Ca and P.

The organ, tissue, or fluid chosen for analysis varies with the element, but estimations of whole blood, plasma or serum trace element, or enzyme concentrations have wide applicability and do not, of course, require sacrifice of the animal. The levels of certain mineral elements in hair or wool, urine, and even in milk are also of value in the detection of deficiency or toxicity states, although individual variability can be very high and external contamination provides problems for trace element status evaluation.

Whole blood or blood serum or plasma is widely used for studies in mineral nutrition. Values significantly and consistently above or below "normal" concentrations or ranges provide suggestive but not conclusive evidence of a dietary excess or deficiency of particular minerals (Underwood, 1981). Precautions must be taken in interpreting blood mineral data collected or prepared in less than optimum conditions. Factors responsible for elevations of serum or plasma minerals include stress, exercise, hemolysis, temperature and serum separation time (Fick et al., 1979). These factors have often been difficult to control in studies in Latin America and Africa and have resulted in high serum P concentrations compared to extremely low levels of forage P (McDowell et al., 1984).

Because of the many factors that cause variation in mineral content of hair, hair analyses are not likely to be precise indicators of the mineral status of animals. While concentrations of Ca, P, and Cu in hair are not affected by dietary intake, Zn and Se contents of hair may reflect dietary intake, as well as toxic consumptions of Pb, As, and possibly Cd (Combs et al., 1982).

Enzymes and radioactive isotopes have been used to detect mineral deficiencies. Because of the essential role of many minerals in enzymes (i.e., serum ceruloplasmin for Cu and glutathione peroxidase for Se), enzyme activity has been developed into a useful measure in certain mineral deficiencies. Perhaps this approach will have far wider application in the future (Miller and Stake, 1974). The diagnostic possibilities of in vitro uptake of ^{75}Se by ovine red blood cells (Wright and Bell, 1963) and of ^{65}Zn by porcine red blood cells (Berry et al., 1966) have been inversely correlated with the dietary levels of Se and Zn, respectively. However, this technique appears to have limited sensitivity. In a recent investigation of ^{65}Zn uptake by erythrocytes in rats and sheep, a marked species difference was found, and little encouragement was given to the use of the technique for the detection of early or mild Zn deficiency (Chesters and Will, 1978).

V. RESPONSE TO SUPPLEMENTATION

Animals that show deficiency signs and that improve upon supplementation when unsupplemented animals do not, provide confirmation of other diagnostic procedures. Supplementation studies, however, are costly in time and resources if conducted with adequate control and assessment (Egan, 1975). Since each investigation of a suspected subclinical or marginal deficiency is experimental in nature, control groups of animals must be run in common with treated groups but denied access to the supplement. Where separate, uniform paddocks are unavailable for experimental treatments, injections or methods of drenching can be used.

VI. ANALYSES MOST INDICATIVE OF MINERAL STATUS

Since mineral analyses are complicated and expensive, it is important to select and analyze the minimum number of plant and animal tissues (or fluids) that are more indicative of mineral status of ruminants. Table 15.1 illustrates analyses of considerable value in assessing specific mineral deficiencies and toxicities. The word "critical" is used in Table 15.1 to note a concentration in forages below (or above with excesses) what is considered the requirement. This assumes the expected consumption of dry matter. Critical animal tissue concentrations are levels below or above values associated with specific clinical signs as reported in the literature.

Fifteen mineral elements of major importance have been selected for a brief discussion on methods of diagnosis of the status in the grazing ruminant. Excellent reviews on this subject are by the NCMN (1973), Miller and Stake (1974), Egan (1975), Underwood (1979), and Underwood (1981).

A. Calcium and Phosphorus

Visual signs of borderline Ca and P deficiencies are not easily distinguishable from other deficiencies. Inadequate intake of Ca will cause weakened bones, slow growth, low milk production, and tetany (convulsions) in severe deficiencies. Signs of P deficiency are not easily recognized except in severe cases when fragile bones, general weakness, weight loss, emaciation, stiffness, reduced milk production, lowered reproduction, and chewing of wood, rocks, bones, and other objects may be noticed. Abnormal chewing of objects ("pica") may occur, however, with other dietary deficiencies as well. Bone chewing may lead to death as a result of botulism.

One of the earliest biochemical measurements of P deficiency is a reduction in serum inorganic P (Underwood, 1981). Values consistently below 4.5 mg/100 ml in cattle and sheep are an indication of P deficiency. However, the NCMN (1973) did not consider serum inorganic P to be sufficiently sensitive to recommend it in diagnosing problems with cattle since forage analyses give earlier and more detailed information. Even though serum Ca does decline with deficiency, especially in some species and ages of animals, the homeostatic or physiological mechanisms regulating it are more effective than for P or for most other minerals (Underwood, 1981). Normal serum Ca concentration is 9–12 mg/100 ml.

TABLE 15.1
Diagnosis of Specific Mineral Deficiencies or Toxicities in Cattle

Element	Dairy cow[a,b]	Beef cattle[a,c]	Sheep[a]	Tissue	Critical levels[d,e,f]
A. Deficiency					
Calcium, %	0.54	0.18–0.53	0.21–0.52	Bone (fat free)	24.5%
				Bone ash	37.6%
				Plasma	8 mg/100 ml
Magnesium, %	0.20	0.05–0.25	0.04–0.08	Serum	1–2 mg/100 ml
				Urine	2–10 mg/100 ml
Phosphorus, %	0.38	0.18–0.37	0.16–0.37	Bone (fat free)	11.5%
				Bone ash	17.6%
				Plasma	4.5 mg/100 ml
Potassium, %	0.80	0.65	0.50		
Sodium, %	0.18	0.08	0.04–0.10	Saliva	100–200 mg/ml
Sulfur, %	0.20	0.10	0.14–0.26		
Cobalt, ppm	0.10	0.10	0.1	Liver	0.05–0.07 ppm
Copper, ppm	10	8	5.0	Liver	25–75 ppm
				Serum	0.65 µg/ml
Iodine, ppm	0.50	0.5	0.1–0.8	Milk	300 µg/day
Iron, ppm	50	50	30–50	Hemoglobin	10 g/100 ml
				Transferrin	13–15% saturation

	Dairy cow[a,b]	Beef cattle[a,c]	Sheep[a]		
Manganese, ppm	40	40	20–40	Liver	6 ppm
Selenium, ppm	0.1	0.20	0.1	Liver	0.25 ppm
				Serum	0.03 µg/ml
				Hair or wool	0.25 ppm
Zinc, ppm	40	30	35–50	Serum	0.6–0.8 µg/ml

B. Toxic concentration

Element	Dairy cow[a,b]	Beef cattle[a,c]	Sheep[a]	Tissue	Critical levels[d,e,f]
Copper, ppm	80	115	8–25	Liver	700 ppm
Fluorine, ppm	30	20–100	60–200	Bone	4500–5500 ppm
Manganese, ppm	1000	1000		Hair	70 ppm
Molybdenum, ppm	6	6	5–20	Liver	4 ppm
Selenium, ppm	5	5	>2.0	Liver	5–15 ppm
				Hair	10 ppm
Zinc, ppm	500	500	1000		

[a] Requirements below which a deficiency occurs.
[b] Recommendations for lactating dairy cows (500 kg) giving 17–23 kg of milk (NRC, 1978).
[c] Recommendations for growing, fattening steers, and heifers (NRC, 1984).
[d] References for critical levels are found in the following reviews: McDowell (1976), NRC (1980), Mtimuni (1982), McDowell et al. (1984).
[e] Nonmineral assays for the following elements are sensitive diagnostic techniques: cobalt (vitamin B_{12}), iodine (free thyroxine), copper (ceruloplasmin), and selenium (glutathione peroxidase).
[f] Soil concentrations suggesting deficiencies are as follows: calcium (0.35 meq/100 g), potassium (0.15 meq/100 g), magnesium (0.07 meq/100 g), phosphorus (10 ppm), cobalt (0.1 ppm), copper (0.6 ppm), manganese (19 ppm), and zinc (2 ppm).

Serum P is a good indicator of P status of ruminants only if stress factors, hemolysis, temperature, and serum separation time can be strictly controlled. Due to the limitations of serum P, Cohen (1973) concluded that bone parameters (i.e., Ca, P, and ash) provide a more reliable method of assessing Ca and P status of cattle. Collection of bone biopsy samples is being widely used as a survey technique to locate mineral deficiencies in tropical regions (McDowell et al., 1984).

B. Magnesium

Hypomagnesemic tetany (grass tetany or staggers) is a metabolic disturbance most commonly occurring in adult cows and ewes, especially those that are lactating heavily and grazing lush grass pastures. It is manifested by irritability, tetany, and convulsions, followed by death. In many instances, animals on pasture are found dead without illness having been observed.

The clinical signs of tetany are caused by inadequate Mg in serum and other extracellular fluids. Generally, lactating cows exhibiting tetany will have serum Mg levels below 1.0 mg/100 ml (Underwood, 1981). Dutch workers (NCMN, 1973) concluded that daily urinary Mg excretion is an earlier indicator of Mg supply than serum concentrations, with 2–10 mg/100 ml indicating inadequacy. The chance of hypomagnesemia increases as the Mg content of forage falls and as the product of K and crude protein contents of herbage increase (through reduced availability of dietary Mg) (NCMN, 1973).

C. Potassium

Potassium deficiency is characterized by nonspecific signs such as slow growth, reduced feed, and water intake, lowered feed efficiency, muscular weakness, nervous disorders, stiffness, and emaciation. Evaluation of K deficiency is difficult. Low serum K analyses have some diagnostic value for establishing deficiencies, but these may also be caused by malnutrition, negative N balance, gastrointestinal losses, and endocrine malfunction. Reduced feed consumption appears to be an early sign of inadequate dietary K. Because reliable evaluations of a K deficiency are not available, dietary K concentration appears to be the best indicator of K status.

D. Sodium

The initial sign of Na deficiency is a craving for salt, demonstrated by the avid licking of wood, soil, and sweat from other animals, and by drinking water. A prolonged deficiency causes loss of appetite, decreased growth, unthrifty appearance, reduced milk production, and loss of weight. Because of its rapid reaction to deficiency long before clinical signs appear, the best criterion for assessment of Na status is the concentration of Na and K in saliva. Deficiency causes a fall in Na and a rise in K. Skydsgaard (1968) suggested the normal Na:K ratio in saliva to be from 17:1 to 25:1 and suggested that if it is between 10:1 and 15:1, Na deficiency can be suspected.

E. Sulfur

Signs of sulfur (S) deficiency have been described as loss of weight, weakness, lacrimation, dullness, depressed milk production, and death. In a S deficiency, microbial protein synthesis is reduced, and the animal shows signs of protein malnutrition. A lack of S also results in a microbial population that does not utilize lactate; therefore, lactate accumulates in the rumen, blood, and urine. It is difficult to diagnose a deficiency, especially a borderline one. Serum sulfate levels have been suggested as an indicator of S deficiency, but blood lactate and dietary S levels may be the most reliable indicators of S status.

F. Cobalt

Clinical signs of Co deficiency in grazing ruminants are not specific. Deficient animals show a normocytic, normochromic anemia with a concomitant loss of appetite, retarded growth, general emaciation, rough hair coat, and loss of milk production.

Both soil (extracted with acetic acid, 2.5%) and forage Co concentrations of less than 0.1 ppm are considered low. The levels of Co in the livers of sheep and cattle are sufficiently responsive to changes in Co intake to have value in the detection of Co deficiency, with liver vitamin B_{12} an even more reliable criterion. Values of 0.10 μg vitamin B_{12}/g wet weight or less are "clearly diagnostic of Co deficiency disease" (Underwood, 1979). Subnormal plasma glucose, vitamin B_{12}, and elevated blood pyruvate levels are good indicators of Co deficiency in ruminants, together with loss of appetite (MacPherson *et al.*, 1976). While herbage and tissue analyses are helpful in diagnosing the deficiency, the definite proof is the prompt improvement in feed intake following the feeding of Co.

G. Copper and Molybdenum

Copper deficiency in cattle is characterized by poor growth, anemia, bone fragility, loss of hair color, diarrhea, and myocardial fibrosis. Milk production and body condition are poor, fertility is low, and calves may show congenital rickets. With sheep, demyelination of certain tracts in the fetal and neonatal nervous system results in incoordination, immobilization, blindness, and death, with the disease known as swayback or enzootic ataxia. With sheep, bone fragility may occur, and wool loses pigmentation and crimp (a condition known as steely wool).

In some regions, excessive dietary Mo and S induce a Cu deficiency (conditioned deficiency). Frank cases of Mo toxicity occur on pastures containing an excess of the element and are characterized by a profuse scouring.

The determination of Cu in the diet has limited diagnostic value and can, in fact, be seriously misleading unless other elements with which Cu interacts, including Mo and S in particular, are determined also. The criteria most widely used for Cu deficiency are the concentrations of Cu in liver and blood (Underwood, 1981). Plasma Cu can indicate a deficiency but does not reflect higher "marginal safety" liver storage (Hartmans, 1974). Copper status of cattle and sheep can be readily ascertained from serum ceruloplasmin activity, a Cu-containing enzyme. A high percentage of the plasma Cu exists as ceruloplasmin, with high correlations between serum Cu and ceruloplasmin activity reported (Miller and Stake, 1974).

Chronic Cu toxicity signs in ruminants include suddenly depressed appetite, jaundice, blood in urine, sudden debility, and death (NCMN, 1973). Copper content in the liver of poisoned cattle and sheep is always in excess of 700 ppm (NCMN, 1973).

H. Fluorine

The most sensitive index of the toxic fluoride effect is the mottling, staining, and excessive wearing of the permanent teeth formed during the time of excessive fluoride ingestion. Lameness caused by bony exostoses may result.

The remarkable capacity of the bone to sequester fluoride can be gauged from the fact that F concentrations in compact bone below 4500 ppm are considered to be innocuous, with a saturation point on the order of 15,000–20,000 ppm (Suttie *et al.*, 1958). Fluoride concentrations of tail bones represents a valuable means of detecting bovine fluorosis (Suttie, 1967). In cows, 20–30 ppm in the urine is considered to indicate borderline F toxicity and over 35 ppm of F to be indicative of systemic signs of toxicity (Shupe *et al.*, 1963). Carlson (1966) reports that 0.2 ppm of F can be considered a critical plasma concentration in young cattle.

I. Iodine

A simple I deficiency results in an enlargement of the thyroid gland (goiter), with newborn more likely to be affected than adults. Calves or lambs may be stillborn, weak and hairless, with irregular or suppressed estrus, or fetal development may be arrested with death, resorption, or abortion. Bulls may exhibit a decline in libido and a deterioration in semen quality.

Goitrogenic substances occur frequently and increase the animals' requirement for I. Consequently, the I content of the diet has limited value in diagnosing a deficiency problem. Serum protein-bound-I values in adult cattle less than 3–4 μg/dl or total serum I values less than 5–10 μg/dl may indicate an inadequate I intake. Milk I levels less than 10 μg/dl are associated with inadequate intake (Hemken et al., 1972).

J. Iron

Iron deficiency is rarely of practical concern for grazing livestock except when blood is lost from parasitic infestation or disease. Deficiency results in a hypochromic, microcytic anemia with low serum Fe, increased total serum Fe binding capacity, and a decreased transferrin saturation (Underwood, 1977). Low hemoglobin and hematocrit values are not sensitive indicators of early Fe deficiency stages because they do not occur until storage is depleted severely. Many consider percent saturation of transferrin to be the most practical means of detecting Fe deficiency in its early stages (Underwood, 1977).

K. Manganese

Manganese deficiency in ruminants results in skeletal abnormalities, delayed estrus, reduced fertility, abortions, and deformed young. Calves have deformed legs with "over-knuckling" and enlarged joints, and grow poorly. Deficient heifers are slower to exhibit estrus and to conceive.

The Mn content of most body tissues is remarkably resistant to change with low intake (Underwood, 1977). However, some reduction occurs in liver, bones, and hair. McDowell et al. (1984) concluded that a Mn deficiency can best be detected by the combination of liver (<6 ppm) and forage (<25 ppm) analyses, and a toxicity is suspected when hair samples contain over 70 ppm Mn.

L. Selenium

The predominant Se deficiency disease in young ruminants is nutritional muscular dystrophy or white muscle disease. Affected lambs or calves exhibit difficulty in standing. White muscle disease can be diagnosed by gross and histological examination of the affected muscles. However, it is possible to measure serum glutamic-oxalacetic transaminase (SGOT) which is markedly elevated in calves and lambs with white muscle disease. The units of the enzyme per milliliter ranged from 295 to 3,460 for lambs and calves with white muscle disease, and only 22–191 in normal animals (Blincoe and Dye, 1958).

The Se concentrations in the tissue of animals reflect the dietary Se level over a wide range from deficient to toxic intakes. The kidney and the liver are the most sensitive indicators of the Se status of the animal, and the Se concentrations in these organs can provide valuable diagnostic criteria. The most widely used assessment of Se status is blood Se concentrations. Low blood Se is always found in Se-deficient conditions. A direct relationship between blood glutathione peroxidase (a Se-containing enzyme) activity and Se concentrations has been established. Since Se is deposited in all the tissues of the body, except fat, of animals consuming seleniferous feeds, high concentrations of the element provide indisputable evidence of an excessive intake.

M. Zinc

Early effects of Zn deficiency include reduced feed intake, growth rate and feed efficiency, and skin disorders. Clinical signs include alopecia, general dermatitis of the neck and head, listlessness, reduced testicular growth, swollen feet, and wounds that fail to heal properly.

Under experimental conditions, many biochemical changes have been identified in severely Zn-deficient animals (Miller and Stake, 1974). Those with the most promising diagnostic value are the Zn concentrations in plasma, hair, and bone, and alkaline phosphatase content of plasma or other tissues (Blackmon et al., 1967). Serum or plasma Zn is greatly and quickly reduced in animals fed a severely deficient diet (Miller and Stake, 1974).

VII. A MAPPING TECHNIQUE FOR DETERMINING MINERAL DEFICIENCIES AND TOXICITIES

Mineral deficiencies or toxicities in grazing livestock can be predicted by use of a systematic mapping survey technique or regional reconnaissance. Analyzed Se and Co levels of U.S. forages have been related to Se- and Co-responsive diseases (Kubota et al., 1967; Kubota, 1968). Similar mapping techiques based on forage analyses have been undertaken for Ca and P in Brazil (Gavillon and Quadros, 1970) and Se in Venezuela (Jaffe et al., 1969). Egan (1975) reported that the sampling and analysis of stream bed sediments have revealed areas of hitherto unsuspected Mo-induced Cu deficiency in sheep and cattle, Mn deficiency in cattle, and Co deficiency in sheep. Cobalt and/or Cu deficiencies of grazing ruminants have also been established in specific Brazilian regions as a result of low liver concentrations of these elements (Tokarnia and Döbereiner, 1973). Deficiencies of P were likewise established in Venezuela (Chicco and French, 1959) and in Panama (Chicco, 1972), on the basis of low serum P levels.

In Uganda, the mineral status of dairy farms were established from analyses of local pastures (Long et al., 1972). Boyazoglu (1973) analyzed liver samples for trace elements and identified mineral deficiencies in 10 regions of South Africa and adjoining territories. Liver Cu concentrations from the Sudan (Tartour, 1975) and South Africa (Van Niekerk, 1978) have been used to detect areas of Cu deficiency.

Since 1974, the University of Florida, with support from the Agency for International Development, has been engaged in cooperative mineral research with institutions in Latin America, Africa, and Southeast Asia. The purpose of this research has been to locate mineral deficiencies or excesses for grazing livestock by use of a systematic mapping technique that analyzes plant and animal tissues and by observing the biological response from mineral supplements. A systematic mapping technique using forage and animal tissue analyses has been employed in large areas of Colombia, Costa Rica, and Venezuela (McDowell et al., 1984).

REFERENCES

Berry, R. K., Bell, M. C., and Wright, P. L. (1966). *J. Nutr.* **88**, 284–290.
Blackmon, D. M., Miller, W. J., and Morton, J. D. (1967). *Vet Med. Small Anim. Clin.* **62**, 265–270.
Blincoe, C., and Dye, W. B. (1958). *J. Anim. Sci.* **17**, 224–226.
Boyazoglu, P. A. (1973). *South Afr. J. Anim. Sci.* **3**, 149–152.
Carlson, J. R. (1966). Ph.D. thesis. Univ. of Wisconsin, Madison.
Chapman, H. L., Jr., Cox, D. H., Haines, C. H., and Davis, G. K. (1963). *J. Anim. Sci.* **22**, 733–737.
Chesters, J. K., and Will, M. (1978). *Trace Elem. Metab. Man Anim. Proc. Int. Symp. 3rd*, p. 211.
Chicco, C. F. (1972). "Estudio de la Nutrición Mineral del Ganado de la Región Occidental de Panama." Project UNDP/SF No. 323, David, Panama.
Chicco, C. F., and French, M. H. (1959). *Agron. Trop.* **9** (2), 41–62.

Cohen, R. D. H. (1973). *Aust. J. Exp. Agric. Anim. Husb.* **13**, 5–8.
Combs, D. K., Goodrich, R. D., and Meiske, J. C. (1982). *J. Anim. Sci.* **54**, 391–398.
Conrad, J. H. (1978). "Proceedings Latin American Symposium on Mineral Nutrition Research with Grazing Ruminants" (J. H. Conrad and L. R. McDowell, eds.), pp. 143–148. Univ. of Florida, Gainesville.
Conrad, J. H., Sousa, J. C., Mendes, M. O., Blue, W. G., and McDowell, L. R. (1980). *In* "World Conference on Animal Production IV" (L. S. Verde and A. Fernandez, eds.), pp. 48–53. Buenos Aires.
Egan, A. R. (1975). "Trace Elements in Soil-Plant-Animal Systems" (D. J. Nicholas and A. R. Egan, eds.), pp. 371–389. Academic Press, New York.
Fick, K. R., McDowell, L. R., Miles, P. H., Wilkinson, N. S., Funk, J. D., and Conrad, J. H. (1979). "Methods of Mineral Analysis for Plant and Animal Tissues," 2nd ed. Univ. of Florida, Gainesville.
Gavillon, O., and Quadros, A. T. (1970). "Calcium and Phosphorus in Native Pastures in Rio Grande do Sul: Comparisons of Deficiencies in Spring and Summer." Depto. da Produccion Animal Boletin Tecnico No. 17, Porto Alegre, Brazil.
Gitter, M. C., Boarer, D. A., Howard, D. A., and Herbert, C. N. (1975). *Trop. Anim. Health Prod.* **7**, 95–104.
Hartmans, J. (1970). *Trace Elem. Metab. Anim. Proc.* WAAP/IBP *Int. Symp. 1969*, pp. 441–445.
Hartmans, J. (1974). *Trace Elem. Metab. Anim. Proc. Int. Symp. 2nd,* pp. 261–273.
Healy, W. B. (1973). *In* "Chemistry and Biochemistry," Vol. 1, pp. 567–588. Academic Press, New York.
Hemken, R. W., Vandersall, J. H., Oskarsson, M. A., and Fryman, L. R. (1972). *J. Dairy Sci.* **55**, 931–934.
Jaffe, W. G., Chavez, J. R., and Mondragón, M. C. (1969). *Arch. Latinoam. Nutr.* **19** (3), 299–307.
Kiatoko, M., McDowell, L. R., Bertrand, J. E., Chapman, H. L., Pate, F. M., Martin, F. G., and Conrad, J. H. (1982). *J. Anim. Sci.* **55**, 18–37.
Kubota, J. (1968). *Soil Sci.* **106**, 122–130.
Kubota, J., Allaway, W. H., Carter, D. L., Cary, E. E., and Lazar, V. A. (1967). *J. Agric. Food Chem.* **15**, 448–453.
Lattuer, J. P. (1962). "Cobalt Deficiencies and Sub-deficiencies in Ruminants." Center d'Information du Cobalt, Brussels.
Little, D. A. (1972). *Aust. Vet. J.* **48**, 668–670.
Long, M. I. E., Marshall, B., Ndyanabo, W. K., and Thornton, D. D. (1972). *Trop Agric. (Trinidad)* **49**, 227–234.
MacPherson, A., Moon, F. E., and Voss, R. C. (1976). *Br. Vet. J.* **132**, 294–308.
McDowell, L. R. (1976). *In* "Beef Cattle Production in Developing Countries" (A. J. Smith, ed.), pp. 216–241. Univ. of Edinburgh Press, Edinburgh.
McDowell, L. R., Bauer, B., Galdo, E., Koger, M., Loosli, J. K., and Conrad, J. H. (1982). *J. Anim. Sci.* **55**, 964–970.
McDowell, L. R., Conrad, J. H., and Ellis, G. L. (1984). *In* "Symposium on Herbivore Nutrition in Sub-Tropics and Tropics—Problems and Prospects" (F. M. C. Gilchrist and R. I. Mackie, eds.), pp. 67–88. Craighall, South Africa.
Miller, W. J., and Stake, P. E. (1974). *In* "Proceedings Georgia Nutrition Conference for Feed Industry," pp. 25–43. Univ. of Georgia, Atlanta.
Mitchell, R. L. (1963). *J. R. Agric. Soc. Engl.* **124**, 75–86.
Mtimuni, J. P. (1982). Ph.D. dissertation. Univ. of Florida, Gainesville.

National Research Council (NRC) (1978). "Nutrient Requirements of Dairy Cattle," 5th rev. ed. Nutrient Requirements of Domestic Animals, No. 3. Natl. Acad. Sci., Washington, D.C.
National Research Council (NRC) (1980). "Mineral Tolerance of Domestic Animals." Natl. Acad. Sci., Washington, D.C.
National Research Council (NRC) (1984). "Nutrient Requirements of Beef Cattle," 6th rev. ed. Nutrient Requirements of Domestic Animals, No. 4. Natl. Acad. Sci., Washington, D.C.
Netherlands Committee on Mineral Nutrition (NCMN) (1973). "Tracing and Treating Mineral Disorders in Dairy Cattle." Centre for Agricultural Publishing, Wageningen, The Netherlands.
Shupe, J. L., Harris, L. E., Greenwood, D. A., Butcher, J. E., and Nielsen, H. M. (1963). *Am. J. Vet. Res.* **24,** 300–305.
Skydsgaard, J. M. (1968). *Nutr. Abstr. Rev.* **38,** 411.
Sutmöller, P., Vahia de Abreu, Ant., Van der Grift, J., and Sombroek, W. G. (1966). "Mineral Imbalances in Cattle in the Amazon Valley." Royal Trop. Inst., Dept. Agric. Res. Communication No. 53, Amsterdam.
Suttie, J. W. (1967). *Am. J. Vet. Res.* **28,** 709–712.
Suttie, J. M., Phillips, P. H., and Miller, R. F. (1958). *J. Nutr.* **65,** 293–304.
Tartour, G. (1975). *Trop. Anim. Health Prod.* **7,** 87–94.
Tejada, R., McDowell, L. R., Martin, F. G., and Conrad, J. H. (1985). *Nutr. Rep. Int.* (in press).
Tokarnia, C. H., and Döbereiner, J. (1973). *Pesqui. Agropecu. Bras. Ser. Vet.* **8,** 1–6.
Tokarnia, C. H., Döbereiner, J., Canella, C. F. C., and Damaso, M. N. R. (1961). *Arq. Inst. Biol. Anim.* **4,** 195–202.
Ullrey, D. E. (1983). *In* "Proceedings of the International Minerals Conference," pp. 41–56. International Minerals and Chemical Corporation, St. Petersburg, Florida.
Underwood, E. J. (1977). "Trace Elements in Human and Animal Nutrition," 4th ed. Academic Press, New York.
Underwood, E. J. (1979). *In* "Proceedings of the Florida Nutrition Conference," pp. 203–230. Univ. of Florida, Gainesville.
Underwood, E. J. (1981). "The Mineral Nutrition of Livestock." Commonwealth Agricultural Bureaux, London.
Van der Veen, R. R. (1973). *South Afr. Med. J.* **47,** 344–347.
Van Niekerk, B. D. H. (1978). *In* "Proceedings of the Latin American Symposium on Mineral Nutrition Research with Grazing Ruminants" (J. H. Conrad and L. R. McDowell, eds.), pp. 194–200. Univ. of Florida, Gainesville.
Wright, P. L., and Bell, M. C. (1963). *Proc. Soc. Exp. Biol. Med.* **114,** 379–382.

16

Incidence of Nutrient Deficiencies and Excesses in Tropical Regions and Beneficial Results of Mineral Supplementation

L. R. McDOWELL

Department of Animal Science
University of Florida
Gainesville, Florida

I.	Introduction.	359
II.	Geographical Distribution of Nutritional Deficiencies and Toxicities.	360
III.	Energy–Protein Deficiencies in Ruminants.	361
IV.	Incidence of Mineral Deficiencies and Toxicities.	364
V.	Mineral Supplementation Results.	369
VI.	Disease Conditions Related to Minerals.	372
VII.	Seasonal Needs for Supplemental Minerals.	376
VIII.	Economic Benefits from Mineral Supplementation.	377
	References	378

I. INTRODUCTION

Undernutrition is commonly accepted to be the most important limitation to ruminant livestock production in tropical countries. If any one of the essential nutrients is lacking, growth stops, reproduction may cease, and, eventually, the animals will die, often as a result of complications with diseases from lowered resistance to infections (Loosli, 1974). For grazing livestock, forages are and will continue to be the major source of the essential nutrients of energy, protein, vitamins, and minerals. Because of microbial synthesis and normal abundance in plants, only vitamin A of the vitamins is generally considered important as a dietary component for grazing livestock. Vitamin E, under conditions of low selenium (Se) intake, can sometimes be important since it is known to have a sparing effect on the requirement of this mineral. Conditions favoring vitamin deficiencies are discussed in Chapter 18 and water limitations in Chapter 3 of this volume.

II. GEOGRAPHICAL DISTRIBUTION OF NUTRITIONAL DEFICIENCIES AND TOXICITIES

Nutritional disorders including deficiencies, toxicities, and imbalances are severely inhibiting grazing livestock in developing tropical countries and are of more significant consequence than are infectious diseases. Deficiencies or toxicities of certain nutrients are limited to specific world regions, but deficiencies of other nutrients can in no way be related to geographical locations.

It is quite obvious that nonsalinous water is limited in many regions of the world. In desert areas of the world there is often a shortage of water. Likewise, high quality water in some world regions is often a problem during certain months of the year or may be plentiful some years and absent in others. The subject of drought feeding of livestock in northern Nigeria and other countries of South Africa has been reviewed (Loosli, 1974).

In general protein, energy, and vitamin deficiencies can less accurately be ascribed to geographical locations. Geographical locations that will not support sufficient plant growth continually or seasonally because of adverse climatic conditions (i.e. lack of adequate moisture, low temperatures) obviously will be deficient in total feedstuffs. Inadequate supplies of energy and/or protein are usually the major nutritional limitations to livestock production in many countries of the world. In all latitudes, there are at times deficiencies of energy, protein, and vitamins due to the serious variations in feed supplies. These nutrients can be deficient in all world regions and under all climatic and environmental conditions. For these nutrients and also frequently with water and minerals, the determining factor as to the incidence of a deficiency may be the feeding and management capabilities of the livestock producer.

III. ENERGY-PROTEIN DEFICIENCIES IN RUMINANTS

Many instances of energy and protein deficiency arise from the poor quality herbage, which, unfortunately, is often the only feed available to ruminants during a long dry season. For grazing livestock, meeting the energy—protein requirements depends on (1) the quantity of pasture available, (2) the energy–protein concentrations of forages, (3) intakes of the particular forages, and (4) the requirement of particular classes of animals at a given level of production. Forages can be produced in abundance, particularly in warm, humid areas, but often they are low in protein and digestible energy for reasonable levels of animal performance. The availability of energy and protein as measured by their apparent digestion by sheep and cattle declines with advancing forage maturity (Butterworth, 1967). As they advance beyond a few weeks growth, most tropical forages have a characteristically high lignin content, which influences both digestibility and the amount the animals will eat. Lignin must be considered as the primary structural inhibitor of quality in tropical grasses within a given species (Moore and Mott, 1973). These researchers noted that tropical forages have a lower maximum intake and digestibility, and this may be due to higher cell wall contents (CWC) in tropical forages at comparable stages of growth.

Low digestible energy and protein content of the diet imposes a severe physical restriction on the amount of feed an animal is able to consume. Data by Smith (1962) from mature *Hyparrhenia* pasture showed that *Bos indicus* steers eat dry matter (DM) equivalent to 1.2% of their body weight when the herbage contained 50% digestible organic matter (DOM), but as the dry season progressed intake of forage fell to 0.8% of body weight when DOM dropped to 38%.

Many publications have supported the conclusion that tropical grasses are inherently of low quality compared to temperate grasses. Crude protein is often the main limiting nutrient for livestock in the tropics, with approximately 7% crude protein as the minimum level required for positive nitrogen balance in mature grazing animals (Milford and Haydock, 1965). Cohen (1975) notes that tropical forages are largely devoid of legumes and have a low crude protein content, a factor that is almost invariably associated with low organic matter digestibility. Additions of legumes to a grass pasture improved the quality of the feed, especially during the dry season when the protein content in mature grass (standing hay) is extremely low. Even a small percentage of legume in the sward increases roughage intake by ruminants. The nitrogen (N) content and digestibility of tropical legumes do not decline as rapidly with age as those of tropical grasses, and dietary N can be maintained above the critical level for allowing optimum forage intake (Stobbs, 1975).

Many investigators have supported the conclusion that tropical grasses are inherently of low quality compared to temperate grasses (Moore and Mott, 1973). Table 16.1 compares distributions of proximate analysis components between tropical Latin American and temperate U.S.–Canadian forages (McDowell *et al.*, 1977). Almost 25% of both the Latin American and U.S.–Canadian forage entries contained 7% crude protein or less. Tropical Latin American forages tended to have higher ash and crude fiber but lower nitrogen-free extract (NFE) than did U.S.–Canadian forages. Approximately 53% of the Latin American and 41% of the U.S.–Canadian forage entries contained over 10% ash. Likewise, approximately 54% of the Latin American and 41% of the U.S.–Canadian forage entries contained over 30% crude fiber. Crude fiber content of grasses commonly constitutes some 30–36% of the dry matter in Caribbean (Devendra and Gohl, 1970) and Nigerian (Oyenuga, 1957) grasses. Previously, French (1961) concluded that tropical grasses were higher in fiber than were temperate grasses, thus explaining their lower quality. A much higher proportion of U.S.–Canadian forages contained over 45% NFE than did Latin American forages (62.5 and 46.2%, respectively). Lower NFE and higher crude fiber and ash concentration for tropical forages would thus indicate a lower available energy than for temperate forage species.

The effect of climate on forage structure and quality may contribute to the differences between temperate and tropical grasses. Minson and McLeod (1970) demonstrated high negative correlations between the DM digestibility of temperate and tropical forages and the mean temperature during growth ($r = -0.76$), evaporation ($r = -0.64$), and both temperature and evaporation ($r = -0.83$). They concluded that it was these unavoidable aspects of climate that gave rise to the much lower digestibility of tropical grasses than temperate grasses. Deinum (1966) noted the effects of light, temperature, and water supply on forage quality and suggested that nutrient composition of a plant was largely determined by the climate in which it grew. Wilson and Ford (1971) made simultaneous comparisons of the effect of temperature on two tropical grasses (*Panicum* and *Setaria*) and two *Lolium perenne* cultivars. As temperature increased from 15.6/10 to 26.7/21°C (day/night temperatures), there was an increase in CWC in all grasses and a decrease in *in vitro* DM digestion. At a given temperature, tropical grasses had higher CWC, by 10 percentage units, and lower *in vitro* digestion, by 12 percentage units, than did temperate grasses.

TABLE 16.1

Proximate Analysis Comparisons between Latin American and United States–Canadian Forages (Dry Matter Basis)[a]

Location	No. of entries	Distribution of crude protein means (%)			
		0–7.0	7.1–10.0	10.1–15.0	>15.0
Latin America	1993	24.7	23.9	22.4	29.0
U.S.–Canada	1729	24.9	21.3	21.7	32.1

Location	No. of entries	Distribution of ether extract means (%)			
		0–3.0	3.1–5.0	5.1–7.0	>7.0
Latin America	1348	75.7	17.0	5.0	2.4
U.S.–Canada	1523	64.0	26.2	6.5	3.3

Location	No. of entries	Distribution of crude fiber means (%)			
		0–20.0	20.1–30.0	30.1–40.0	>40.0
Latin America	1450	9.7	36.1	46.8	7.4
U.S.–Canada	1652	20.0	39.4	37.2	3.5

Location	No. of entries	Distribution of ash means (%)			
		0–5.0	5.1–10.0	10.1–15.0	>15.0
Latin America	1564	5.4	41.9	43.2	9.5
U.S.–Canada	1685	5.9	53.3	29.0	11.8

Location	No. of entries	Distribution of nitrogen-free extract means (%)			
		0–35.0	35.1–45.0	45.1–55.0	>55.0
Latin America	1404	9.9	44.1	37.8	8.3
U.S.–Canada	1619	3.2	34.3	51.6	10.9

[a] Modified from McDowell et al. (1977).

Two excellent reviews (Lamond, 1970; Topps, 1976) on undernutrition and reproductive failure in cattle for various diverse world regions highlight some of the more important facets of this topic. The most devastating economic consequence of energy and/or protein deficiency in the lactating cow is frequently a failure to conceive during the normal breeding season, which coincides approximately with the third month of lactation. Ward (1968) has shown that, with few exceptions, lactating cows do not calve 2 years in succession under Rhodesian conditions. Young (1968) observed the same type of calving histories on native pastures in northern Australia. The effects of lactation on pregnancy were striking in the llanos of Colombia (Stonaker, 1975), since only 9–13% of the nursing cows and heifers were pregnant versus 51–54% of the dry cows and heifers in the 40 llanos farms studied.

Wiltbank et al. (1969) have indicated the basic necessity of a sufficient level of energy before conception can take place. In South Africa, Smuts and Marias (1940) and Louw and Van der Wath (1943) concluded that digestible energy value of winter grass veld was so poor that it was almost valueless as a source of energy for sheep and that it was futile to supplement such grazing with protein alone. Pieterse and Preller (1965) produced statistically and economically significant results by feeding 1.5 kg maize grain per day to grazing cattle on summer veld. Bishop and Kotze (1965) demonstrated that recalving percentages and weaning weights could be considerably improved by feeding 1.75 kg of maize grain per cow per day for a period of about 2 months following calving in spring. This response was attributed to an inadequate intake of energy by cows grazing lush, but short spring pasture.

Other studies from Southern Africa and Australia have shown that energy-rich but protein-deficient feeds not only give poor but even negative responses in animal production (Van Niekerk, 1978; Winks et al., 1970). Some experiments show that energy-rich but protein-deficient supplements such as molasses or maize meal can actually accelerate weight loss due to depressed pasture intake if fed on protein-deficient pastures. Van Niekerk (1978) concluded that energy is not always the first limiting nutrient for grazing livestock in the dry season and that there is little point in supplementing with energy feeds unless more important nutrient deficiencies have first been corrected.

IV. INCIDENCE OF MINERAL DEFICIENCIES AND TOXICITIES

Mineral deficiencies and imbalances for herbivores are reported from almost all tropical regions of the world. Wasting diseases, loss of hair,

16. Incidence of Nutrient Deficiencies and Excesses

depigmented hair, skin disorders, noninfectious abortion, diarrhea, anemia, loss of appetite, bone abnormalities, tetany, low fertility, and pica are clinical signs often suggestive of mineral deficiencies throughout the world. The extent to which a lack of sufficient energy and protein is responsible for these clinical signs is still largely unanswered. However, numerous investigators have observed that livestock sometimes deteriorate in spite of an apparent adequate feed supply (Sutmöller et al., 1966). Ruminants grazing forages in a severe cobalt- or copper-deficient area are even more limited by lack of these elements than either that of energy or protein.

Mineral nutrition disorders range from acute mineral deficiency or toxicity diseases, characterized by well marked clinical signs and pathological changes, to mild and transient conditions difficult to diagnose and expressed as a vague unthriftiness or unsatisfactory growth and reproduction. The latter assumes great importance because they occur over large areas and affect a large number of animals. Mineral deficiency signs can be confusing as the observed conditions can involve more than one mineral and be combined with the effects of energy–protein deficiencies, various types of parasitism, toxic plants, and infectious diseases. Fig. 16.1 illustrates the clinical sign of a broken hip bone in Argentina that could result from a phosphorus (P) and/or copper (Cu) deficiency. A wasting condition known as "Coquera" (Fig. 16.2) in Peru is likely a result of one or more mineral deficiencies but could, on the contrary, involve lack of other nutrients or be the result of a toxic substance or an infectious disease.

Allman and Hamilton (1949) gathered information from various parts of the world on locations of livestock nutritional deficiencies. Russell and Duncan (1956) and Underwood (1981) have reported selected world locations of mineral deficiencies and toxicities. Information on mineral deficiencies and excesses specifically for livestock in Latin America was updated by Phillips (1956) and De Alba (1971). Table 16.2 lists reports of mineral deficiencies or toxicities for grazing livestock in Africa, Latin America, and Asian tropical countries. The majority of Table 16.2 combines more recent reviews on the reported incidence of mineral deficiencies or toxicities in the developing tropical countries of the world and specifically Latin America. The numerous references to support Table 16.2 for reported incidences of deficiencies and toxicities are listed elsewhere (McDowell, 1976; Fick et al., 1978; McDowell et al., 1984). An additional tropical country not listed in Table 16.2 is Australia where reported mineral deficiencies for grazing livestock are P, sodium (Na), sulfur (S), cobalt (Co), Cu, and Se (Stobbs and Minson, 1980).

Reports of mineral deficiencies or excesses include both confirmed as well as highly suspected geographical areas of mineral deficiencies and

Fig. 16.1. Broken hip bone, the result of P and/or Cu deficiency. (Courtesy of Bernardo Jorge Carrillo, Centro de Investigaciones en Ciencias Veterinarias (C.I.C.V.), INTA, Castelar, Argentina.)

toxicities for ruminants. Listing countries constitutes a very generalized approach with important geographical omissions inevitable, but it does indicate the scope of the problem. The term "deficient or toxic area" is used to denote a region in which deficiencies or toxicities can be readily found, but it does not imply that all plants or animals in that area are, in fact, in a deficient or toxic state. The extent of affected areas is not generally appreciated, and it is inevitable that reports of mineral inadequacies will greatly increase as more tropical countries undertake mineral research and, thereby, improve their methods of detection. Some countries reported few if any mineral deficiencies or excesses since they lack facilities and capabilities to confirm these problem regions.

The mineral most likely deficient for grazing cattle is P, followed by Cu and Co. Deficiencies of Na and iodine (I) are equally widespread, as are lack of Co and Cu. However, under most circumstances, deficiencies of Co and Cu generally are more detrimental to ruminant production than are either Na or I. Only since 1975 has the severity and wide extent of Se deficiency been established, due primarily to improvement of detection techniques. Toxicity of Se, molybdenum (Mo), and flourine (F) are widespread throughout the tropical countries of the world.

A wide range of mineral deficiencies and excesses has been established in many tropical countries on the basis of forage analysis (see Chapter 8 of this volume). A summary of mineral concentration of 2615 forage samples included in the 1974 "Latin American Tables of Feed Composition" indicated that mineral deficiencies were severe and widespread. Based on mineral requirements for grazing beef cattle, the percentage of forage samples deficient were as follows: calcium (Ca), 31%; P, 73%; Na, 60%; magnesium (Mg), 35%; Co, 43%; Cu, 47%; Fe, 24%; manganese (Mn), 21%; and zinc (Zn), 75%. Molybdenum was over 3 ppm in 14% of the samples. Both I and Se were known to be widely deficient, based on other

Fig. 16.2. A disease condition in Peru referred to as "coquera." The etiology of this wasting disease is unknown.

TABLE 16.2

Geographical Locations of Mineral Deficiencies or Toxicities of Ruminants in Tropical Countries of Latin America, Africa, and Asia[a]

Required elements	
Calcium	Argentina, Bolivia, Brazil, Colombia, Costa Rica, El Salvador, Guatemala, Guyana, India, Malawi, Mexico, Panama, Peru, Philippines, Senegal, Surinam, Uganda, Venezuela, Zaire
Magnesium	Argentina, Brazil, Chile, Colombia, Costa Rica, Guatemala, Guyana, Haiti, Honduras, Jamaica, Kenya, Malawi, Peru, Surinam, Trinidad, Uganda, South Africa, Uruguay, Venezuela
Phosphorus	Antigua, Argentina, Bolivia, Botswana, Brazil, Ceylon, Chile, Colombia, Costa Rica, Cuba, Dominican Republic, Ecuador, El Salvador, Egypt, Ghana, Guatemala, Guyana, Haiti, Honduras, India, Indonesia, Jamaica, Kenya, Malagasy Republic, Malawi, Malaysia, Mexico, Nicaragua, Nigeria, Panama, Paraguay, Peru, Philippines, Puerto Rico, Senegal, Somalia, South Africa, Surinam, Swaziland, Tanzania, Trinidad, Uganda, Uruguay, Venezuela, Zaire, Zimbabwe
Potassium	Brazil, Haiti, Nigeria, Panama, Swaziland, Uganda, Venezuela
Sodium	Bolivia, Brazil, Chad, Colombia, Dominican Republic, Guatemala, Kenya, Malawi, New Guinea, Nigeria, Panama, Philippines, Senegal, Somalia, South Africa, Surinam, Swaziland, Thailand, Uganda, Uruguay, Venezuela, Zimbabwe
Sulfur	Brazil, Colombia, Ecuador, Uganda
Cobalt	Argentina, Brazil, Colombia, Costa Rica, Cuba, Egypt, El Salvador, Guyana, Haiti, India, Indonesia, Katanga, Kenya, Malaysia, Mexico, Nicaragua, Northern Africa, Peru, Philippines, South Africa, Surinam, Uganda, Uruguay, Zaire
Copper (or molybdenum toxicity)	Argentina, Bolivia, Brazil, Colombia, Costa Rica, Cuba, Dominican Republic, Ecuador, El Salvador, Ethiopia, Guatemala, Guyana, Haiti, Honduras, India, Indonesia, Kenya, Malaysia, Malawi, Mexico, Panama, Peru, Philippines, Senegal, South Africa, Sudan, Surinam, Swaziland, Tanzania, Trinidad, Uruguay, Venezuela, Zaire, Zimbabwe.
Iodine	Worldwide.
Iron	Brazil, Costa Rica, India, Panama.
Manganese	Argentina, Brazil, Burma, Costa Rica, Panama, South Africa, Uganda.
Selenium	Bahamas, Bolivia, Brazil, Colombia, Costa Rica, Dominican Republic, Ecuador, Guyana, Honduras, Indonesia, Malawi, Mexico, Paraguay, Peru, South Africa, Swaziland, Thailand, Uganda, Uruguay, Venezuela.
Zinc	Argentina, Bolivia, Brazil, Colombia, Costa Rica, Dominican Republic, Ecuador, El Salvador, Guatemala, Guyana, India, Indonesia, Kenya, Malawi, Mexico, Panama, Peru, Philippines, Puerto Rico, South Africa, Sudan, Swaziland, Uganda, Uruguay, Venezuela.

TABLE 16.2 (Continued)

Toxic elements	
Fluorine	Algeria, Argentina, Ecuador, Guyana, India, Mexico, Morocco, Saudi Arabia, South Africa, Tanzania, Tunesia.
Manganese	Brazil, Costa Rica, Indonesia, Peru, Surinam.
Selenium	Argentina, Brazil, Central African Republic, Chad, Chile, Colombia, Ecuador, Honduras, India, Iran, Kenya, Madagascar, Mexico, Nigeria, Northern Africa, Peru, Puerto Rico, South Africa, Sudan, Upper Volta, Venezuela.

[a] McDowell (1976), Fick et al. (1978), McDowell et al. (1984), Mtimuni (1982).

criteria, but few analyses were available (McDowell et al., 1974; 1977). Table 16.3 presents forage protein and mineral analyses data representing 2813 samples collected from seven provinces in Costa Rica during both the wet and dry seasons. Average concentrations for Ca, P, Cu, and Zn were low during both seasons in relation to cattle requirements. Significant differences existed between provinces with protein, P, and Cu concentrations lower during the dry season.

V. MINERAL SUPPLEMENTATION RESULTS

Positive results of mineral supplementation for specific minerals have been noted in Chapters 9 through 13 of this volume. Numerous reports of improved weight gains by mineral-supplemented cattle have been summarized for various world regions (Tokarnia and Döbereiner, 1973; Cohen, 1975; Fick et al., 1978; McDowell et al., 1984). Results from cattle mineral supplementation trials in Bolivia, Brazil, and Peru illustrate growth benefits from minerals. For the year end results from the Beni of Bolivia (McDowell et al., 1982), bonemeal-supplemented weaned animals gained 96 kg versus 81 kg for controls. From Peru, cattle grazing native pasture and supplemented with P had daily gains of 0.59 kg versus 0.27 kg for the controls (Echevarria et al., 1977). From the state of Goias, Brazil, cattle receiving common salt gained only 53 gm/animal/day versus 90, 137, and 112 gm for animals receiving P, P + Co + Cu + I, and P + Co + Cu + I + Zn + Fe + Mn, respectively (E. Lopes, unpublished data). From Indonesia (Yates, 1983), sheep offered *Leucaena leucocephala* (leucaena) increased from 31 to 74 gm/day (experiment 1) and from 40 to 86 gm/day (experiment 2) when provided a trace mineral supplement, compared with controls.

TABLE 16.3

Mineral Analysis of Costa Rican Forages Collected during the Wet and Dry Seasons[a]

Province	No. samples	Protein (%)	Calcium (%)	Phosphorus (%)	Magnesium (%)	Potassium (%)	Iron (ppm)	Copper (ppm)	Manganese (ppm)	Zinc (ppm)
A. Wet season										
San José	81	18.0	0.21	0.25	0.18	2.6	197	12.5	57	29.8
Alajuela	367	10.0	0.81	0.14	0.13	1.5	320	5.9	59	16.4
Cartago	58	16.4	0.24	0.20	0.17	2.6	338	11.1	76	28.4
Heredia	36	13.7	0.31	0.24	0.21	2.3	225	8.0	182	31.1
Guanacaste	539	7.6	0.23	0.15	0.11	1.2	175	3.1	21	9.8
Puntarenas	228	8.2	0.23	0.16	0.11	1.1	169	2.8	43	11.6
Limón	159	9.8	0.14	0.15	0.13	1.2	1103	11.3	220	28.7
Average		12.0	0.16	0.18	0.16	1.8	361	7.8	94	22.3
B. Dry season										
San José	84	16.0	0.24	0.10	0.28	4.1	228	10.3	73	32.9
Alajuela	391	9.0	0.20	0.10	0.18	2.0	313	6.6	84	20.6
Cartago	65	14.9	0.20	0.13	0.21	3.4	278	7.5	142	33.3
Heredia	20	7.5	0.20	0.15	0.18	1.4	125	5.0	104	16.0
Guanacaste	521	4.1	0.27	0.08	0.14	1.1	213	3.2	40	23.5
Puntarenas	192	5.0	0.25	0.10	0.14	1.4	259	3.4	89	36.7
Limón	62	8.4	0.21	0.11	0.24	3.1	524	7.2	199	30.8
Average		9.3	0.22	0.11	0.20	2.4	277	6.2	104	27.7

[a] Unpublished data of mineral research programs of Facultad de Agronomia, Universidad de Costa Rica. Program leaders include Hernán Fonseca, Emilio Vargas, Carlos Campabadal, and Jorge Sanchez. All analysis on a dry matter basis.

A Colombian study was completed that included data on mineral supplementation and animal performance for selected ranches [Centro Internacional de Agricultura Tropical (CIAT) 1979]. Miles (unpublished data) summarized this information to divide 16 ranches on the basis of high or low P supplementation (Table 16.4). The impact of mineral supplementation on reducing abortions and calf deaths is impressive, with a 61% higher weight gain for those with higher intakes.

Although growth responses from mineral supplementation is important, the most devastating economic result of mineral deficiencies is reproductive failure, with mineral supplementation dramatically increasing fertility levels in grazing cattle from many parts of the world. Increased reproductive performance due to mineral supplementation is illustrated in Table 16.5 for 17 locations in Latin America, Africa, and Asia. Averaging the 17 reports together resulted in a mean calving percentage of 52.6% for animals receiving salt only versus 75.6% for those receiving additional supplemental minerals. The specific mineral or minerals responsible for increasing reproductive performance in the given experiments is for some experiments unclear, however, P most likely contributed greatest to this improvement.

During a 4-year study the importance of mineral supplementation on overall production of cattle in the llanos region of Colombia was very dramatic (Miles and McDowell, 1983) (Table 16.6). Multiplication of the weaning percentage by the weaning weight gave 88.7 kg of calf produced per cow with complete minerals compared with 44.8 kg with salt alone.

TABLE 16.4

Effect of Phosphorus Supplementation on Production Performance of 16 Ranches in the Llanos of Colombia

Rate of phosphorus supplementation	No. of farms	Abortions (%)[b]	Calf mortality[c]	Yearly gain (kg)
High group[a]	8	9.7	2.7	83.2
Low group	8	18.8	11.6	51.6

[a] Most ranchers used a prepared mix that also contained calcium, sodium, iodine, and cobalt in relatively adequate quantities while copper and zinc were at such low levels to likely have little favorable effect.

[b] Abortions are as a percentage of total pregnancies.

[c] Calf deaths during first month as a percentage of total births.

TABLE 16.5

Latin American, African, and Asian Studies on Effects of Mineral Supplementation on Increased Calving Percentages[a]

Country	Control[b]	Control + mineral supplement	References
Bolivia	67.5	80.0[c]	Bauer (unpublished data)
Bolivia	73.8	86.4[d]	Bauer et al. (1981)
Brazil	55.0	77.0[e]	Conrad and Mendes (1965)
Brazil	49.0	72.0[c]	Guimaraes and Nascimento (1971)
Brazil	25.6	47.3[c]	Grunert and Santiago (1969)
Colombia	50.0	84.0[e]	Stonaker (1975)
Panama	62.2	68.8[f]	Rios Arauz (1972)
Panama	42.0	80.0[c]	Poultney (personal communication)
Peru	25.0	75.0[g]	Echevarria et al. (1974)
Philippines	57.0	79.0[e]	Calub and Amril (1979)
Philippines	76.0	80–82[e]	Nocom (personal communication)
South Africa	51.0	80.0[c]	Theiler et al. (1924, 1928)
Thailand	49.0	67.0[c]	Tumwasorn (1981)
Uruguay	48.0	64.0[c]	De Leon Lora (1963)
Uruguay	86.9	96.4[c]	Schiersmann (1965)
Uruguay	50.0	75.0[h]	Pittaluga et al. (1980)
Uruguay	27.0	70.0[d]	Arroyo and Mauer (1982)

[a] Modified from McDowell et al. (1984).
[b] Control animals received only common salt (NaCl).
[c] Bone meal.
[d] Bone phosphate.
[e] Complete mineral mixture.
[f] Dicalcium phosphate + triple superphosphate.
[g] Dicalcium phosphate + copper sulfate.
[h] A combination of various treatments including bone meal and complete mineral mixture and phosphorus fertilization.

VI. DISEASE CONDITIONS RELATED TO MINERALS

In addition to specific mineral deficiency or toxicity diseases in tropical countries, there are a number of "wasting" type diseases that respond to minerals, but the exact etiology is unknown. Two disease conditions "cara inchada" and "secadera" have been major deterrents to cow–calf production in vast areas of Brazil and Colombia, respectively. Although details of the etiology of these two conditions remain to be completely identified, both of these conditions have responded in a major way to adequate and balanced trace mineral supplementation.

A periodontal disease widely known as "cara inchada" (swollen face)

TABLE 16.6

Four-year Colombian Study Evaluating Supplemental Minerals[a]

Item	Salt only	Complete minerals[b]
Abortions, %	9.3	0.75
Death losses, birth to weaning, %	22.6	10.5
Calves weaned, %	38.4	60.4
Weaning weight (9 months), kg	117	147
Gain of growing cattle (572 days), kg	86	141
Average gain per day, gm	150	247
Kg/year/cow[c]	44.8	88.7

[a] Evaluation of CIAT (1977) report; Miles and McDowell (1983).

[b] Evaluation of the complete mineral supplement indicated adequate concentrations of most minerals but suboptimum levels of Zn and Cu, with no added Se and S.

[c] Weaned calf percentage multiplied by weaning weight.

affects 10–20% of the young cattle in certain regions of central Brazil and to a lesser extent in Bolivia. Clinical symptoms indicate a periodontitis in nursing calves as young as 30 days of age, followed by loosening and shedding of the teeth (premolar and molar) and enlargement of the maxillary and, less frequently, mandibular bones (Fig. 16.3). These oral lesions are accompanied by emaciation, severe diarrhea, acromotrichia, and retarded growth. Gingivitis appears to be aggravated by forage consumption, and reduced integrity of the periodontal ligament permits forage impaction in this damaged tissue, which results in an alveolas pyorrhea and chronic ossifying periostitis. Normal mastication is impeded, which results in unproductive animals and death from starvation. Controlled experiments and extensive field trials indicate that low Cu and Zn, in addition to the Cu–Mo–S interrelationship, are involved in periodontitis of Brazilian cattle (Camargo et al., 1981).

A wasting disease called "secadera" is probably the most common cattle disease in the eastern plains of Colombia and is of much greater significance than foot and mouth disease. In a study of 37 ranches, "secadera" was found on 32% of the ranches in the wet season and 42% of the ranches in the dry season (Corrier et al., 1978). Contributing factors include the interaction between nutritional deficiencies and infectious diseases. The former due to poor, highly leached, acid soils and extended dry period, and the latter are due to hemoparasitic diseases including anaplasmosis, babesiosis, and trypanosomiases. "Secadera" has been described as progressive inanition in which there is evidence of failure in

 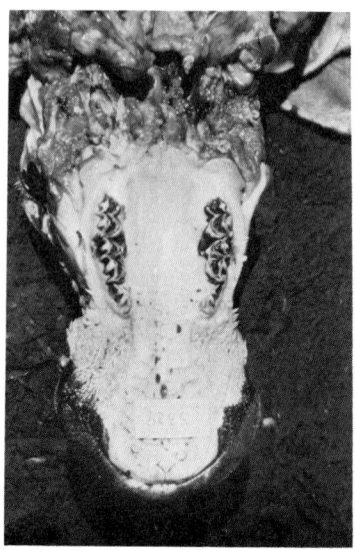

Fig. 16.3. Ten-month-old Zebu (Gir) (left) in the region of Jaciara, Mato Grosso, Brazil, affected by "cara inchada." Bi- or unilateral swelling of the maxillary bones in the advanced stage of the periodontal disease is the reason for its popular name that means "swollen face". The 2-month-old Zebu calf (right) from the region of Torixoreu in the Araguaia river valley, Mato Grosso, did not yet show any swelling of its face. However, deep, mostly symmetrical lesions at the site of the Papilla interdentalis lingualis between the maxillary Pd_3 and Pd_4, characterize the progressing periodontal disease. (Courtesy of Jürgen Döbereiner, EMBRAPA, Rio de Janeiro, Brazil.)

intermediary metabolism. Pathogenic sign of incipient "secadera" in lightskinned animals is a marked darkening, made more visible by brittle, dry hair that tends to curl (Fig. 16.4). Later the skin thickens and loses pliability, and there is a rapid and continuing loss of weight. Approximately 50% of the cases develop muscular weakness with 12% showing incoordination and ataxia involving the rear legs (Mullenax, 1982). Zinc deficiency appears to be a contributing factor, but the most satisfactory treatment has been adequate consumption of a mineral supplement well fortified with Cu, S, Se, and Zn (Miles and McDowell, 1983). Use of a high quality mineral supplement has completely eliminated the condition. In a study of the mineral status of 10 ranches in the eastern plains of Colombia where "secadera" is prevalent, the most widespread forage mineral deficiencies were Ca, P, Na, Cu, and Zn (Vargas et al., 1984).

An additional disease condition in the llanos of Venezuela is known as "borrachera" a type of falling disease (Fig. 16.5). Animals with this condition have a very low tissue status of Cu. However, the etiology of this

Fig. 16.4. A wasting disease ("secadera") of cattle in the llanos of Colombia. Both a young and mature animal (bottom) with the condition. Animals (top) are characterized by an emaciated condition in spite of good-quality available forage.

Fig. 16.5. A condition known as "borrachera," a falling disease, found in the llanos of Venezuela may be associated with Cu deficiency. The toxic plant *Arrabidaea bilabiata* has also been associated with this condition.

condition has yet to be established due to lack of controlled supplementation experiments.

VII. SEASONAL NEEDS FOR SUPPLEMENTAL MINERALS

Since tropical forages contain less minerals during the dry season, one might expect more deficiencies during this time, but numerous reports, including those from Kenya, Colombia, and South Africa, noted specific mineral deficiencies were prevalent during the wet season (McDowell, 1976). Grazing cattle were more prone to develop Co or P deficiencies, and the clinical signs were severest after the rains when pastures were green and plentiful. From Africa, Van Niekerk (1978) noted that the beneficial effect of P was primarily during the wet season, although the P content in the grass was at its highest. Research from a 2-year experiment in Colombia determined the seasonal need for minerals (Laredo, 1979). No benefit was derived by feeding minerals all year long versus during the 5-month rainy season.

Phosphorus supplements would not be beneficial while animals are in a process of losing weight. In reviewing a number of mineral supplementation studies, Van Niekerk (1978) cited some studies showing that P if given as the sole supplement actually accentuated winter weight loss in both cattle and sheep. From South Africa is a report that P supplementation alone during winter (or dry season) is detrimental to animals on open range when they are in a negative growth pattern (G. N. Louw, personal communication). On the contrary, Van Schalkwyk and Lombard (1969) demonstrated a significant P carryover effect of P supplemented during the dry season (reviewed by Van Niekerk, 1978). These results show that cattle fed P during the dry season are apparently able to build up P reserves that benefit the animal during the subsequent period of rapid growth.

Increased incidence of mineral deficiencies during the wet season is less related to forage mineral concentration than to the greatly increased requirements for these elements by the grazing animal. During the wet season, livestock gain weight rapidly since energy and protein supplies are adequate, and thus the mineral requirements are high, while during the dry season, inadequate protein and energy result in animals losing weight, which lowers mineral requirements.

There are notable exceptions as to season of the year when mineral supplementation is most critical. In the wet llanos of Venezuela, northern Colombia, and Bolivia, as the water recedes in the dry season, cattle enter the lowlands to graze a great variety of species. For this reason, in these areas, breeding and calving are more frequent during the dry season than in the rainy season (Stonaker, 1975). Under these conditions, incidence of mineral deficiencies would not be expected to be more prevalent during the wet season. Van Niekerk (1978) reports incidence of Zn deficiency for grazing cattle of South Africa in the dry but not the wet season.

VIII. ECONOMIC BENEFITS FROM MINERAL SUPPLEMENTATION

The economic benefits of mineral supplementation would seem obvious; however, the majority of grazing livestock producers do not regularly provide minerals with the exception of common salt. Miles and McDowell (1983) report that sales of mineral supplements for cattle are at a level that seems to indicate that the vast majority of the estimated 3 million head of cattle in the llanos of Colombia do not receive mineral supplements. Supplement sales figures have indicated that 50–80% of llanos cattle in Colombia do not receive mineral supplementation.

A large number of experiments have shown dramatic production benefits from providing mineral supplements that obviously result in high benefit/cost ratios. Cunha (1983) illustrated the low cost insurance of adding trace minerals to salt sold in the United States, the additional yearly cost over salt being 25 cents per head for beef cattle and 53.5 cents for dairy cows. Economical return on mineral investment has been at least 2–1 in some Latin American studies (Conrad, 1976). Using production data from the CIAT herd system in Colombia (CIAT, 1974) and mineral supplement cost and gross sales of beef, W. H. Miles (unpublished data) calculated the cost/benefit ratio of feeding minerals. A dramatic return of 15.6 pesos for each peso invested in minerals resulted over animals receiving only common salt. From Mato Grosso, Brazil (Sousa et al., 1983), steers gained 40 kg during 336 days when supplemented with only salt compared to 144 kg when fed a complete mineral supplement. The cost/benefit ratio of this mineral supplementation was shown to be 1:26.

When there is no information available on mineral status for specific regions, then complete ("shotgun") mineral mixtures are warranted. However, with additional information on likely limiting minerals, more economical mixes can be formulated. From Colombia, 10 cattle experiments were designed to evaluate feeding complete commercial supplements versus supplements formulated to contain a minimum number of minerals that had previously been established as deficient on the basis of forage and animal tissue analyses (Laredo, 1980). The specifically tailored mineral supplements that contained only required minerals produced equal production responses at one-half the cost.

Determination of mineral deficient areas in different countries and correction of these disorders by economically feasible mineral supplementation programs adapted to local conditions, along with improvement of the energy and protein supply will contribute to raising the nutritional level and, therefore, the productivity of tropical livestock.

REFERENCES

Allman, R. T., and Hamilton, T. S. (1949). "Nutritional deficiencies in livestock," FAO Agricultural Studies No. 5. Washington, D.C.

Arroyo, G., and Mauer, E. (1972). B.S. thesis. Univ. of the Republic, Montevideo, Uruguay.

Bauer, B., Galdo, E., McDowell, L. R., Koger, M., Loosli, J. K., and Conrad, J. H. (1981). *Trace Elem. Metab. Man and Anim. Proc. Int. Symp. 4th,* pp. 50–53.

Bishop, E. J. B., and Kotze, J. J. J. (1965). *Farming S. Afr.* **41,** 6.

Butterworth, M. H. (1967). *Nutr. Abstr. Rev.* **37,** 349–368.

Calub, A., and Amril, M. A. (1979). *Philipp. Farmers J.,* Dec., p. 50.

16. Incidence of Nutrient Deficiencies and Excesses

Camargo, W. V. A., Veiga, J. S., and Conrad, J. H. (1981). *Trace Elem. Metab. Man Proc. Int. Symp. 4th,* pp. 47–49.
Centro Internacional de Agricultura Tropical (CIAT) (1974). "Centro Internacional de Agricultura Tropical Annual Report." CIAT, Cali, Colombia.
Centro Internacional de Agricultura Tropical (CIAT) (1977). "Centro Internacional de Agricultura Tropical Annual Report." CIAT, Cali, Colombia.
Centro Internacional de Agricultura Tropical (CIAT) (1979). "Pastos Tropicales." CIAT, Cali, Colombia.
Cohen, R. D. G. (1975). *World Rev. Anim. Prod.* **11**(2), 27–43.
Conrad, J. H. (1976). "Ruminant Livestock Production System." Georgetown, Guyana.
Conrad, J. H., and Mendes, M. O. (1965). "Report of Escritório Fécnico de Agricultura." Revista dos Criodores, Brazil.
Corrier, D. E., Cortez, J. M., Aycardy, E. R., Wells, E. A., Bohorguez, M., and Salazar, J. J. (1978). *Br. Vet. J.* **134**, 101–107.
Cunha, T. J. (1983). "Salt and Trace Minerals for Livestock, Poultry and Other Animals." Salt Institute, Alexandria, Virginia.
De Alba, J. (1971). "Feeding of Livestock in Latin America," 2nd ed. La Prensa Medica Mexicana, Mexico.
Deinum, B. (1966). *Proc. Int. Grassl. Congr., 10th,"* pp. 415–418.
De León, Lora, L. A. (1963). M. S. thesis. Instituto Interamericano de Ciencias Agrícolas, Turrialba, Costa Rica.
Devendra, C., and Gohl, B. I. (1970). *Trop. Agric. (Trinidad)* **47**, 335–342.
Echevarria, M., Valdivia, R., Barúa, J., Santhirasegaram, K., and Campos, L. (1974). *ALPA Memoria* **9,** R19 (abstr.).
Echevarria, M., Riesco, A., Morales, V., del Valle, O., and Garcia, M. (1977). "Asociación Latinoamericana de Produccíon Animal (ALPA) Proceedings," pp. R-53 (abstr.). Havana, Cuba.
Fick, K. R., McDowell, L. R., and Houser, R. H. (1978). *In* "Proceedings Latin American Symposium on Mineral Nutrition Research with Grazing Ruminants," pp. 149–163. Univ. of Florida, Gainesville.
French, M. H. (1961). *Turrialba* **11,** 78–84.
Grunert, E., and Santiago, C. (1969). *Brasilien Zuchthyg.* **4,** 65.
Guimarães, J. M. A., and do Nascimento, C. N. B. (1971). *Estudos sobre Bovinos,* 1(2), 37–51.
Lamond, D. R. (1970). *Animal Breeding Abstr.* **38,** 354–372.
Laredo, M. A. (1979). "Programa de Nutricion Animal. Informe de Progreso 1979." Instituto Colombiano Agropecuario, Bogota, Colombia.
Laredo, M. A. (1980). "Programa de Nutricion Animal. Informe de Progreso 1980." Instituto Colombiano Agropecuario, Bogota, Colombia.
Loosli, J. K. (1974). *Proc. Nigerian Soc. Anim. Prod.* **1,** 74–82.
Louw, J. G., and Van der Wath, J. G. (1943). *Onderstepoort J. Vet. Sci. Anim. Ind.* **18,** 177–190.
McDowell, L. R. (1976). *In* "Beef Cattle Production in Developing Countries" (A. J. Smith, ed.), pp. 216–241. Univ. of Edinburgh Press, Edinburgh.
McDowell, L. R., and Conrad, J. H. (1977). *World Anim. Rev.* **24,** 24–33.
McDowell, L. R., Conrad, J. H., Thomas, J. E., and Harris, L. E. (1974). "Latin American Tables of Feed Composition." Univ. of Florida, Gainesville.
McDowell, L. R., Conrad, J. H., Thomas, J. E., Harris, L. E., and Fick, K. R. (1977). *Trop. Anim. Prod.* **2,** 273–279.

McDowell, L. R., Bauer, B., Galdo, E., Koger, M., Loosli, J. K., and Conrad, J. H. (1982). *J. Anim. Sci.* **55**, 964–970.
McDowell, L. R., Conrad, J. H., and Ellis, G. L. (1984). *In* "Symposium on Herbivore Nutrition in Sub-Tropics and Tropics—Problems and Prospects" (F. M. C. Gilchrist and R. I. Mackie, eds.), pp. 67–88. Craighall, South Africa.
Miles, W. H., and McDowell, L. R. (1983). *World Anim. Rev.* **46**, 2–10.
Milford, R., and Haydock, K. P. H. (1965). *Aust. J. Exp. Agr. Anim. Husb.* **5**, 13–17.
Minson, D. J., and McLeod, M. N. (1970). *Proc. Int. Grassl. Congr., 11th*, pp. 719–722.
Moore, J. E., and Mott, G. O. (1973). *In* "Anti-quality Components of Forages," pp. 53–96. Crop Sci. Soc. Am. Special Publication No. 4, Madison, Wisconsin.
Mtimuni, J. P. (1982). Ph. D. thesis. Univ. of Florida, Gainesville.
Mullenax, C. (1982). *Bovine Pract.* **35**, 16.
Oyenuga, V. A. (1957). *Emp. J. Exp. Agric.* **25**, 237–255.
Phillips, R. W. (1956). "Recent Developments Affecting Livestock Production in the Americas," pp. 83–98. FAO Agriculture Development Paper No. 55, Washington, D.C.
Pieterse, P. J. S., and Preller, J. H. (1965). *Proc. South Afr. Soc. Anim. Prod.* **4**, 123–126.
Pittaluga, O., Allegri, M., Bombs, M., and Riet, F. (1980). *Invest. Agronómicas* **1**, 42–45.
Ríos Araúz, S. (1972). M.S. thesis Instituto Interamericano de Ciencias Agrícolas, Turrialba, Costa Rica.
Russell, F. C., and Duncan, D. L. (1956). "Minerals in Pasture: Deficiencies and Excesses in Relation to Animal Health." *Technical Communication No. 15*. Commonwealth Bureau of Animal Nutrition, Rowett Institute, Aberdeen, Scotland.
Schiersmann, G. C. S. (1965). M.S. thesis. Instituto Interamericano de Ciencias Agrícolas. La Estanzuela, Colonia, Uruguay.
Smith, C. A. (1962). *J. Agric. Sci.* **58**, 173–178.
Smuts, D. B., and Marias, J. S. C. (1940). *Onderstepoort J. Vet. Sci. Anim. Ind.* **15**, 187–196.
Sousa, J. C. de, Gomes, R. F. C., Rezende, A. M., Rosa, I. V., Cardoso, E. G., Gomes, A., Costa, F. P., Oliveira, A. R. de, Coelho, N. L., and Curvo, J. B. E. (1983). *Pesqui. Agropecu. Bras.* **18**(3), 311–318.
Stobbs, T. H. (1975). *World Rev. Anim. Prod.* **11**(2), 58–65.
Stobbs, T. H., and Minson, D. J. (1980). *In* "Digestive Physiology and Nutrition of Ruminants" (D. C. Church, ed.), 3rd ed., pp. 257–277. O & B Books, Corvallis, Oregon.
Stonaker, H. H. (1975). *J. Anim. Sci.* **41**, 1218–1227.
Sutmöller, P., Vahia de Abreu, A., Van der Grift, J., and Sombroek, W. G. (1966). "Mineral Imbalances in Cattle in the Amazon Valley," *Royal Tropical Institute Communication No. 53*. Department of Agriculture Research, Amsterdam.
Theiler, A., Green, H. H., and Du Toit, P. J. (1924). *Union S. Afr. Dept. of Agr. Pam.* **8**, 460–504.
Theiler, A., Green, H. H., and Du Toit, P. J. (1928). *J. Agr. Sci.* **18**, 369–371.
Tokarnia, C. H., and Döbereiner, J. (1973). *Pesqui. Agropequ. Bras. Ser. Vet.* **8**(Suppl.) 1–6.
Topps, J. H. (1976). *In* "Beef Cattle Production in Developing Countries" (A. J. Smith, ed.), pp. 204–215. Univ. of Edinburgh, Edinburgh.
Tumwasorn, S. (1981). *In* "Proceedings of Second Seminar on Mineral Nutrition in Thailand." (P. Vijchulata, ed.), pp. 40–59. Kasetart University, Bangkok.
Underwood, E. J. (1981). "The Mineral Nutrition of Livestock." Commonwealth Agricultural Bureaux, London.
Van Niekerk, B. D. H. (1978). *In* "Proceedings Latin American Symposium on Mineral

Nutrition Research with Grazing Ruminants" (J. H. Conrad and L. R. McDowell, eds.), pp. 194–200. Univ. of Florida, Gainesville.
Van Schalkwyk, A., and Lombard, P. E. (1969). *Agioanimalia* **1,** 45.
Vargas D., R., McDowell, L. R., Conrad, J. H., Martin, F. G., Buergelt, C., and Ellis, G. L. (1984). *Trop. Anim. Prod.* **9,** 103–113.
Ward, H. K. (1968). *Rhod. J. Agric. Res.* **6,** 93–101.
Wilson, J. R., and Ford, C. W. (1971). *Aust. J. Agric. Res.* **22,** 563–571.
Wiltbank, J. N., Kasson, C. W., and Ingalls, J. E. (1969). *J. Anim. Sci.* **29,** 602–605.
Winks, L., Alexander, G. I., and Lynch, D. (1970). *Proc. Aust. Soc. Anim. Prod.* **8,** 34–38.
Yates, N. G. (1983). *Trop. Anim. Prod.* **8,** 50–52.
Young, J. S. (1968). *Aust. Vet. J.* **44,** 350–357.

17

Free-Choice Mineral Supplementation and Methods of Mineral Evaluation

L. R. McDOWELL

Department of Animal Science
University of Florida
Gainesville, Florida

I.	Introduction.	383
II.	Methods of Providing Minerals to Grazing Livestock.	384
III.	Free-Choice Mineral Supplementation.	385
IV.	Factors Affecting Mineral Consumption.	386
	A. Soil Fertility and Forage Type Consumed.	386
	B. Available Energy–Protein Supplements.	387
	C. Individual Requirements.	387
	D. Salt Content of Drinking Water.	387
	E. Palatability of Mineral Mixture.	387
	F. Availability of Fresh Mineral Supplies.	389
	G. Physical Form of Minerals.	392
V.	Selecting a Free-Choice Mineral Supplement.	395
VI.	Information Required for Mineral Supplement Formulation.	396
	A. Requirements.	397
	B. Biological Availability.	397
	C. Intakes of Mineral Supplement and Dry Matter.	397
	D. Element Concentration in Mineral Mixture.	399
VII.	Calculations Required for Mineral Supplement Formulation.	400
VIII.	Mineral Supplement Evaluation.	402
	References.	406

I. INTRODUCTION

Grazing livestock usually do not receive mineral supplementation, except for common salt, and must depend almost exclusively upon forages for their requirements. Only rarely, however, can tropical forages completely satisfy all mineral requirements. As an illustration, borderline or deficient levels of certain elements were noted for many forages in the 1974 "Latin American Tables of Feed Composition" (McDowell et al., 1977). These minerals and the percentage of the forage samples deficient were: cobalt (Co), 43%; copper (Cu), 47%; magnesium (Mg), 35%; phosphorus (P), 73%; sodium (Na), 60%; and zinc (Zn), 75%.

For many classes of livestock including swine, poultry, feedlot cattle, and dairy cows, mineral supplements are incorporated into concentrate rations, which generally insures that animals are receiving required minerals. However, for grazing livestock to which concentrate feeds cannot be economically fed, it is necessary to rely on self-feeding of mineral supplements. Even though it is found that some animals will either overconsume or underconsume mineral supplements, usually it is the most practical way of supplying mineral needs under grazing conditions.

II. METHODS OF PROVIDING MINERALS TO GRAZING LIVESTOCK

Indirect methods of providing minerals to grazing cattle include use of mineral-containing fertilizers, altering soil pH, and encouraging growth of specific pasture species. A number of reports have indicated that increasing soil pH influences forage mineral uptake, thereby potentially causing deficiencies of Cu and Co and excesses of selenium (Se) and molybdenum (Mo). Taking advantage of large variations in mineral content of different plant species growing on the same soil can be used to promote or discourage availability of specific minerals to grazing livestock. Underwood (1981) reports that the indirect approach as a means of controlling mineral deficiencies is not without its problems arising from the great complexity of soil–plant–mineral interrelations and difficulties related to erratic climate and cost.

Where economic and climatic considerations are favorable, fertilizer treatment of the soil is an effective means of improving both the yield and the mineral composition of herbage. Recent Australian research (Underwood, 1981) has shown that not only does superphosphate fertilizer increase herbage P but results in improved palatability and digestibility of the forage. Increased mineral content of forages through fertilization has an additional advantage of assuring a more uniform mineral consumption, since all animals would be consuming higher quantities of minerals in the forage. The major problem with free-choice mineral supplements is that not all animals in a herd will consume adequate quantities. Unless there are definite forage yield increases that can be utilized effectively by grazing herbivores, use of mineral-containing fertilizers is economically prohibitive. Direct administration of minerals to cattle in water, mineral licks, mixtures, and drenches, rumen preparations (i.e., Co pellets, copper oxide needles, and glass bullets containing various trace elements), and injections are generally the most economical methods of supplementation. Benefits and disadvantages of mineral supplementation methods are discussed by Underwood (1981).

III. FREE-CHOICE MINERAL SUPPLEMENTATION

Voluntary consumption of individual minerals or mineral mixtures by animals is referred to as free-choice feeding. Pamp et al. (1976) reviewed the literature on free-choice feeding of minerals to livestock. This practice of feeding minerals free choice to ruminants has been used for many years to supply needed minerals, but this practice is often based on the erroneous assumption that the animal knows which minerals are needed and how much of each mineral is required.

Early reports from South African researchers (Theiler et al., 1924; Green, 1925) described P-deficient cattle with depraved appetites chewing on bones. Since bones were a good source of P, the belief was held that animals have the ability to select feeds that contain minerals lacking in the diet. Additional early studies with ruminants fed P-deficient diets indicated that cows and lambs may consume sufficient P free choice to meet their requirements (Becker et al., 1933; Dew et al., 1954; Bohstedt, 1957). Becker and co-workers (1933) further showed that serum P was increased to normal by consumption of bonemeal, and the greater the intake of P from feed sources, the less bonemeal was eaten. Stoddard and Mickleson (1961) stated that free-choice feeding of mineral supplements is recommended to allow for variation in the mineral requirements of individual animals. These workers concluded that cattle have the ability to consume minerals in amounts needed to meet requirements, if a palatable mineral source is available.

Arnold (1964) stated that much evidence in the literature shows that most mammals exhibit little nutritional wisdom and that animals will select a palatable but poor quality diet in preference to an unpalatable, nutritious diet, even to the point of death. Gordon et al. (1954) had earlier measured the preferences of P-deficient cattle and sheep for supplemental calcium carbonate alone or combined with an equal part of dicalcium phosphate. The animals failed to consume enough of the P-containing supplement to prevent aphosphorosis. Experiments were carried out using lactating dairy cows to determine if sufficient dicalcium phosphate would be consumed to meet Ca and P requirements (Coppock et al., 1972, 1976). Under conditions of low Ca or P intake, cows did not consume enough dicalcium phosphate free choice to meet requirements or to correct the deficiencies. Therefore, it was concluded that lactating dairy cows had no, or only very limited, appetite for Ca or P. Burghardi et al. (1982) reported that although consumption of free-choice calcium carbonate was greater for lambs fed calcium (Ca) deficient diets than for controls, daily gains and feed efficiencies of control animals were superior to those receiving the Ca-deficient diet.

Another approach to providing free-choice minerals is the use of a "cafeteria style" mineral feeder, which offers the animal a choice of as many as 10 or more minerals and vitamins, versus the practice of providing one or two individual minerals or complete supplements. The general conclusions from two studies evaluating the "cafeteria style" provision of free-choice minerals were that dairy cows do not consume sufficient amounts of minerals to meet requirements and that acceptability rather than appetite or craving for minerals influences free-choice consumption (Hutjens and Young, 1976; Muller et al., 1977). Maller (1967) presented evidence that domestication has produced an animal that is more responsive to the sensory qualities of feed than to nutritive value. Thus, the ability to select needed nutrients may have been lost through domestication.

IV. FACTORS AFFECTING MINERAL CONSUMPTION

The average daily intake of free-choice mineral mixtures by grazing livestock is highly variable. Coppock et al. (1972) measured individual daily consumption of dicalcium phosphate by 69 lactating dairy cows in a 22-week trial. Individual variation in mineral consumption was large, with a range from 0 to over 1000 g per head daily. Factors that affect the consumption of mineral mixtures have been cited by Cunha et al. (1964), Cunha (1980), and Coppock (1970).

A. Soil Fertility and Forage Type Consumed

Usually, the higher the level of soil fertility, the lower the consumption of minerals. Barrows (1977) reports that for cattle, salt, Ca, P, and Mg each appear to be consumed in relation to the content of the particular element in the grass. A number of reports have shown that cattle on native range consume more mineral supplement than those cattle on improved pastures. Cattle on low-quality or overgrazed pastures consume more mineral supplement.

B. Available Energy–Protein Supplements

The kind and level of protein–energy supplementation will influence mineral intake. Protein and energy supplements that likewise provide minerals will decrease both the need and desire for free-choice minerals.

C. Individual Requirements

Growth rate, percentage of calf crop, and milk production influence mineral needs. Added requirements of gestation and lactation increase mineral needs and, thereby, consumption. Barrows (1977) reported that consumption of mineral elements tended to decline as cows increased in age.

D. Salt Content of Drinking Water

Naturally high salt concentration of drinking water decreases mineral supplement intake. Livestock have a natural craving for salt (Fig. 17.1). However, if that desire is fulfilled from drinking water high in salt, grazing livestock will consume less or none at all of a free-choice mineral mixture based on salt. Where naturally occurring salt content of water is high, mineral supplements cannot be based on salt and should be reformulated with other palatability stimulators such as cottonseed meal and molasses.

E. Palatability of Mineral Mixture

As previously mentioned, research has shown that cattle have no particular desire for the majority of minerals, with the exception of common salt. In a review on salt appetite, Denton (1967) noted that all mammals have the ability to taste salt, and there is a universal liking for salt. Becker et al. (1944) noted that the attitude of cattle toward salt in a mineral supplement is inversely related to the amount of salt present in feeds and water. Common salt, because of its palatability, is a valuable "carrier" of other minerals. If mixtures contain 30–40% salt or more, they are generally consumed on a free-choice basis in sufficient quantities to supply supplementary needs of other minerals.

Dew et al. (1954) allowed dairy cows free access to combinations of sodium chloride and steamed bonemeal. When sodium chloride was furnished in a separate container, bonemeal consumption dropped to an

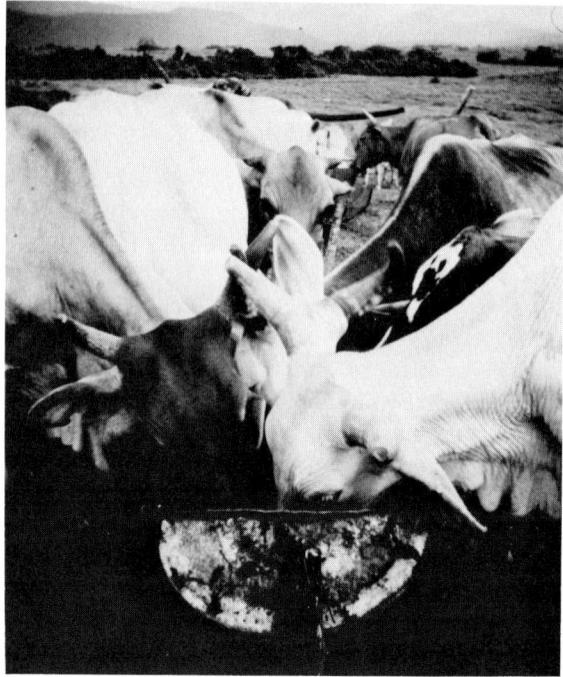

Fig. 17.1. Mineral feeder containing only common salt in Indonesia.

average of 1 g per cow daily. However, when bonemeal and sodium chloride were mixed together, bonemeal consumption increased eightfold. Many reports testify to the beneficial effects of bonemeal in free-choice supplements. Processing methods for bonemeal and other supplements affect both the nutritive value of the products and also palatability and, consequently, consumption. Improperly processed bonemeals can emit an unpleasant odor, which reduces consumption. Also, the danger of botulism, foot and mouth disease, and other disease conditions can be transmitted from inadequately processed bonemeal.

A P source other than bonemeal that is relatively palatable is monosodium phosphate. Coppock *et al.* (1972) reports that dicalcium phosphate was preferred to defluorinated phosphate by dairy cattle fed three different ration regimes. These researchers concluded that cattle preferred an acid supplement (pH 3.5), such as dicalcium phosphate, to an alkaline supplement (pH 8.5), such as defluorinated phosphate.

The concept of consuming only palatable supplements is well illustrated when Mg-deficient cattle are provided with a free-choice supply of supplementary Mg, such as magnesium oxide. Cattle will end up dying of grass tetany rather than consume this unpalatable source of Mg. However, when even high concentrations of magnesium oxide (i.e., 25%) are combined with palatable ingredients, grass tetany is prevented due to adequate consumption of the mixture.

Palatability and appetite stimulators such as cottonseed meal, dried molasses, dried yeast culture, and fat help achieve more uniform, herd-wide consumption. Some of these products not only give the supplement a dust-free, moist, and free-flowing character, but also provide energy and protein. Ingredients that increase palatability must be used in moderation, or they will cause overconsumption.

F. Availability of Fresh Mineral Supplies

Previous diet or access to mineral supplements is a factor affecting short-term consumption of minerals. When animals are not allowed access to minerals for long periods of time, they may become so voracious that they often injure each other in attempting to reach salt. Under these conditions, they will consume 2–10 times the normal daily quantities of minerals until their appetite is satisfied.

Various types of mineral feeders are illustrated in Figs. 17.1 through 17.9. Rainproof mineral boxes help increase mineral intake by preventing

Fig. 17.2. Wind-controlled mineral feeder constructed of galvanized metal with a fiberglass-coated wooden stationary two-compartment tray. These are designed to keep the closed side to the wind to eliminate waste from wind and rain. Because holes develop in the ground around permanently installed feeders, the feeder stand may be made portable. (Courtesy of C. E. Fenton, Arcadia, Florida.)

Fig. 17.3. Mineral feeder at Pichilingue Experiment Station in Ecuador.

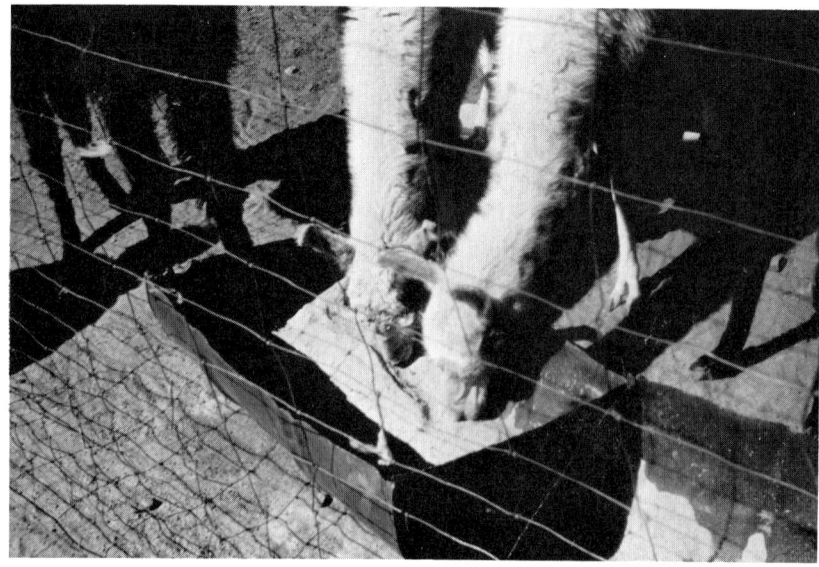

Fig. 17.4. Bolivian llamas at mineral feeder.

Fig. 17.5. Mineral feeder in Malaysia.

caking, molding, and blowing away during windy weather. The choice of palatability or appetite stimulators is important when considering the keeping value of a supplement. Cornmeal is a good appetite stimulator when included in a mineral mixture but is more easily fermentable than a proteinaceous product such as cottonseed meal. The use of 20–40% salt prevents molding and blowing.

Mineral feeders will be used more frequently by livestock if they are located near water tanks, shaded loafing areas, back rubbers, and areas of

Fig. 17.6. Mineral feeder on village farm in Thailand.

Fig. 17.7. Empty mineral feeder in the Colombian llanos.

best grazing. Mineral feeders should be constructed low enough so that calves can also consume minerals. They should be located on dry ground accessible to trucks for checking and servicing throughout the year. Feeders should be spaced at intervals of less than one-half mile and be adequate in number for the stocking capacity of the pasture. One suggestion is to have approximately one mineral feeder per 50 head of livestock. Less minerals are consumed if grazing livestock must travel long distances to the mineral box.

In some tropical regions with vast grazing areas, there are great difficulties in locating feeders so that animals have constant access to minerals. This is a particular problem where animals graze over large areas with no central location for drinking water. Also, in regions that seasonally flood, locating mineral feeders above the water level is sometimes a problem (Fig. 17.10).

G. Physical Form of Minerals

Mineral consumption is often 10% less when provided in block (Figure 17.11) versus loose form. Mineral blocks can be developed on the basis of

Fig. 17.8. Mineral feeder in Pulcallpa, Peru. Feeder is constructed of durable products for long life and designed to minimize effects of rain. (Courtesy of Mariano Echevarría, IVITA, UNMSM, Pulcallpa, Peru.)

Fig. 17.9. Mineral feeder in the Dominican Republic designed for use by calves.

Fig. 17.10. Cattle in the Venezuelan llanos. Photo illustrates the difficulty of providing minerals during the wet season, which is characterized by flooding.

Fig. 17.11. Salt block in a mineral feeder in the Dominican Republic.

degree of hardness to take into consideration rainfall, humidity and other environmental conditions. Rain will dissolve a block to soft causing mineral lose, and yet, livestock experience difficulty consuming enough hard block to fulfill mineral requirements. If the animals remain only a limited time in the vicinity of mineral blocks, then excessive block hardness will result in reduced mineral consumption.

V. SELECTING A FREE-CHOICE MINERAL SUPPLEMENT

Even though grazing livestock have been found not to balance their mineral needs perfectly when consuming a free-choice mixture, there is usually no other practical way of supplying mineral needs under grazing conditions. As a low cost insurance to provide adequate mineral nutrition, "complete" mineral supplements should be available free choice to grazing livestock (Cunha et al., 1964). A "complete" mineral mixture usually includes salt, a low fluoride–phosphorus source, Ca, Co, Cu, iodine (I), manganese (Mn), and Zn. Selenium, Mg, potassium (K), sulfur (S), iron (Fe), or additional elements can also be incorporated into a mineral supplement or can be included at a later date as new information suggests a need. In the case of Mg, an oral supplement would only be of value during the seasonal occurrences of grass tetany (Allcroft, 1961). Calcium, Cu, or Se, when in excess, can be more detrimental to ruminant production than any benefit derived by providing a mineral supplement. In regions where high forage Mo predominates, 3- to 5-times the Cu content in mineral mixtures is needed to counteract Mo toxicity (Cunha et al., 1964). Thus, the exact level of Cu to use in counteracting Mo toxicity is a complex problem and should be worked out for each area. Table 17.1 lists the characteristics of a "good" (complete or "shotgun") mineral supplement.

In relation to feeding minerals, a number of so called "authorities" feel there is no justification for the use of "shotgun" (complete) free-choice mineral mixtures that are designed to cover a wide range of environments and feeding regimens and that contain a margin of safety as an insurance against deficiency. These "authorities" feel that "shotgun" mixtures are economically wasteful and can also be harmful. This author is in disagreement with this viewpoint regarding "shotgun" mixtures for cattle. There is little danger of toxicity or excessive cost in relation to the high probability of increased production rates for cattle from administering a complete "shotgun" free-choice mineral mixture following the guidelines in Table 17.1. Copper and Se added at recommended levels would be the minerals of most concern for toxicity. However, cattle, contrary to sheep, are

TABLE 17.1

Characteristics of a "Good" Complete Free-Choice Cattle Mineral Supplement

An acceptable complete cattle mineral supplement should be as follows:
1. Final mixture containing a minimum of 6–8% total P. In areas where forages are consistently lower than 0.20% P, mineral supplements in the 8–10% phosphorus range are preferred.
2. Calcium–phosphorus ratio, not substantially over 2:1.
3. Provide a significant proportion (i.e., about 50%) of the trace mineral requirements of Co, Cu, I, Mn, and Zn. In known trace-mineral-deficient regions, 100% of specific trace minerals should be provided.
4. Composed of high-quality mineral salts that provide the best biologically available forms of each mineral element, and avoidance or minimal inclusion of mineral salts containing toxic elements. As an example, phosphates containing high F should be either avoided or formulated so that breeding cattle would receive no more than 30–50 ppm F in the total diet. Fertilizer or untreated phosphates could be used to a limited extent for feedlot cattle.
5. Formulated to be sufficiently palatable to allow close to adequate consumption in relation to requirements.
6. Backed by a reputable manufacturer with quality-control guarantees as to accuracy of mineral-supplement label.
7. An acceptable particle size that will allow adequate mixing without smaller size particles settling out.
8. Formulated for the area involved, the level of animal productivity, the environment (temperature, humidity, etc.) in which it will be fed, and is as economical as possible in providing the mineral elements used.

much less sensitive to Cu toxicity, and inorganic forms of Se (i.e., sodium selenite) are not well utilized by livestock when administered in excess of the requirements. In conclusion, it is best to formulate free-choice mixtures on the basis of analyses or other available data. However, when no information on mineral status is known for a given region, a free-choice complete ("shotgun") mineral supplement is definitely warranted, following the recommendations given in Table 17.1.

VI. INFORMATION REQUIRED FOR MINERAL SUPPLEMENT FORMULATION

In order to evaluate a mineral supplement for ruminants, it is necessary to have an approximation of (1) requirements of the target animals for the essential nutrients, which includes the age of the animals involved, stage of current production or reproduction cycle, and intended purpose for which the animals are being fed; (2) relative biological availability of the minerals in the sources from which they will be provided; (3) approximate daily intake per head of the mineral mixture and total dry matter that is anticipated for the target animals; and (4) concentration of the essential nutrients in the mineral mixture (Houser *et al.,* 1978).

A. Requirements

Although precise figures for mineral requirements of different classes of animals under varying conditions are still not known, there are enough research data available from which some general conclusions may be drawn as to dietary recommendations (National Research Council, 1984). These requirements should be used only as a rough guide since the needs of individual animals may differ from averages. It is recognized that most of the requirements were not established with Zebu cattle nor with cattle living under tropical conditions. It is also recognized that with the introduction of crossbreeding and of exotic breeds of livestock, the growth rates have been increased, with a consequent increase in mineral requirements. In spite of these shortcomings, most researchers agree that this is the best information we have and that these requirements should be used as a guide until more precise data are available.

B. Biological Availability

Biological availability of a mineral element implies the availability of that element to some organism for its use. The bioavailability and percentage of mineral elements in some sources commonly used in mineral supplements are shown in Table 17.2. These variations in bioavailability of sources must be taken into consideration when evaluating or formulating a mineral supplement.

C. Intakes of Mineral Supplement and Dry Matter

The many factors influencing mineral supplement consumption have been previously discussed, with the general conclusion that palatability of the supplement affects intake more than does physiological needs. In formulating mineral mixes, estimating the possible need must coincide with adequate intake. A number of reports conclude that grazing livestock do not always consume mineral mixtures well under tropical conditions. Rios (1974) has presented data showing a wide monthly variation in the consumption of mineral mixtures by Venezuelan cattle. I. Rosa (unpublished data), working on ranches in Mato Grosso, Brazil, has also encountered very low voluntary intake of mineral mixtures. Further research is

TABLE 17.2

Percent of Mineral Element in Some Sources Commonly Used in Mineral Supplements and Relative Bioavailability[a]

Element	Source compound	% of element in compound	Bio-availability
Calcium	Steamed bonemeal	29.0(23–37)	High
	Defluorinated rock phosphate	29.2(19.9–35.7)	Intermediate
	Calcium carbonate	40.0	Intermediate
	Soft phosphate	18.0	Low
	Ground limestone	38.5	Intermediate
	Dolomitic limestone	22.3	Intermediate
	Monocalcium phosphate	16.2	High
	Tricalcium phosphate	31.0–34.0	—
	Dicalcium phosphate	23.2	High
	Hay sources	23.3	Low
Cobalt	Cobalt carbonate	46.0–55.0	—[b]
	Cobalt sulfate	21.0	—[b]
	Cobalt chloride	24.7	—[b]
Copper	Cupric sulfate	25.0	High
	Cupric carbonate	53.0	Intermediate
	Cupric chloride	37.2	High
	Cupric oxide	80.0	Low
	Cupric nitrate	33.9	Intermediate
Iodine	Calcium iodate	63.5	High
	Ethylenediamine dihydriodide	80.0	High[c]
	Potassium iodide, stabilized	69.0	High
	Cuprous iodide	66.6	High
Iron	Iron oxide	46.0–60.0	Unavailable
	Ferrous sulfate	20.0–30.0	High
	Ferrous carbonate	36.0–42.0	Low[d]
Magnesium	Magnesium carbonate	21.0–28.0	High
	Magnesium chloride	12.0	High
	Magnesium oxide	54.0–60.0	High
	Magnesium sulfate	9.8–17.0	High
	Potassium and magnesium sulfate	11.0	High
Manganese	Manganous sulfate	27.0	High
	Manganous oxide	52.0–62.0	High
Phosphorus	Defluorinated rock phosphate	13.3(8.7–21.0)	Intermediate
	Calcium phosphate	18.6–21.0	High
	Dicalcium phosphate	18.5	Intermediate
	Tricalcium phosphate	18.0	—
	Phosphoric acid	23.0–25.0	High
	Sodium phosphate	21.0–25.0	High
	Potassium phosphate	22.8	—
	Soft phosphate	9.0	Low
	Steamed bonemeal	12.6(8–18)	High

TABLE 17.2 (*Continued*)

Element	Source compound	% of element in compound	Bio-availability
Potassium	Potassium chloride	50.0	High
	Potassium sulfate	41.0	High
	Potassium and magnesium sulfate	18.0	High
Selenium	Sodium selenate	40.0	High
	Sodium selenite	45.6	High
Sulfur	Calcium sulfate (gypsum)	12.0–20.1	Low
	Potassium sulfate	28.0	High
	Potassium and magnesium sulfate	22.0	High
	Sodium sulfate	10.0	Intermediate
	Anhydrous sodium sulfate	22.0	—
	Sulfur, flowers of	96.0	Low
Zinc	Zinc carbonate	52.0	High
	Zinc chloride	48.0	Intermediate
	Zinc sulfate	22.0–36.0	High
	Zinc oxide	46.0–73.0	High

[a] From Ellis *et al.* (1983).
[b] Critical tests not done, but source effective.
[c] Some liberation of free iodine when mixed with trace minerals.
[d] Some samples are fairly high in availability—but not as available as ferrous sulfate.

needed to determine the pattern of mineral consumption for livestock on tropical pastures. When evaluating mineral supplements where consumption is not known, researchers commonly start with an intake figure of 50 g/day and adjust this figure according to local conditions.

It is virtually impossible to know the total dry matter consumption of cattle on pasture. However, this is essential since requirements are based on intakes of dry matter. The quality of a pasture will to a great degree determine intake. Although 2% body weight is considered a rough estimate of forage dry matter intake by cattle, they may eat much less if the forage is of poor quality. Actual dry matter consumption often becomes a matter of judgment on the part of the researcher or rancher. For grazing mature cattle, often daily dry matter consumption is between 7 and 10 kg.

D. Element Concentration in Mineral Mixture

After an evaluation has been made as to the biological availability of the elements to be supplied by the mineral mixture, and a judgment has been made as to the approximate daily intake of mineral mixture and of total dry matter, the concentration of each element in the mineral mixture can be used to calculate the amount of each element that will be furnished per animal, expressed as a percentage or parts per million (ppm) of the total dry matter intake. The concentration in the total diet of each element furnished by the mineral mixture can be compared to the total requirements for that element in order to determine if a significant amount is being furnished by the supplement. It is difficult to determine what constitutes a significant portion of the requirement for each mineral that should be supplied by the mineral mixture, but it is generally believed the figure should be 25–50% for the trace elements. In zones known to have a trace mineral deficiency, 100% of these elements should be provided.

VII. CALCULATIONS REQUIRED FOR MINERAL SUPPLEMENT FORMULATION

To determine the amount of mineral supplied to the animal, the calculations are made as follows:

$$\frac{(\% \text{ element in mineral mixture}) \times \text{daily intake of mineral mix (g)}}{\text{total daily dry matter intake (g)}} \times 100$$
$$= \% \text{ element in total from mineral mixture}$$

If, for example,

Copper in mineral mixture (%) = 0.12
Daily intake of mineral mixture (g) = 50
Total daily intake of dry matter (g) = 10,000

then:

$$\frac{0.0012 \times 50}{10,000} \times 100 = 0.0006\% \text{ or 6 ppm}$$

Note: To convert percentage to ppm, move the decimal four places to the right. If approximately 10 ppm is considered the allowance for Cu, then 60% of the Cu requirement would thus be supplied by this particular mixture.

To calculate the percentage of the element in the final mixture, use the following formula:

$$\frac{\text{Amount of mineral mix} \times \% \text{ element in mineral mix}}{\text{total amount}} \times 100$$
$$= \% \text{ of element in diluted mix}$$

If, for example, the recommendation is for feeding 1/2 kg mix per 2 kg salt and the % of Ca in the mixture is 18.38%, then:

$$\frac{500 \text{ g} \times 0.1838}{2500 \text{ g}} \times 100 = 3.68\% \text{ Ca in final mixture}$$

Table 17.3 illustrates the estimated trace mineral requirements and percentages of each mineral required in a mixture to meet either 50 or 100% of the requirement. These figures are based on the assumption of a daily mineral consumption of 50 g. With less consumption, the mineral supplement should contain a higher percentage of each mineral. Likewise, a lower intake of dry matter would reduce the percentage of minerals required in the mixture. Each producer should determine mineral consumption for his herd and change products if higher consumption rates are required (i.e., increase the cottonseed meal level of the mineral mixture from 5 to 10%).

An additional calculation needed to formulate mineral supplements is to account for the percentage of the desired mineral in an available compound. As an example, if 0.20% Cu is required to meet the requirement,

TABLE 17.3

Percentage of Trace Minerals Required in an Adequate Mineral Supplement[a,b]

Element	Estimated maximum requirement (ppm)	Percentage of minerals in mixture for the following percent of the requirement[a]	
		50%	100%
Cobalt	0.1	0.001	0.002
Copper	10	0.10	0.20
Iodine	0.8	0.008	0.016
Manganese	25	0.25	0.50
Zinc	50	0.50	1.0
Iron	50	0.50	1.0
Selenium	0.2	0.002	0.004

[a] From McDowell et al. (1983).
[b] This assumes an average consumption of 50 g/day of mineral mixture for cattle and 10 kg of total dry feed per animal daily.

how much cupric carbonate (which is 53.0% Cu) is required? The calculations are made as follows:

$$\frac{\%\text{ element desired in mixture}}{\%\text{ element in available compound}}$$

If, for example,

% Cu required = 0.20
% Cu in cupric carbonate = 53.0

then:

$$\frac{0.20}{53.0} \times 100 = 0.377\%\text{ cupric carbonate required}$$

VIII. MINERAL SUPPLEMENT EVALUATION

Table 17.4 presents an example of mineral allowances and proportions supplied by a typical mineral mixture offered free choice on range or pasture. Many of the minerals were not provided in significant quantities in relation to requirements. Problems concerned with mineral supplementation programs in diverse tropical regions have been summarized (McDowell and Conrad, 1977) and include: (1) insufficient chemical analyses and biological data to determine which minerals are required and in what quantities, (2) lack of mineral consumption data needed for formulating supplements, (3) inaccurate and/or unreliable information on mineral ingredient labels, (4) supplements that contain inadequate amounts or imbalances, (5) standardized mineral mixtures that are inflexible for diverse ecological regions (i.e., supplements containing Se distributed in a Se-toxic region), (6) farmers not supplying mixtures as recommended by the manufacturer (i.e., mineral mixtures diluted 10:1 and 100:1 with additional salt, and (7) difficulties involved with transportation, storage, and cost of mineral supplements.

Some of the information necessary to evaluate mineral supplements can be found printed on the bag or a tag securely attached to the bag or, in the case of bulk shipments, attached to the invoice or other papers involved in the sale. This information is sometimes incorrect and is expressed in different ways, making it difficult for the layman to determine what is being purchased and if it is adequate for the purpose intended. Also, many developing countries do not have quality control standards. Therefore, there may be no analysis on the mineral bags, or the analysis may be inaccurate or fraudulent.

TABLE 17.4

Evaluation of a Mineral Mixture Sold in Latin America[a,b]

Name of mixture: A
Country: X
Estimated consumption of mineral mix: 50 g/day
Recommendation for feeding: 1/2 kg minerals for each 2 kg of salt
Estimated consumption of DM[c]: 10 kg of DM per day
Compounds in mix: Dicalcium phosphate, sodium chloride, magnesium oxide, calcium carbonate, minor elements
Ca–P: 1.2–1

	\multicolumn{12}{c}{Composition (%)}											
	Ca	P	Mg	K	Na	Fe	S	Cu	Co	Mn	Mo	Se	Zn	I
Declared	20.60	15.30	1.08	—	2.40	0.10	—	0.05	0.003	0.1	0.001	0.001	0.24	0.003
Analyses	18.38	14.86	1.07	0.072	2.30	0.22	—	0.0084	0.001	0.206	—	0.00001	0.934	—
Diluted	3.68	2.97	0.21	0.0143	31.90	0.04	—	0.0017	0.0002	0.041	—	0.000002	0.1868	—

Mineral	Dietary allowance (ppm)	Mineral in mixture (%)	Amount from mineral mix (ppm)	Allowance from mineral mix (%)
Cu	10	0.0017	0.085	0.85
Co	0.1	0.00022	0.011	11.00
Fe	80	0.043	2.15	2.70
I	0.8	0.003	0.15	18.75
Mn	25	0.0412	2.06	8.24
Se	0.1	0.000002	0.0001	0.001
Zn	50	0.187	9.34	18.68

[a] From McDowell et al. (1983).
[b] Evaluator comments: Low P in final mixture; ratio of Ca:P satisfactory but low quantities; bioavailability of macroelements satisfactory; sources or minor elements not listed; percentage daily allowance of minor elements is low.
[c] Dry matter, DM.

TABLE 17.5

Relationship between the Mineral Content of Mineral Supplements Sold in Latin America and Laboratory Analyses[a,b]

	Composition (%)							
	Mexico		Ecuador		Peru		Paraguay	
Mineral	Tag[b]	Lab	Tag	Lab	Tag	Lab	Tag	Lab
Ca	12.00	3.42	30.00	21.77	20.60	18.38	16.10	13.99
P	20.40	1.16	20.00	14.62	15.30	14.86	21.10	16.11
Mg	0.0038	0.153	—	1.15	1.08	1.07	1.59	1.38
K	—	0.756	—	0.23	—	0.072	0.076	0.076
Na	0.026	21.21	0.16	2.05	2.40	2.30	—	0.049
Fe	0.16	0.21	—	0.11	0.10	0.22	1.80	1.87
Cu	0.016	0.0035	—	0.87	0.05	0.008	0.072	0.214
Co	0.0016	0.0006	—	—	0.003	0.001	0.197	0.022
Mn	0.48	0.103	—	0.006	0.10	0.21	0.482	0.177
Mo		0.0003	—	0.00009	0.001	—	—	0.0003
Zn	0.0008	0.0056	—	0.015	0.24	0.93	1.06	1.15

[a] From McDowell et al. (1983).
[b] Stated tag composition compared to actual laboratory analyses.

The concentration of elements in the mineral mixture, in most cases, is supplied by the manufacturer on the mineral tag. Making a judgment from such a tag assumes, of course, that the quality control of the supplement has been such that the mixture actually contains the amount stated on the tag. Unfortunately, this has not proven to be the case in many instances. Analyses of mineral mixtures collected throughout Latin America have often shown little relationship between the amount of elements listed on the tag and those actually found in the supplement. Examples from four different countries are shown in Table 17.5 (McDowell et al., 1983).

Responsible firms that manufacture and sell high-quality mineral supplements provide a great service to individual farmers. However, there are companies that are responsible for exaggerated claims of advertising, and some that produce inferior products that are of little value, or worse, those likely to be of detriment to animal production. Table 17.6 provides an example of an inferior mineral mixture available in Latin America. This particular mineral supplement is recommended for cattle, sheep, pigs, and chickens. It is impossible to adequately meet requirements of both ruminants and monogastric animals with the same mixture. This imbalanced mineral mixture, which is extremely high in Ca (29.4%) and low in P (1.8%), would likely be more detrimental to grazing cattle than having no access to supplemental minerals, and may actually produce a P deficiency.

Investigations from Ecuador have illustrated the problem of unreliability of some companies that sell mineral supplements (Fausto Rivera, personal communication). Fifty products claiming to contain minerals were analyzed in relation to stated guarantees. Approximately 10% were reliable, with the majority having inaccurate information on mineral ingredient labels, and others providing inadequate amounts or mineral imbalances. A known practice which demonstrates an extreme fraudulent practice of one particular supplier was to buy a relatively good product from the marketplace, mix it with 20–30% soil, and then resell the resulting mixture under a new brand name.

Some mineral product manufacturers do not put the percentage of individual mineral elements on the tag but rather the compound and its percentage or a combination of both. In order to evaluate the mixture, calculations must be made to determine the individual mineral content.

TABLE 17.6

An Example of an Inferior Mineral Mixture Available in Latin America[a,b,c]

Element	Dietary allowance	Mineral mixture (%)	Amount provided from mineral mix	Allowance for mineral mixture (%)
Sodium Chloride	0.50%	20.00	0.10%	20.0
Calcium	0.30%	29.44	0.147%	49.1
Phosphorus	0.25%	1.80	0.009%	3.6
Magnesium	2000 ppm	3.2	0.016%	8.0
Iron	100 ppm	0.88	44 ppm	44.0
Zinc	50 ppm	0.02	1 ppm	2.0
Cobalt	0.1 ppm	0.002	0.1 ppm	100
Iodine	0.80 ppm	0.001	0.05 ppm	6.25
Copper	10 ppm	0.015	0.75 ppm	7.5
Manganese	25 ppm	0.075	3.75 ppm	15.0
Selenium	0.1 ppm	0.0005	0.025 ppm	25.0

[a] From McDowell et al. (1983).

[b] Mineral mixture is recommended for cattle, sheep, pigs, and chickens. It is assumed that mineral consumption will average approximately 0.5% of the total dietary intake. This is based on an estimated intake of 50 g of mineral mixture for cattle and 10 kg of total dry feed per head daily.

[c] Criticisms of mineral mixture are as follows: (1) Mixture extremely low in P and exceptionally high in Ca. The Ca:P ratio is 16.4:1. (2) The supplement does not provide a significant proportion (i.e., 50%) of the trace mineral requirements of Cu, I, Mn, and Zn. (3) The majority of the Fe is from ferric oxide, an unavailable form of this element. (4) Since this diet contains 29.4% Ca and only 20% salt (NaCl), it is likely to be of low palatability.

TABLE 17.7

Latin American Mineral Mixture with Recommendations to Feed 2 kg per 100 kg of Salt[a]

Element	Analysis of mixture to be mixed with (%)	Analysis of total mixture mixed 1:50 with salt (%)	Animal requirements provided by final mixture[b] (%)
Calcium	15.59	0.31	0.39
Phosphorus	13.65	0.27;	0.45
Magnesium	1.6	0.03	0.07
Potassium	0.057	0.028	0.02
Sodium	0.17	38.56	128.00
Iron	1.6	0.03	1.88
Copper	0.049	0.0096	0.48
Cobalt	0.0095	0.00019	9.50
Manganese	0.18	0.0035	0.70
Molybdenum	0.0004	0.000008	0.40
Selenium	0.00002	0.0000004	0.02
Zinc	0.11	0.0022	0.22
Iodine	—	0.00080	6.70

[a] From Ellis et al. (1983).
[b] The final mineral mixture was to be self-fed to cattle.

Another common occurrence that complicates the evaluation of mineral supplements is that the manufacturer prints on the tag the concentration of the elements and then recommends that the mineral mixture be mixed with a certain quantity of salt. The percentage of elements in the final mixture is not printed on the tag, and it is left to the evaluator to calculate the final mixture, which often dilutes the original mixture to the point that an insignificant quantity of some of the mineral elements are provided. The mixture shown in Table 17.7 collected in Latin America illustrates this point (Ellis et al., 1983).

REFERENCES

Allcroft, R. (1961). *Vet. Rec.* **73**, 1255–1266.
Arnold, G. W. (1964). *Proc. Aust. Soc. Anim. Prod.* **5**, 258–271.
Barrows, G. T. (1977). *Anim. Nutr. Health* **32**, 12–14.
Becker, R. B., Neal, W. N., and Shealy, A. L. (1933). *Bull—Fla. Agric. Exp. Stn.* **699**.
Becker, R. B., Davis, G. K., Kirk, W. G., Glasscock, R. S., Dix A. P. T., and Pace, J. E. (1944). *Bull—Fla. Agric. Exp. Stn.* **401**.
Bohstedt, G. (1957). "Minerals for livestock". Wisconsin Extension Circular 297, Madison, Wisconsin.

Burghardi, S. R., Goodrich, R. D., Meiske, J. C., Thonney, M. L., Theuninck, D. H., Kahlon, T. S., Pamp, D. E., and Krajem, K. (1982). *J. Anim. Sci.* **54**, 410–418.
Coppock, C. E. (1970). *Proc.—Cornell Nutr. Conf. Feed Manuf.* pp. 29–35.
Coppock, C. E., Everett, R. W., and Merrill, W. G. (1972). *J. Dairy Sci.* **55**, 245–256.
Coppock, C. E., Everett, R. W., and Belyea, R. L. (1976). *J. Dairy Sci.* **59**, 571–580.
Cunha, T. J. (1980). *Anim. Nutr. Health* **35**(3), 11, 29.
Cunha, T. J., Shirley, R. L., Chapman, H. L., Jr., Ammerman, C. B., Davis, G. K., Kirk, W. G., and Hentges, J. F. (1964). *Bull—Fla. Agric. Exp. Stn.* **683**.
Denton, D. A. (1967). *In* "Handbook of Physiology, Alimentary Canal," Section 6: Vol. 1. Am. Physiol. Soc., Washington, D.C.
Dew, M. I., Stoddard, G. E., and Bateman, G. Q. (1954). *Utah Farm Home Sci.* **15**, 36.
Ellis, G. L., McDowell, L. R., and Conrad, J. H. (1983). *In* "Proceedings 17th Annual Conference on Livestock and Poultry in Latin America," pp. B41–53. Univ. of Florida, Gainesville.
Gordon, J. G., Tribe, D. E., and Graham, T. C. (1954). *Br. J. Anim. Behav.* **2**, 72–74.
Green, H. H. (1925). *Physiol. Rev.* **5**, 336–348.
Houser, R. H., McDowell, L. R., and Fick, K. R. (1978). *In* "Proceedings Latin American Symposium on Mineral Nutrition Research with Grazing Ruminants" (J. H. Conrad and L. R. McDowell, eds.), pp. 173–180. Univ. of Florida, Gainesville.
Hutjens, M. F., and Young, C. W. (1976). *In* "Proceedings 71st American Science Association Annual Meeting". p. 30.
McDowell, L. R., and Conrad, J. H. (1977). *World Anim. Rev.* **24**, 24–33.
McDowell, L. R., Conrad, J. H., Thomas, J. E., Harris, L. E., and Fick, K. R. (1977). *Trop. Anim. Prod.* **2**, 273–279.
McDowell, L. R., Conrad, J. H., Ellis, G. L., and Loosli, J. K. (1983). "Minerals for Grazing Ruminants in Tropical Regions." Bulletin, Univ. of Florida, Gainesville.
Maller, O. (1967). *In* "The Chemical Senses and Nutrition" (M. R. Kare and O. Maller, eds). Johns Hopkins Press, Baltimore.
Muller, L. D., Schaffer, L. V., Ham, L. C., and Owens, M. J. (1977). *J. Dairy Sci.* **60**, 1574–1582.
National Research Council (NRC). (1984). "Nutrient Requirements of Beef Cattle," 6th rev. ed. Nutrient Requirements of Domestic Animals, No. 4. Natl. Acad. Sci., Washington, D.C.
Pamp, D. E., Goodrich, R. D., and Meiske, J. C. (1976). *World Rev. Anim. Prod.* **12**, 13–18.
Rios, J. E. (1974). *Agron. Trop.* **24**, 227–234.
Stoddard, G. E., and Mickelson, C. H. (1961). *Utah Farm Home Sci.* **22**, 103, 111–113.
Theiler, A., Green, H. H., and Du Toit, P. J. (1924). *Union S. Afr. J. Dep. Agric.* **8**, 460–504.
Underwood, E. J. (1981). "The Mineral Nutrition of Livestock." Commonwealth Agricultural Bureaux, London.

18

Vitamin Nutrition for Ruminants

L. R. McDOWELL

Department of Animal Science
University of Florida
Gainesville, Florida

I. Introduction	409
II. Vitamin A	410
A. Requirements	410
B. Function and Deficiency	411
C. Body Stores of Vitamin A	415
D. Natural Sources of Vitamin A	416
E. Vitamin A Toxicity	417
F. β-Carotene Function Independent of Vitamin A	417
G. Vitamin A Supplementation	418
III. Vitamin D	419
IV. Vitamin E	421
V. Vitamin K	422
VI. B-Complex Vitamins	423
A. Niacin	423
B. Thiamine	424
C. Choline	426
VII. Providing Vitamin Supplements	427
References	428

I. INTRODUCTION

Besides a source of energy, protein, and minerals, the ruminant requires in its diet, or must obtain from microorganisms present in the digestive tract, a number of highly specific nutrients, the vitamins.

Vitamins that can be added to animal diets are divided into two groups: (1) the fat-soluble vitamins, which are vitamins A, D, E, and K; and (2) the water-soluble or B-complex vitamins, which are thiamine, riboflavin, niacin, pantothenic acid, vitamin B_6, biotin, folacin, vitamin B_{12}, and choline. Vitamin C is synthesized in the tissues of ruminants, and, therefore, it is assumed dietary sources are not required. A reasonable assumption is that ruminants, at the tissue level, require the same vitamins as monogastric animals. Similarity of requirements has been shown for the young ruminant before development of the rumen (usually 6–8 weeks of age). Deficiencies of thiamin, riboflavin, vitamin B_6, pantothenic acid, biotin, niacin, and vitamin B_{12} have all been produced experimentally in young ruminants prior to the development of the rumen (Miller, 1979).

Milk received by nursing ruminants is an excellent source of the water-soluble vitamins. Also, milk is generally a good source of vitamins A and E, but quantities are affected by dietary concentrations of these vitamins. Rumen microorganisms normally synthesize B vitamins and vitamin K in sufficient quantities to meet the requirements of grazing ruminants. However, newer information suggests that under certain situations the ruminant animal may be benefited by thiamin, niacin, and choline supplementation.

In practical feeding of grazing ruminants, under most circumstances, vitamin deficiencies should not be a major problem. Due to synthesis of vitamins D, K, and the B vitamins by grazing ruminants, only supplemental vitamin A, and possibly vitamin E, may be required. The present review will emphasize the nutrition of vitamin A for grazing ruminants with reference made to circumstances where ruminants may respond to supplementation of vitamins D, E, K, thiamin, niacin, and choline.

II. VITAMIN A

A. Requirements

Vitamin A is considered to be the most critical vitamin for grazing ruminants. The National Research Council (NRC) subcommittes have established requirements for growing ruminants that approximate 2200 IU per kilogram of diet for both beef and dairy cattle (NRC, 1984 and 1978) and 588–1962 IU for sheep (NRC, 1975) (see Chapter 2 of this volume). Requirements range from 1.5- to 3-times greater for mature animals than for growing animals. For beef cattle, growing and finishing cattle require 2200 IU of vitamin A per kilogram of dry diet, pregnant heifers and cows require 2794 IU per kilogram, and lactating cows and bulls require 3894 IU per kilogram (Perry, 1980). During the last trimester of pregnancy additional vitamin A is required (NRC, 1978).

The vitamin A requirements for grazing ruminants are naturally met by the provitamin A (carotene) in feedstuffs. A number of carotenoid compounds exist in plants, with β carotene the most active, and other carotenoids varying from 0 to 57% of the potency of β carotene. The carotenoids can be converted to physiologically active vitamin A in varying degrees in the wall of the small intestine.

One milligram of β carotene in the diet is equivalent to approximately 400 IU of vitamin A for beef cattle (NRC, 1984), dairy cattle (NRC, 1978), and goats (NRC, 1981) and 400–700 IU for sheep (NRC, 1975). Some factors that influence the rate at which carotenoids are converted to vitamin A are type of carotenoid, breed of animal, individual differences in animals, and level of carotene intake (NRC, 1984). Stress conditions, such as extremely hot weather, viral infections, and altered thyroid function have also been suggested as causes for reduced carotene to vitamin A conversion.

Additional factors may possibly affect the metabolism and increase the requirements of vitamin A including free nitrates in feeds, inadequate protein, a zinc (Zn) deficiency, and low dietary phosphorus (P) (Harris, 1975). Parasitism may also harm the intestinal wall and decrease carotene-to-vitamin A conversion. Considerable work and controversy has been reported on the relationship between nitrates and vitamin A nutrition. In a review of this subject by Rumsey (1975), it was concluded that although nitrates can be shown to have an adverse effect on vitamin A in vitro, this does not appear to translate into a significant effect under most feeding conditions.

B. Function and Deficiency

Vitamin A plays many critical roles in the animal body, with a deficiency causing at least four different and probably physiologically distinct lesions. These are (1) loss of vision due to a failure of rhodopsin formation in the retina, (2) defects in bone growth, (3) defects in reproduction (i.e., failure of spermatogenesis in the male and resorption of the fetus in the female animal, and (4) defects in growth and differentiation of epithelial tissues. Numerous studies have also demonstrated increased frequency and severity of infection in vitamin A–deficient animals. Lack of vitamin A results in decreased antibody production and impaired cell-mediated immune processes against infective agents (Davis and Sell, 1983).

Clinical signs can be specific for vitamin A deficiency, or only general signs are observed including loss of appetite, loss of weight, unthrifty appearance (Fig. 18.1), thick nasal discharge, and reduced fertility. The normal epithelium in various locations throughout the body becomes replaced by a stratified, keratinized epithelium when vitamin A is deficient.

Fig. 18.1. This vitamin A deficient calf is emaciated and shows evidence of diarrhea. The calf also shows excessive lacrimation and nasal discharges characteristic of the deficiency. [From Church (1971).]

This effect has been noted in the respiratory, alimentary, reproductive, and genitourinary tracts, as well as in the eye. If vitamin A is lacking, ruminants may get pinkeye, or other illnesses related to the mucous membranes. This keratinization lowers the resistance of the epithelial tissues to the entrance of infectious organisms. Thus respiratory troubles, such as colds and sinus infections, tend to be more severe in vitamin A deficiency. Likewise, in relation to maintaining a healthy epithelium, vitamin A and β carotene have been shown to play an important role in reducing the incidence and severity of mastitis in dairy cows (Chew *et al.,* 1982). A diet adequate in vitamin A is necessary to help maintain the normal powers of resistance, but additional intakes will not increase resistance to infections that enter through the epithelium.

The only physiological function of vitamin A that has been clearly defined on a biochemical basis is its role in vision. Vitamin A is combined with a protein rhodopsin, in visual purple. This compound breaks down in the physiological process of sight as a result of photochemical reaction. A deficiency, in terms of the needs for the resynthesis of visual purple, results in night blindness (nyctalopia), which is a clinical sign in all animals. In a severe vitamin A deficiency, characteristic changes occur in the eye including excessive watering, keratitis, softening and cloudiness of the cornea, and development of xerophthalmia that is characterized by drying of the conjunctiva. Copious lacrimation (rather than xeropthalmia) is the most prominent clinical sign of vitamin A deficiency involving the eye in cattle (Maynard *et al.*, 1979). Clinical sign of eye involvement as a result of vitamin A deficiency is illustrated in Fig. 18.2. Blindness may follow eye infections caused by vitamin A deficiency.

Vitamin A is concerned in the normal development of bone. The bones are altered in shape during growth, with the teeth also affected. A failure of the spine and some other bones to develop normally results in pressure on the nerves and in their degeneration. For example, a blindness in calves results from constriction of the optic nerve caused by a narrowing of the bone canal through which it passes (Maynard *et al.*, 1979). Bone changes may also be responsible for muscle incoordination and other nervous symptoms shown by vitamine A-deficient cattle and sheep.

In an advanced vitamin A deficiency the cerebrospinal fluid pressure is

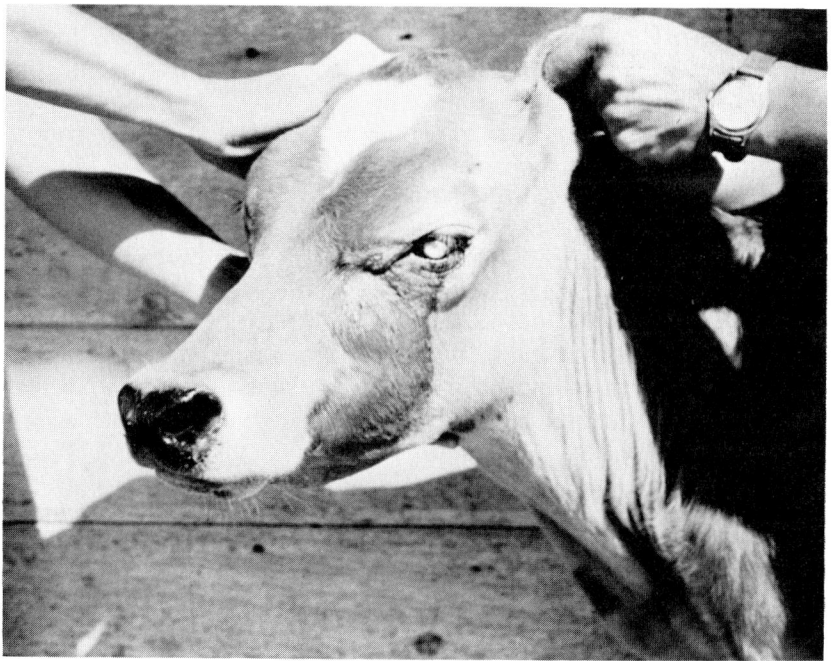

Fig. 18.2. Calf in the Philippines (south of Manila) showing a vitamin A deficiency characterized by copious lacrimation and blindness. The 6-month-old animal had been fed reconstituted skim milk powder and poor-quality bleached hay (practically devoid of carotene). (Courtesy of J. K. Loosli, Univ. of Florida, Gainesville.)

elevated and may result in a staggering gait and convulsive seizures, probably caused by the increased cerebrospinal fluid pressure (NRC, 1978). Vitamin A deficiency lowers reproductive efficiency in both males and females. In cattle, key indications of the deficiency are shortened pregnancies, either as abortions or reduced gestation length; a high incidence of retained placenta; and the birth of dead, weak, incoordinated, or permanently blind calves caused by bone abnormalities in the optic foramen that constricts the optic nerve (Miller, 1979). Severe and intermittent diarrhea at advanced stages of deficiency is characteristic. In finishing cattle, generalized edema may occur, with signs of lameness in the hock and knee joints and swelling in the brisket area (NRC, 1984).

A number of methods are available to evaluate the vitamin A status of livestock including production response, liver–vitamin A stores, plasma vitamin A, and cerebrospinal fluid pressure. For beef cattle, Perry et al. (1962) suggested that plasma vitamin A is probably the most accurate indicator of a borderline deficiency, with a level of less than 40 μg of vitamin A per 100 ml of blood serum indicating deficiency. For dairy cattle, liver–vitamin A values below 1.0 ppm is indicative of a critical deficiency (NRC, 1978). Feedlot performance of beef cattle was good as long as liver–vitamin A stores were above 2 ppm (Kohlmeier and Burroughs, 1970). In the feeding trials of Perry et al. (1967), cattle showed a positive response to supplemental vitamin A when their liver concentrations were 3–4 ppm.

Tropical geographical locations where vitamin A deficiency might be expected would be regions where there are extensive dry seasons. Many tropical regions in Africa, Asia, and Latin America routinely have dry seasons 6 months or longer in duration. Under conditions in India, it has been reported that fair grazing is possible for only 3 months of the year (Ray, 1963). With the cessation of the monsoon rains, the grasses mature rapidly with a large drop in their carotene content. The value has been found to decrease from 100 to 200 mg per kilogram on a dry matter basis during midrainy season, to as low as 0.5–1 mg per kilogram during the dry season. As a result, clinical signs including night blindness, blindness in newborn calves, and birth of weak calves have been reported from all parts of India.

Research has indicated that Zn is required for mobilization of vitamin A from liver stores and that vitamin A deficiency has resulted for grazing cattle due to a shortage of this mineral (Guerin, 1981). Research from tropical northern Australia indicated that a 12% annual cattle mortality was due in part to a slow release of liver vitamin A. Guerin (1981) indicated that high-calcium (Ca) and low-Zn forage concentrations apparently contributed to this slow liver–vitamin A release. Since tropical forages have been shown to be low in Zn (see Chapter 13 of this volume), conditioned vitamin A deficiencies may be resulting even though liver–vitamin A values indicate adequate concentrations of this vitamin.

C. Body Stores of Vitamin A

Vitamin A is stored in liver and fat of animals during times when intake exceeds requirements. Although most of the vitamin A reserves are in the liver, when carotene intake is high, some is stored in fat. Yellow body fat is associated with carotene, with species and breed of ruminants strongly influencing the amount of yellow fat. During periods of low carotene supplies in the diet, this stored vitamin A can be mobilized and utilized without signs of a vitamin A deficiency. At birth, the ruminant usually does not have sufficient vitamin A reserves to provide for its needs for any substantial time. Accordingly, it is important that young ruminants receive colostrum, which generally is rich in vitamin A, within a few days after birth. If the cow has received a diet low in vitamin A activity, the newborn calf is likely to be susceptible to a vitamin A deficiency because the body reserves are low and the colostrum will have a subnormal content. (Miller, 1979).

Grazing livestock with access to green, high-quality pastures can store sufficient vitamin A in the liver to provide for periods of low intake during the winter or dry season. How long stored vitamin A will be adequate during periods of low carotene intake is an unanswered question, but perhaps 4–6 months. Cattle grazing good pasture will have 30–80 ppm of liver vitamin A (Rumsey, 1975). Cattle entering the feedlot with 20–40 ppm will have adequate liver stores for 3–4 months (Perry et al., 1967). The intramuscular injection of emulsified vitamin A at the rate of 1 million IU apparently provides sufficient vitamin A to prevent deficiency signs for 2–4 months in growing or breeding beef cattle (NRC, 1984).

About 200 days are required to entirely deplete the vitamin A stores in the livers of ewe lambs previously pastured on green feed (NRC, 1975). Because of this storage, sheep that graze on green forage during the normal growing season are able to do reasonably well on a low-carotene diet of dry feed for periods of 4–6 months. The tendency of the goat to search out palatable green plant parts ensures it an advantage over other ruminant species. (NRC, 1981).

Florida beef cattle finished on Roselawn St. Augustine grass during the summer did not require supplemental vitamin A for production; however, 25,000 IU per animal daily increased weight gains of cattle by approximately 10% when cattle were pastured during the winter (Chapman et al., 1964). In a study of nine cattle ranches from four regions in Florida, forage carotene and, consequently, liver vitamin A were lower during the winter season (Kiatoko *et al.*, 1982). In this study, cattle in the most northern region, and hence with less total grazing days, had significantly lower liver vitamin A than those in the other three regions, and approached critical levels.

D. Natural Sources of Vitamin A

Provitamin A carotenoids, mainly β carotene in green feeds, are the principal source of vitamin A for grazing livestock. All green parts of growing plants are rich in carotene and, therefore, have a high vitamin A value. Good pasture always provides a liberal supply, and the kind of pasture plant, whether grass or legume, appears to be of minor importance. At maturity, however, leaves contain much more than stems, and, thus, legume hay is much richer in vitamin content than is grass hay (Maynard *et al.*, 1979). Mature, overly ripe plant materials have greatly decreased quantities of carotene. Hays that are cut in the bloom stage or earlier and cured without exposure to rain or to too much sun retain a considerable proportion of their carotene content, but those that are cut in the seed stage and exposed to rain and to the sun for extended periods of time lose carotene almost entirely. Green-hay curing in the swath may lose one-half of its vitamin A activity in one day's exposure to the sunlight and perhaps lose practically all of it if it is exposed to rain as well as sunlight.

The carotene content of dried or sun-cured forages decreases upon storage, with the rate of such destruction depending upon factors such as temperature, exposure to air and sunlight, and length of storage. Vitamin A and carotene destruction also occur due to processing of feeds with steam and pressure, or when mixed with certain oxidizing materials such as minerals (Scott, 1972).

Aside from yellow corn and its by-products, practically all of the concentrates used in feeding animals are devoid of vitamin A value, or nearly so. In addition, yellow corn contains a high proportion of non-β-carotenoids (i.e., cryptoxanthin, lutein, and zeacarotene) that vary from almost no vitamin A value up to 57% of that of β carotene. There is evidence that yellow corn may lose carotene rapidly during storage. For instance a hybrid, high in carotene, lost about one-half of its carotene in 8 months storage at 25°C and about three-quarters in 3 years. Less carotene was lost during storage at 7°C (Quackenbush, 1963).

A marked discrepancy exists between the carotene content of corn silage and the vitamin A status of ruminants fed corn silage. On the average, corn silage carotenes were found to be about two-thirds as effective as β carotene for maintaining liver stores in rats (Rumsey, 1975).

More mature silages were not able to sustain liver–vitamin A stores in beef steers, particularly if the ensiled corn plant was finely chopped. Miller *et al.* (1969) have reported that ethanol, sometimes found in corn silage as a product of fermentation, may reduce liver–vitamin A stores as much as 26% by increasing mobilization of vitamin A from the liver.

Wing (1969) reported carotene digestibility in plants to be greater during the warmer months than during the winter months. Variations were found in the digestibility of carotenes in plants due to year, species of plant, dry matter content, and form of forage. Carotene digestibility was somewhat lower in silages than in pastures or hay.

E. Vitamin A Toxicity

In general, the possibility of vitamin toxicities for livestock is remote. Of all vitamins, vitamin A is most likely to be provided in toxic concentrations to livestock. Clinical signs associated with hypervitaminosis A include acceleration of bone formation, intramuscular and subcutaneous hemorrhaging, thickening of skin, mucous cell formation in keratinized membranes, and reproductive failure (McDowell, 1984).

Extended vitamin A toxicity also causes decreased growth, lower feed consumption, enlarged liver, heart, and kidney, and elevated heart rate (Hazzard *et al.*, 1964). The dietary level at which damage occurs in cattle varies among affected tissues, with some changes observed in bones when as little as 60,000 IU of vitamin A per 45.5 kg of body weight per day is given (Hazzard *et al.*, 1964). This level is only about 30 times the requirement. In contrast, weight gains were depressed above 400,000 IU of vitamin A per 45.5 kg of body weight.

The efficiency of carotene conversion to vitamin A declines progressively with increasing intakes. This appears to be a natural "homeostatic control mechanism" that protects grazing livestock from any harmful effects due to the great abundance of carotene present in high-quality, fresh forages when they are the major feed for long periods of time (Miller, 1979). Likewise, the comparatively rapid disposal of very high levels of stored vitamin A is a protective mechanism. On a practical basis, toxicity is more easily caused by vitamin A than by carotene. Even so, vitamin A toxicity is not a practical problem, except when unreasonably large amounts are given accidentally (Miller, 1979).

F. β-Carotene Function Independent of Vitamin A

Since 1978, a number of studies have indicated that β-carotene has a function independent of vitamin A in dairy cattle (Lotthammer, 1978). Dairy cattle receiving extra β carotene have had a higher intensity of estrus, increased conception rates, and reduced frequency of follicular cysts than do controls. The corpus luteum of the cow has higher β-carotene concentrations than does any other organ, and it has been suggested that β carotene has a specific effect on reproduction. Other researchers have found no effect of β-carotene supplementation on the fertility of dairy cattle.

G. Vitamin A Supplementation

Ruminants offered mainly pastures or fed good-quality hay have little need for additional vitamin A supplement. The amounts of carotene in fresh, green forages is very high relative to the dietary requirements. When grazing livestock receive a modest amount of fresh, green pasture forage, there is little likelihood of a deficiency. Likewise, with a substantial amount of good silage made from green forage, or with liberal feeding of fresh hay with a good green color, a deficiency will not occur. However, low-quality forages and weathered leached hay may contain very little available carotene.

Practical feeding conditions where supplemental vitamin A may be needed include (1) feeding of poor quality forage with little or no green color, (2) diets composed primarily of concentrates and no green pasture, (3) feeding mainly corn silage and a concentrate mixture low in vitamin A activity (Jordan *et al.*, 1963), (4) young calves fed milk from cows on a low intake of vitamin A or carotene, (5) when calves are fed relatively little whole milk or colostrum (NRC, 1978), and (6) when the background of purchased cattle is unknown and when they appear unthrifty, body stores of vitamin A might be suboptimal. (Perry, 1980).

The use of concentrate feeds in place of forages is probably the largest single factor that has increased the need for vitamin A supplements in ruminant diets. Inefficient utilization of corn carotene and the destruction of carotene and vitamin A in the rumen are reasons why the total daily vitamin A requirement is usually added to high-concentrate diets as a supplement, regardless of what is present in the diet (Rumsey, 1975).

Vitamin A may be supplemented (1) as part of a concentrate or liquid supplement, (2) as part of a free-choice mineral mixture, and (3) as an injectable product. The most effective means to provide vitamin A to feedlot animals is inclusion with the concentrate mixtures, which will provide uniform consumption of the vitamin. For grazing livestock vitamin A can be provided as part of a free-choice mineral mixture, with the limitation being unknown consumption by individual animals and destruction of the vitamin with time. Vitamin A is easily destroyed with stability in a mineral mix affected by abrasion, moisture, and trace element metals particularly copper (Cu). Stabilized and protectively coated (or beaded) forms of vitamin A will slow down destruction of the vitamin, but for highest potency fresh supplies of the mixture should be available on a regular basis. A reliable method of providing recommended quantities of vitamin A to grazing animals is by injection. Intramuscular injection of emulsified vitamin A at a level of 1 million IU will provide sufficient vitamin A to prevent deficiency signs for about 3 months in growing or breeding beef cattle (Perry, 1980).

The decision to add vitamin A to the diet should be based mainly on whether or not a deficiency could be a practical problem. As with most nutrients, a borderline deficiency is much more likely than is a severe deficiency. Based on the positive results that may be derived and taking into account that vitamin A supplementation is inexpensive and no toxicity problems have been reported when given at recommended levels, it seems beneficial to supplement vitamin A at all times when ruminants are not grazing or receiving green pastures or roughages.

III. VITAMIN D

The estimated dietary requirements of vitamin D for ruminants varies from 160 to 600 IU per kilogram of dry matter (see Chapter 2 of this volume). Vitamin D has been used in massive doses either in the feed (Hibbs and Conrad, 1976) or as injections (Payne and Manston, 1967), as a method of preventing milk fever. The role of massive doses of vitamin D in reducing the incidence of milk fever apparently is not a nutritional effect but a pharmacological effect (Miller, 1970). Effective control of milk fever by appropriate dietary Ca and P concentration has been discussed (see Chapter 9 of this volume).

Vitamin D is essential for absorption and metabolism of Ca and P. In its absence, or at low levels, normal bone development is impaired. Soft, irregular shaped leg and rib bones resulting from a vitamin D deficiency results in rickets. Clinical signs of the deficiency are decreased appetite and growth rate, digestive disturbances, stiffness in gait, labored breathing, irritability, weakness, and, occasionally, tetany and convulsions. Later, enlargement of the joints, slight arching of the back, bowing of the legs, and the erosion of the joint surfaces cause difficulty in locomotion (NRC, 1984). In older ruminants, a vitamin D deficiency causes osteomalacia characterized by reabsorption of the mineral from the already formed bone. For dairy cattle, milk production may be decreased and estrus inhibited by inadequate vitamin D (NRC, 1978). If vitamin D is marginally deficient, the dietary requirement for Ca and P are increased.

Rickets and osteomalacia can also be caused by a deficiency of Ca or P, or by an unbalanced ratio of these minerals (see Chapter 9 of this volume).

Vitamin D is available to animals both through the diet and as a result of exposure to sunlight. Ultraviolet radiation from sunlight acts on ergosterol, a plant sterol, and on 7-dehydrocholesterol, a sterol of animal origin, to produce compounds having antirachitic activity (i.e., vitamin D_2 and D_3, respectively). Thus, sun-cured hays are excellent sources of vitamin D, but, in silages, the content is dependent on the amount of sun drying plus the content of Vitamin D in dead leaves. Animals exposed to sunlight can obtain their requirement directly from irradiation of 7-dehydrocholesterol in the skin.

Recent research indicates that the dietary forms of vitamin D, D_2, and D_3 are not the forms used in tissues but are converted to more active forms including 25-hydroxy D_3, 1,25-dihydroxy D_3, and 24,25 dihydroxy D_3 (DeLuca, 1979). The formation of the vitamin D metabolites is regulated by feedback mechanisms with hormonal involvement including the parathyroid hormones and calcitonin.

When grazing livestock have normal exposure to direct sunlight or are fed normal amounts of sun-cured forage, little chance for vitamin D deficiency exists. However, seasons of minimum sunlight, artificially cured forages, sheep with full fleece, feedlot animals without access to sunlight or sun-cured forages, and high producing dairy cows are considered situations that require dietary supplementation.

Very high levels of vitamin D cause high blood plasma Ca, deposition of Ca in many soft tissues including the heart and arteries, and other pathological changes that can become sufficiently severe to cause death (Miller, 1979). Vitamin D toxicity can be caused by excessive supplementation of the vitamin or by a plant-induced calcinosis. Grazing animals in several parts of the world develop calcinosis, a disease characterized by the deposition of Ca salts in soft tissues (Morris, 1982). During the development of the disease, destruction of connective tissues occurs, and this precedes mineralization in which magnesium (Mg) is involved, as well as Ca and P. The physical deterioration accompanying these changes means that the value of carcasses is greatly reduced, and fertility and milk production are adversely affected. The ingestion of the leaves of the shrub *Solanum malacoxylon* by grazing animals causes enzootic calcinosis in Argentina and Brazil where the disease is referred to as "enteque seco" and "espichamento", respectively. The calcinogenic factor in *S. malacoxylon* is $1,25(OH)_2D_3$, an active hormonal form of vitamin D_3, which is present as a glycoside. Enzootic calcinosis of unknown etiology exists in various parts of the world. In some fields in Argentina, between 10 and

30% of cattle show clinical signs of "enteque seco" and *S. malacoxylon* is now regarded as one of the most important poisonous plants of that country (Morris, 1982).

IV. VITAMIN E

Vitamin E functions as an antioxident, thereby protecting fats within the membranes from breaking down. Other physiological functions include a role in prostaglandin synthesis, blood clotting, and immunity. Vitamin E is an important vitamin in ruminant nutrition. The body has tremendous potential for storage of vitamin E. Therefore, efforts to study deficiency signs and determine exact requirements have been impeded by such body stores, which can prevent true clinical signs from appearing for long periods of time. At present, there are insufficient data to establish minimum requirements; however, some recommendations are published in NRC references (see Chapter 2 of this volume). The NRC (1984) on beef cattle estimates the requirement for younger calves to range from 15 to 60 IU dl-α-tocopherol per kilogram of dry diet. Although a similar level may be adequate for dairy calves, much higher levels may be required if milk replacers containing oils are fed.

The metabolic role of vitamin E is linked to that of selenium (Se). Some disease conditions associated with low dietary concentrations of Se (i.e., less than 0.1 ppm) are analogous to those of vitamin E deficiency. In some instances, disease syndromes have responded fully to vitamin E. In others, either Se is more or equally as effective as vitamin E. White muscle disease (WMD), also known as nutritional muscular dystrophy, a serious muscle degeneration disease in young ruminants, is due to an Se or vitamin E deficiency, or both.

White muscle disease is seen in young nursing ruminants and is characterized by generalized weakness, stiffness, and deterioration of muscles, with affected animals having difficulty standing (see Chapter 12 of this volume). In calves, the musculature of the tongue may be affected, therefore, prohibiting sucking (NRC, 1978). Often death occurs suddenly from heart failure as a result of severe damage to the heart muscle. In milder cases with calves where the chief clinical signs are stiffness and difficulty standing, dramatic, rapid, improvement can result with vitamin E–Se injections. Many cattlemen make it a practice to inject newborn calves intramuscularly with a combination of vitamin E and Se. For dystrophic lambs, an oral therapeutic dose of 500 mg of dl-α-tocopherol followed by a 100 mg on alternate days, until recovery is successful (Rumsey, 1975).

Most preventive preparations for WMD in ruminants contain a combination of both vitamin E and Se.

Other than for young ruminants, severe vitamin E deficiency is uncommon, and the effects have not been fully described for the adult species. Attempts to establish a practical role for vitamin E in reproduction of both males and females have been largely unsuccessful (NRC, 1978). The relationship to reproduction is of special interest since early rat research demonstrated that reproductive failure was a key feature of vitamin E deficiency. Four generations of female and male dairy cattle were fed low vitamin E diets. Although growth, reproduction, and milk production were normal, several cattle died suddenly of apparent heart failure between 21 months and 5 years of age. From a different aspect of reproduction for dairy cattle, Harrison *et al.* (1984) reported that supplemental vitamin E was required in addition to Se for prevention of retained placenta. Groups administered vitamin E alone, Se alone, and a control group had a retained placenta incidence of 17.5% compared with none for animals receiving both vitamin E and Se.

It has been shown that there is a need for adequate amounts of vitamin E in the ration to prevent oxidation flavors in milk. However, the cost is high because vitamin E is one of the most expensive vitamins, and the efficiency of transfer into milk is less than 2% (NRC, 1978).

Natural diets supply adequate amounts of vitamin E under most conditions. There are eight forms (tocopherols) of vitamin E in feedstuffs, with α-tocopherol the most potent source. Green pasture is an excellent source of α-tocopherol, but high grain diets would be lower in the vitamin. The tocopherols are relatively resistant to heat, but they are readily destroyed by oxidation. Artificial dehydration or processing of forages and grains will reduce the availability of tocopherol. Generally, the amount of vitamin E in natural feedstuffs is reduced during storage. For example, in one study, 80% of the vitamin E was lost in hay making (King *et al.*, 1967). Ensiling or rapid dehydration retains most of the vitamin.

V. VITAMIN K

Vitamin K is required for the formation of prothrombin and other related proteins in the liver, which are essential for normal blood clotting. Vitamin K is widely distributed in forages and is synthesized in the rumen and intestine (Matschiner, 1970). A requirement for supplementation would only be justified when high dicumarol forages, such as molded sweet clover, are fed.

VI. B-Complex Vitamins

After the development of rumen function, B vitamin requirements are usually supplied in adequate amounts by rumen microbial synthesis and by the vitamins normally found in feeds. In regard to B vitamin supplementation, only thiamin, niacin, and choline will be considered, since there is no evidence for additional dietary needs of the remaining B vitamins. These three vitamins would also not normally be needed for grazing ruminants, but rather for ruminants receiving high concentrate diets and low quantities of forage (i.e. feedlot cattle and high-producing dairy cows). Supplementation of vitamin B_{12} is not warranted, since this is synthesized when ruminants received sufficient dietary Co (see Chapter 12 of this volume).

A. Niacin

Niacin is a component of two coenzymes (NAD and NADP) that are important in the production and utilization of energy and the metabolism of carbohydrate, fats, and protein. A decrease in the availability of these coenzymes will depress growth and decrease feed efficiency.

Ruminants had previously been thought to synthesize adequate niacin by microorganisms and/or conversion from the amino acid, tryptophan. However, the synthesizing ability appears to be low, since production responses can be demonstrated in both beef and dairy cattle. Niacin supplementation is especially beneficial to stressed animals, such as beef cattle being adapted to high grain diets or lactating cows that have just calved.

Byers (1979) summarized 14 beef cattle studies demonstrating improved gains and feed efficiency by 9.7 and 10.9%, respectively. Growth from all trials was especially beneficial during the adaptation of cattle to feedlot diets (i.e., during the first 40 days). Feeder calves responded positively in rate and in efficiency of gain in the 29-day adaptation study by gaining an additional 8.3 kg with 70 ppm of niacin added (Byers, 1979). Niacin appears to be effective in enhancing acclimation and adaptation to urea-supplemented diets. Minimal amounts of niacin supplied via microbial synthesis and diet become adequate in the later stages of the feedlot period, whereas earlier they are insufficient to meet demands. Summaries of beef cattle studies over the total feeding period (73–176 days) indicate 50 or 100 ppm of niacin is more effective than are higher levels of 150, 250, or 500 ppm, with respect to gain.

A report has indicated that about 50% of dairy cows in high production herds go through borderline ketosis during early lactation (Emery *et al.*,

1964). Fronk and Schultz (1979) indicated that treating ketotic dairy cows with 12 g doses of nicotinic acid daily had a beneficial effect on the reversal of both subclinical and clinical ketosis. More recent studies indicate 6 g of niacin may be sufficient for this purpose.

Other workers have reported that supplemental niacin increased microbial protein synthesis. Enhanced production of microbial protein might explain increased milk production, weight gain, and feed efficiency observed when urea-containing rations were supplemented with 250–500 ppm niacin (Cunha, 1982). Daily niacin supplementation of 3–6 g per day to early lactation dairy cows has resulted in slight increases of milk production. Jaster *et al.* (1983) showed only a slight increase in milk fat percentage in six commercial dairy herds supplemented with niacin.

In a review, Olentine (1984) has summarized factors affecting the niacin requirements of ruminants as follows:

1. Protein balance—Excess of leucine, arginine, and glycine increase the requirement.
2. High tryptophan content of feeds—As tryptophan content increases, niacin requirements decrease.
3. Energy content—High energy rations require more niacin per unit of feed.
4. Antibiotics—Depending on the product, niacin requirements may increase or decrease.
5. Dietary rancidity—If fat is rancid, niacin requirements are increased.
6. Gastrointestinal synthesis—Niacin is synthesized in the gastric and intestinal regions.
7. Availability of niacin in feedstuffs—Cereal grains and other feedstuffs have varying degrees of niacin availability.

B. Thiamin

Thiamin is involved in intermediary metabolism of carbohydrates as a coenzyme (TPP). In this role, it is associated with many cellular metabolic processes and is involved in central nervous system (CNS) function.

Interest in thiamin centers around the CNS condition known as polioencephalomalacia (PEM) or cerebrocortical necrosis. This condition is characterized by circling, head pressing, blindness, convulsion, and death (Fig. 18.3). Thiamin has been shown to dramatically improve the condition with feeding 150 mg thiamin daily, preventing PEM disorders.

Despite the fact that rumen microbes synthesize thiamin and also feeds, particularly whole grains contain thiamin, deficiencies do develop in rumi-

Fig. 18.3. An animal (top) with polioencephalomalacia, a thiamin deficiency. Feedlot cattle suffering from this condition show dullness and sometimes blindness with a series of nervous disorders such as circling, head pressing, and convulsions. After injections of thiamin, the animal (bottom) was able to stand after 6–8 h. With continued thiamin treatment, in 3–5 days the animal returned to almost normal with slight brain damage. (Courtesy B. Bock, Univ. of Florida, Gainesville.)

nants. Polioencephalomalacia generally occurs in feedlot cattle, frequently about 3 weeks after a ration change. The condition has caused significant economic losses in tropical countries where high levels of molasses are fed and in feedlots where high-grain diets are fed.

Clinical reports of PEM have shown that under high-concentrate feeding systems of beef cattle and lambs, thiaminase, an enzyme that destroys thiamin, may become active in the rumen and cause a thiamin deficiency in animals with functional rumens (Edwin and Lewis, 1971). Thiaminases are found in certain plant species and are produced by some microorganisms believed to be responsible for PEM. Moldy feeds can also contain thiaminases (Davies *et al.*, 1968).

Research suggests that PEM is associated with lactic acid acidosis, with both conditions relating to adaptation to grain diets. Oltjen *et al.* (1962) reported that thiamin in the rumen is decreased by a reduction in rumen pH; a low ruminant pH is characteristic of cattle fed high-concentrate diets. Although little information is available on the direct addition of thiamin to finishing-cattle diets, Brethour (1972) reported that in two trials, a combination of thiamin and sodium carbonate supplement increased feed intake by 5 and daily gain by 8%. In a third trial, thiamin administered alone gave an intermediate response to calves immediately after weaning.

Thiaminase can also be found in a number of plant species. This has been a special problem in Australia where PEM occurs under pasture conditions, apparently being derived from some of the fern species. From Colombia, a wasting disease known as "secadera" has been reported as a thiamin deficiency since the condition has been alleviated with thiamin injections (C. Mullenax, unpublished data). Mullenax (unpublished data) suggests that a fungus associated with native forages contains a thiaminase. On the contrary, Miles and McDowell (1983) report that "secadera" can be successfully controlled with a highly fortified, complete mineral supplement (see Chapter 16 of this volume). Control of this wasting disease may be possible by supplementation with either thiamin or minerals through different mechanisms.

C. Choline

Choline is required for normal growth and metabolism, with one function as a component of lecithins, found in most fats. The amino acid methionine and certain other compounds can at least partially replace choline in the diet. Grains and green forages are good sources of choline. Also, choline is synthesized in the rumen.

Improved performance of feedlot cattle has been related to use of dietary choline, although not consistently in all experiments. Several reports from the states of Washington (Swingle and Dyer, 1970) and Maryland (Rumsey, 1975) have shown increased gains by as much as 6–7% and improved feed efficiency by 2.5–8% for finishing cattle when supplemented with 500–750 ppm choline. Thus, choline, under certain conditions of high-concentrate feeding, may be limiting in the diet. Atkins *et al.* (1983) studied the effects of supplementing choline on roughage adequate diets for dairy cattle and concluded that added choline had no effect on milk production, although slight increases were seen in feed intake and milk fat percentage.

VII. PROVIDING VITAMIN SUPPLEMENTS

If a decision is made to supplement dairy or feedlot ruminants with vitamins, the vitamins can be provided by mixing with concentrate feeds. For the highest potencies of vitamins A, D, and E, fresh feed should be provided on a regular basis, since these vitamins are gradually destroyed in storage. Many studies have shown that vitamins A, D, and E are relatively unstable in mixed feed, especially in the presence of trace minerals (i.e., Cu, iron (Fe), and manganese (Mn) and when stored under hot environmental conditions. As an example, stored concentrate mixtures can lose over 50–60% of the vitamin A value when stored for a year. For feedlot diets, even in the presence of minerals, niacin and choline are stable, but more than 50% of thiamine is destroyed in premixes.

For grazing livestock, the only vitamins that may be required during the winter or dry season are vitamins A and E. These vitamins can be provided as part of a free-choice mineral supplement. Due to the destructive action of minerals, the ester forms of these vitamins (i.e., retinyl acetate and tocopheryl acetate) are used due to a greater stability than that of the pure vitamin. These vitamins are further stabilized by enveloping minute droplets of the vitamins in a stable fat or gelatin, forming small beads, and thus preventing most of the vitamin from coming into contact with oxygen until it is digested in the intestinal tract of the animal (Scott, 1972). Vitamins A and E, even when stablilized and protectively coated (or beaded), will lose potency when included with a free-choice mineral mixture over a period of time.

Destruction of vitamins A and E in concentrate or mineral mixtures can be overcome by the use of injections, which often include vitamins A, D, and E. The use of injectable A, D, and E is quite a common practice but certainly of questionable value when grazing livestock are receiving adequate green roughage.

REFERENCES

Atkins, K. B., Erdman, R. A., and Vandersall, J. H. (1983). *J. Dairy Sci.* **6,** 175 (abstr.).
Brethour, J. R. (1972). *J. Anim. Sci.* **35,** 260 (abstr.).
Byers, F. M. (1979). *Animal Nutr. Health* **35,** 20–22.
Chapman, H. L., Shirley, R. L., Palmer, A. Z., Haines, C. E., Carpenter, J. W., and Cunha, T. J. (1964). *J. Anim. Sci.* **23,** 669–673.
Chew, B. P., Hollen, L. L., Hillers, J. K., and Herlugson, M. L. (1982). *J. Dairy Sci.* **65,** 2111–2118.
Church, D. C. (1971). "Digestive Physiology and Nutrition of Ruminants," Vol. 2, p. 636. D. C. Church and Oregon State Univ. Book Stores, Inc., Corvallis.
Cunha, T. J. (1982). "Niacin in Animal Feeding and Nutrition." National Feed Ingredients Association (NFIA), Fairlawn, New Jersey.
Davies, E. T., Pill, A. H., Austwick, P. K. A. (1968). *Vet. Rec.* **83,** 681.
Davis, C. Y., and Sell, J. L. (1983). *J. Nutr.* **113,** 1914–1919.
DeLuca, H. F. (1979). *Nutr. Rev.* **37,** 161–188.
Edwin, E. E., and Lewis, G. (1971). *J. Dairy Res.* **38,** 79–90.
Emery, R. S., Burg, N., Braur, L. D., and Blank, G. N. (1964). *J. Dairy Sci.* **47,** 1074–1079.
Fronk, T. J., and Schultz, L. H. (1979). *J. Dairy Sci.* **62,** 1804–1807.
Guerin, H. B. (1981). *J. Anim. Sci.* **53,** 758–764.
Harris, B., Jr. (1975). *Feedstuffs* **47**(48), 42–43.
Harrison, J. H., Hancock, D. D., and Conrad, H. R. (1984). *J. Dairy Sci.* **67,** 123–132.
Hazzard, D. G., Woelfel, C. G., Calhoun, M. C., Rousseau, J. E. Jr., Eaton, H. D., Neilsen, S. W., Grey, R. M., and Lucas, J. J. (1964). *J. Dairy Sci.* **47,** 391–401.
Hibbs, J. W., and Conrad, H. R. (1976). *J. Dairy Sci.* **59,** 1944–1946.
Jaster, E. H., Hartwell, G. F., and Hutjens, M. F. (1983). *J. Dairy Sci.* **66,** 1046–1051.
Jordan, H. A., Smith, G. S., Neumann, A. L., Zimmerman, J. E., and Breniman, G. W. (1963). *J. Anim. Sci.* **22,** 738–745.
Kiatoko, M., McDowell, L. R., Bertrand, J. E., Chapman, H. L., Pate, F. M., Martin, F. G., and Conrad, J. H. (1982). *J. Anim. Sci.* **55,** 28–37.
King, R. L., Burrows, F. A., Hemken, R. W., and Bashore, D. L. (1967). *J. Dairy Sci.* **50,** 943–944.
Kohlmeier, R. H., and Burroughs, W. (1970). *J. Anim. Sci.* **30,** 1012–1018.
Lotthammer, K. H. (1978). "Importance of Beta-Carotene for Bovine Fertility." Hoffmann-La Rocho, Inc., Basel, Schweiz, Switzerland.
McDowell, L. R. (1984). "Vitamins. Lecture Notes I." Mimeograph. Univ. of Florida, Gainesville.
Matschiner, J. T. (1970). *J. Nutr.* **100,** 190–192.
Maynard, L. A., Loosli, J. K., Hintz, H. F., and Warner, R. G. (1979). *In* "Animal Nutrition," pp. 283–355. McGraw-Hill, New York.
Miles, W. H., and McDowell, L. R. (1983). *World Anim. Rev.* **46,** 2–10.
Miller, W. J. (1970). *In* "Proceedings Georgia Nutrition Conference for Feed Industry," pp. 32–42. Dept. Anim. Sci., Univ. of Georgia, Athens.
Miller, W. J. (1979). "Dairy Cattle Feeding and Nutrition." Academic Press, New York.

Miller, R. W., Hemken, R. W., Waldo, D. R., and Moore, L. A. (1969). *J. Dairy Sci.* **52,** 1998–2000.
Morris, K. M. (1982). *Vet. Hum. Toxicol.* **24,** 34–48.
National Research Council (NRC) (1975). "Nutrient Requirements of Sheep" 5th rev. ed. Nutrient Requirements of Domestic Animals, No. 5. Natl. Acad. Sci., Washington, D.C.
National Research Council (NRC) (1978). "Nutrient Requirements of Domestic Animals," 5th rev. ed. Nutrient Requirements of Dairy Cattle, No. 3. Natl. Acad. Sci., Washington, D.C.
National Research Council (NRC) (1981). "Nutrient Requirements of Domestic Animals," 1st ed. Nutrient Requirements of Goats. Natl. Acad. Sci., Washington, D.C.
National Research Council (NRC) (1984). "Nutrient Requirements of Beef Cattle" 5th rev. ed. Natl. Acad. Sci., Washington, D.C.
Olentine, C. (1984). *Feed Manage.* **35**(4), 18–24.
Oltjen, R. R., Sirny, R. J., and Tillman, A. D. (1962). *J. Nutr.* **77,** 269–277.
Payne, J. M., and Manston, R. (1967). *Vet. Rec.* **81,** 214–216.
Perry, T. W. (1980). "Beef Cattle Feeding and Nutrition," pp. 24–32. Academic Press, New York.
Perry, T. W., Beeson, W. M., Mohler, M. T., and Smith, W. H. (1962). *J. Anim. Sci.* **21,** 333–339.
Perry, T. W., Beeson, W. M., Smith, W. H., and Mohler, M. T. (1967). *J. Anim. Sci.* **26,** 115–118.
Quackenbush, F. W. (1963). *Cereal Chem.* **40,** 266–269.
Ray, S. M. (1963). *In* "Proceedings of the 1st World Conference on Animal Production" (abstr.), p. 109. European Association of Animal Production, Rome.
Rumsey, T. S. (1975). *Feedstuffs* **47,** 30–34.
Scott, M. L. (1972). "Proceedings Effect of Processing on the Nutritional Value of Feeds" pp. 119–130. Natl. Acad. Sci., Washington, D.C.
Swingle, R. S., and Dyer, I. A. (1970). *J. Anim. Sci.* **31,** 404–408.
Wing, J. M. (1969). *J. Dairy Sci.* **52,** 479–483.

Appendix

A.1 Metric Weights and Measures with Customary Equivalents. 431
A.2 Table of Equivalents. 433

TABLE A.1
Metric Weights and Measures with Customary Equivalents[a,b]

Length		
1 millimeter	=	0.03937 inch
1 centimeter	=	0.3937 inch
1 meter	=	39.37 inches
	=	3.281 feet
	=	1.094 yards
1 kilometer	=	0.6214 mile
Area		
1 square centimeter	=	0.155 square inch
1 square meter	=	1.196 square yards
	=	10.764 square feet
1 hectare (10,000 m^2)	=	2.471 acres
1 square kilometer	=	0.386 square mile
	=	247.1 acres
Capacity or Volume		
1 cubic centimeter	=	0.061 cubic inch
1 cubic meter	=	35.315 cubic feet
	=	1.308 cubic yards
1 milliliter	=	0.0338 fluid ouce (U.S.)
1 liter	=	33.81 fluid ounces (U.S.)
	=	2.1134 pints (U.S.)
	=	1.057 quarts (U.S.)
	=	0.2642 gallon (U.S.)
1 kiloliter	=	264.18 gallons (U.S.)
Weight		
1 gram	=	0.03527 ounce (avdp.)
1 kilogram	=	35.274 ounces (advp.)
	=	2.205 pounds (avdp.)
1 metric ton (1,000 kg.)	=	0.984 ton (long)
	=	1.102 tons (short)

(*continued*)

TABLE A.1 (*Continued*)

	=	2204.6 pounds (avdp.)
Volume per unit area		
1 liter/hectare	=	0.107 gallon (U.S.)/acre
Weight per unit area		
1 kilogram/square centimeter	=	14.22 pounds (avdp.)/square inch
1 kilogram/hectare	=	0.892 pound (avdp.)/acre
Area per unit weight		
1 square centimeter/kilogram	=	0.0703 square inch/pound (avdp.)
Temperature conversion formulas		
Centigrade (Celsius)	=	5/9 (Fahrenheit −32)
Fahrenheit	=	9/5 centigrade (Celsius) +32

[a] Modified from *J. Anim. Sci.* (1966). **25**, 270–271.

[b] When conversions are made the results should be rounded to a meaningful number of digits, relative to the accuracy of original measurements. Values for weights and volumes are based on pure water at 4°C under 760 mm of atmospheric pressure.

TABLE A.2
Table of Equivalents[a]

Parts per million	per-cent	Grams per kilo	Grams per pound	Grams per 100 lb	Grams per ton	Oz per 100 lb	Oz per ton	lbs per 100 lb	lbs per ton
1. 1.0	0.0001	0.001	0.00045	0.0453	0.907	0.0016	0.032	0.0001	0.002
2. 10000.0	1.0	10.0	4.53	453.6	9.072	16.0	320.0	1.0	20.0
3. 1000.0	0.1	1.0	0.45	45.3	907.0	1.60	32.0	0.1	2.0
4. 2200.	0.22	2.2	1.0	100.00	2000.0	3.53	70.6	0.22	4.41
5. 22.0	0.0022	0.022	0.01	1.0	20.0	0.035	0.706	0.0022	0.044
6. 1.10	0.0001	0.0011	0.0005	0.05	1.0	0.0017	0.035	0.00011	0.0022
7. 625.0	0.0625	0.625	0.283	28.3	566.0	1.0	20.0	0.0625	1.25
8. 31.2	0.00312	0.0312	0.0142	1.42	28.3	0.05	1.0	0.00312	0.0625
9. 10000.	1.0	10.0	4.53	453.6	9072.0	16.0	320.0	1.0	20.0
10. 500.0	0.05	0.50	0.227	22.7	453.6	0.8	16.0	0.05	1.0

[a] To find equivalent, locate the horizontal line in which 1.0 units of given the term occurs. The equivalent factor will be found on the horizontal line in the appropriate column. For example—to find parts per million (ppm) of 100 grams per ton, line 6 shows 1.0 grams per ton equivalent to 1.10 ppm 1.10 × 100 = 110 ppm = 100 grams/ton.

Index

A

Aluminum, 50, 184, 318, 320, 326–327, 334
Andropogon gayanus, 75–76
 advantages, 76
 disadvantages, 76
 establishment, 76
 growth habit, 76
 management and utilization, 76
 season of growth, 75
 soil condition, 75
Anhydrous ammonia, 142, 147
Animal traction, 27
Antibiotics, 34, 424
Arsenic, 32, 51, 185, 317–320, 326–328, 330, 334

B

Bagasse, 152
Bahiagrass, *see Paspalum notatum*
Bananas, 151
Bermudagrass, *see Cynodon*
Biuret, 140
Blood meal, 145
Bone calcification, 190–191, 419–420
Boron, 50–52, 108, 318, 328, 330
"Borrachera," 374, 376
Botulism, 202, 347
Brachiaria brizantha, 76–78
 advantages, 77
 disadvantages, 77–78
 establishment, 77
 growth habit, 77
 irrigation, 77
 management and utilization, 77
 season of growth, 76
 soil condition, 77
Brewer's grains, 145–146
Broiler litter, 152–153
Buffers, 34
B vitamins, 260–262, 409–410, 423–427
Bypass protein, 142

C

Cadmium, 52, 306, 320–321, 323–325, 327
Calcinosis, 420
Calcium, 23–25, 28–29, 32–33, 42, 45, 48–50, 54, 94, 96, 152, 166–172, 174–175, 180–181, 189–204, 334, 347–350, 355, 367–368, 371–374, 386, 395–396, 398–399, 401, 403–406, 419–420
 absorption, 191
 deficiency, 193–203, 347–350, 355, 367–368, 371, 374
 forage, 94, 96, 166–172, 174–175, 194, 196–198, 200–201, 355
 metabolism, 190–192
 milk fever, 193–194
 relationship to phosphorus, 192–194, 420
 requirements, 23–25, 50, 192–193
 supplementation, 203–204, 386, 395–396, 401, 403–406
 toxicity, 204
 vitamin D relationship, 191–192, 419–420
"Cara inchada," 372–373, 375
Carotene, *see* Vitamin A

Cassava-leaf meal, 144
Cassava tubers, 150–151
Centrosema pubescens, 91–93
  advantages, 92
  disadvantages, 92–93
  establishment, 92
  growth habit, 92
  irrigation, 92
  management and utilization, 92
  season of growth, 92
  soil condition, 92
Cereal grains, 146, 151
Cerebrocortical necrosis, 424
Chlorine, *see* Salt (NaCl)
Chloris gayana, 78–79
  advantages, 79
  disadvantages, 79
  establishment, 78
  growth habit, 78
  irrigation, 78
  management and utilization, 78–79
  season of growth, 78
  soil condition, 78
Choline, 410, 423, 427
Chromium, 50–51, 317, 328–331, 334
Cobalt, 29–30, 49–51, 81–82, 166–168, 171, 173–176, 259–268, 348–349, 351, 355, 365, 367, 371, 374, 376, 384, 395–396, 401, 403–406
  absorption, 262
  coast disease, 260
  deficiency, 263–266, 351, 355, 365, 367, 371, 374, 376
  forage, 166–168, 171, 173–175, 265–266, 351
  metabolism, 260–262
  phalaris staggers, 262
  requirements, 29–30, 50, 262
  "salt sick," 260
  supplementation, 267–268, 384, 395–396, 401, 403–406
  toxicity, 268
  vitamin B_{12}, 260–261
  wasting disease, 260
Cobalt pellets, 385
Cocoa husks, 153–154
Coffee pulp, 153
Copper, 29–32, 49–52, 82, 84, 91, 166–170, 173, 175, 181, 184–185, 237–251, 334, 348–349, 351–352, 355, 365–368, 373–374, 384, 395–396, 398–406, 419, 427
  absorption, 238
  deficiency, 185, 237, 334, 351–352, 355, 365–368, 373–374
  forage, 166–170, 173, 175, 240–242, 247–249, 251, 352–352
  metabolism, 238–239
  relationship to molybdenum, 237–251, 352, 373
  requirements, 29–32, 50, 239–240
  supplementation, 248–249, 384, 395–396, 398–406
  toxicity, 29, 249–251, 396
Copper oxide needles, 385
Cottonseed meal, 143, 158
Crop residues, 148–150
  treatment of, 149–150
Cynodon, 79–81
  advantages, 80–81
  disadvantages, 81
  establishment, 80
  growth habit, 80
  irrigation, 80
  management and utilization, 80
  season of growth, 79
  soil condition, 79–80

D

Desmodium uncinatum, 93–94
  advantages, 94
  disadvantages, 94
  establishment, 93
  growth habit, 93
  irrigation, 93
  management and utilization, 93–94
  season of growth, 93
  soil condition, 93
Dicumarol, 422
Digestible energy, 24–26
Digitaria decumbens, 81–82
  advantages, 82
  disadvantages, 82
  establishment, 81
  growth habit, 81
  management and utilization, 81–82
  season of growth, 81
  soil condition, 81

Digitgrass, *see Digitaria decumbens*
Dry-season forage improvement, 135–139
  forage quality, 137–139
  forage quantity, 136–137

E

Elephantgrass, *see Pennisetum purpureum*
Energy–protein deficiencies, 62–63, 65–66, 69, 129–130, 137, 360–364
Energy–protein requirements, 22–28, 33, 69
  animal traction, 27–28
  forage intake, 22, 26
  maintenance, 26
  production, 26
  nonprotein nitrogen, 27
Energy requirements, 22-28
Energy supplements, 146–154
  bagasse, 152
  bananas, 151
  broiler litter, 152–153
  cassava tubers, 150–151
  cereal grains, 146–151
  cocoa husks, 153–154
  coffee pulp, 153
  crop residues, 148–150
  hay, 147–148, 154–156
  molasses, 140, 148, 151–152
  rice bran, 153
  rice hulls, 153
  rice polishings, 153
  silage, 148, 154–158
  sugarcane, 144, 146–147
"Enteque seco," 420
"Espichamento," 420

F

Fish meal, 145, 158
Fluorine, 29, 52, 204–210, 328, 348–349, 352, 367, 369
  chemical forms, 208
  essentiality, 204–205
  toxicity, 29, 205–207, 352, 367, 369
Forage, as sources of minerals, 166–176
  biological availability, 166, 169–170
  factors affecting mineral content, 170–176
  grazing selectivity, 170

Forage intake, 22, 32, 62, 170
Forage minerals, affected by, 170–176
  climate, 175–176
  drainage, 172–173
  fertilization, 171–172
  forage maturity, 174–175
  forage species and varities, 173–174
  forage yield, 175–176
  pasture management, 175–176
  soil pH, 172–176
  soils, 171–172
Free-choice minerals, 369, 371–374, 376–378, 385–406

G

Gambagrass, *see Andropogon gayanus*
Glass bullets, 385
Glycine, see Neonotomia wightii
Goiter, 268–269, 271–274, 353
Goitrogens, 32, 270, 353
Gossypol, 143
Grass–legume mixtures, 112–115, 144
  animal performance, 114
  role of grasses, 113
  role of legumes, 113
Grass tetany, 220–228, 350
Grass types, 75–91
Grazing management systems, 115–119
  grazing behavior, 117
  grazing pressure, 117, 155
  grazing selectivity by animal, 118
Greenleaf desmodium, *see Desmodium uncinatum*
Guineagrass, *see Panicum maximum*

H

Hay, 147–158, 154–156
Heat stress, 33–34, 58–70, 125
  acid–base balance, 68–69
  digestion, 64–65
  energy metabolism, 65–66
  feed consumption, 62–63
  forage quality, 64
  hormonal responses, 61–62
  metabolic responses, 61–62
  mineral requirements, 67–68
  physiological responses, 60
  protein requirement, 69

vitamin A requirement, 69–70
water balance, 66–67
water requirement, 66–67
Herbage quality, 119–122
  chemical composition, 119
  digestibility, 120
  forage intake, 32, 121
  palatability, 122
Hormones, effect on nutrient requirements, 34
Hyparrhenia rufa, 82–83
  advantages, 83
  disadvantages, 83
  establishment, 82
  growth habit, 82
  irrigation, 83
  management and utilization, 83
  season of growth, 82
  soil condition, 82

I

Iodine, 51, 268–275, 348–349, 353, 367–368, 371, 395–396, 398–399, 401, 403, 405–406
  absorption, 270
  deficiency, 261, 271–273, 353, 367–368, 371
  goitrogens, 32, 270
  metabolism, 268, 270
  requirements, 270–271
  supplementation, 273–275, 395–396, 398–399, 401, 403, 405–406
  thyroid hormones, 268, 269
  toxicity, 275
Ionophores, 34
Ipil-ipil, *see Leucaena leucocephala*
Iron, 29, 31–32, 49–51, 166–170, 173, 175, 180, 184, 242, 291–297, 334, 348–349, 353, 367–368, 395–396, 398–399, 401, 405–406, 427
  absorption, 292
  deficiency, 294–296, 353, 367–368
  forage, 166–170, 173, 242, 293, 295
  metabolism, 292, 294
  requirements, 29, 31–32, 50, 294
  supplementation, 297, 395–396, 398–399, 401, 403, 405–406
  toxicity, 297

J

Jaragua, *see Hyparrhnia rufa*
Jitirana, *see Centrosema pubescens*

K

Ketosis, 423–424
Kudzu, *see Pueraria phaseoloides*

L

Lasalocid, see Ionophores
Lead, 51–52, 185, 318–320, 322–323, 327, 334
  deficiency, 322–326
  toxicity, 319–322
Legumes, 91–101
Leucaena leucocephala, 94–95
  advantages, 95
  disadvantages, 95
  establishment, 94
  growth habit, 94
  irrigation, 94
  management and utilization, 94–95
  season of growth, 94
  soil condition, 94
Linseed meal, 144
Lithium, 318, 330–321, 328
Livestock production systems, 130–135

M

Macroptilium atropurpureum, 95–96
  advantages, 96
  disadvantages, 96
  establishment, 95–96
  irrigation, 96
  growth habit, 96
  management and utilization, 96
  season of growth, 95
  soil condition, 95
Magnesium, 29–31, 42, 45, 48–50, 52, 54, 79, 108, 166–168, 170–173, 175, 180–181, 214, 220–228, 348–350, 368, 384, 386, 398–399, 403–406
  absorption, 221
  deficiency, 222–226, 350, 368
  forage, 79, 108, 166–173, 175, 222, 226
  grass tetany, 220–228, 350, 395

metabolism, 221
requirements, 28–29, 222
supplementation, 226–228, 350, 384, 386, 398–399, 403–406
toxicity, 228, 367
Manganese, 28–29, 49–51, 53, 166–168, 173–174, 176, 292, 297–304, 348–349, 353, 367–368, 384–385, 395–396, 398–399, 403–406, 427
   absorption, 298–299
   deficiency, 300–302, 353, 367–368
   forage, 166–168, 173–174, 293, 301–302, 304, 353, 384
   metabolism, 298–299
   requirements, 28–29, 50, 299
   supplementation, 303, 384–385, 395–396, 398–399, 403–406
   toxicity, 298, 303–304
Meat and bone meal, 145
Meat meal, 145, 158
Melinis minutiflora, 83–84
   advantages, 84
   disadvantages, 84
   establishment, 83–84
   irrigation, 84
   growth habit, 84
   management and utilization, 84
   season of growth, 83
   soil condition, 83
Mercury, 51, 53, 318, 320–321, 325–326, 334
Metabolizable energy, 24–25, 27, 33
Milk fever, 30, 193–194, 419
Mineral blocks, 392–395
Mineral consumption, factors affecting, 386–395
   energy–protein supplements, 387
   forage type 386–387
   fresh supplies, 389, 391–392
   palatability, 387–389
   physical form, 392, 395
   requirements, 387
   salt content water, 387
   soil fertility, 386–387
Mineral deficiency–toxicity diagnosis,
   calcium and phosphorus, 347–350
   cobalt, 348–349, 351
   copper and molybdenum, 348–349, 351–352

   fluorine, 348–349, 352
   iodine, 348–349, 353
   iron, 348–349, 353
   magnesium, 348, 350
   manganese, 348–349, 353
   potassium, 348, 350
   selenium, 348–349, 354
   sodium, 348–351
   sulfur, 348–349, 351
   zinc, 348–349, 354
Mineral feeders, 388–394
Mineral mapping survey technique, 354–355
Mineral requirements, factors influencing 28–31
   age, 31
   biological availability, 31
   breed and adaptation, 30
   intake and season, 32
   interrelationships, 31
   physiological state, 28
   production level, 28
Mineral sources, 397–398
Mineral status evaluation, 339–355, 367
   analysis
      animal tissues and fluids, 343–346
      water, soil, forage, 341–343
   clinical and pathological, 340–341
   response to supplementation, 346
Mineral supplementation
   evaluation, 377–378, 402–406
   formulation, 396–402
   methods, 384–386
   results, 369–378
   seasonal needs, 376–378
Molasses, 140, 148, 151, 152
Molassesgrass, *see Melinis minutiflora*
Molybdenum, 29, 31–32, 51, 166–168, 173, 176, 180, 237–251, 348–349, 351–352, 355, 367–369, 404, 406
   absorption, 238
   deficiency, 237
   metabolism, 238–239
   relationship to copper, 237–251
   requirements, 29, 240
   toxicity, 29, 238, 240–251, 249–251, 351–352, 355, 404–406
Monensin, *see* Ionophores
Mineral tag (label), 402, 404–406

N

Napiergrass, see *Pennisetum purpureum*
Neonatal ataxia, 241
Neonotonia wightii, 96–97, 137
  advantages, 97
  disadvantages, 97, 137
  establishment, 97
  growth habit, 97
  management and utilization, 97
  season of growth, 96
  soil condition, 96–97
Net energy, 23–24, 28
Niacin, 410, 423–424
Nickel, 51, 53, 317, 328, 330–332, 334
Night blindness, 412
Nitrate, 54
Nitrite, 54
Nonprotein nitrogen (NPN), 27, 139
  anhydrous ammonia, 142
  biuret, 140
  poultry waste, 141
  urea, 27, 139
Nutrition
  relation to disease, 32–33
  relation to parasites, 32–33, 124–125
Nyctalopia, see Night blindness

O

Oilseed meals, 142–143
Osteomalacia, 194, 419–420
Oxalates, 32, 194

P

Pangolagrass, see *Digitaria decumbens*
Panicum maximum, 84–86, 137, 147–148, 156–157
  advantages, 86
  disadvantages, 86
  establishment, 85
  growth habit, 85
  irrigation, 85
  management and utilization, 85
  season of growth, 85
  soil condition, 85
Paragrass, see *Brachiaria brizantha*
Parakeratosis, 304, 307, 354
Parturient paresis, see Milk fever
Paspalum notatum, 86–87
  advantages, 87
  disadvantages, 87
  establishment, 86
  growth habit, 86
  irrigation, 86
  management and utilization, 87
  season of growth, 86
  soil condition, 86
Pasto rodes, see *Chloris gayana*
Pasture maintenance, 104–109
  fertilization, 105–106
  recycling of nutrients, 107
  renovation practices, 108
  weed control, 108
Pasture type and grazing management, 109–112
  native, 109
  pasture supplementation, 111
  sown and improved, 110
Pasture use, 122–125, 130–135
  calf rearing, 124
  cow-calf operations, 122
  dry cows, 125
  finishing, 123
  milk production, 124
Peanut meal, 143
Pennisetum purpureum, 87–89
  advantages, 89
  disadvantages, 89
  establishment, 88
  growth habit, 88
  irrigation, 88
  management and utilization, 88–89
  season of growth, 88
  soil condition, 88
Phosphorus, 23–25, 27–28, 30–33, 42, 49–50, 83, 88, 93–94, 98, 100, 105–107, 121, 152, 166–172, 175, 180–181, 184, 189–204, 334, 347–350, 355, 365–368, 371–372, 374, 376–377, 384–385, 395–396, 398–399, 403–406, 419–420
  absorption, 191
  deficiency, 193–203, 334, 347–350, 355, 365–368, 371, 374, 376, 377
  fertility, 201–202, 371–372
  forage, 83, 88, 93–94, 98, 100, 105–107, 121, 166–172, 175, 194, 196–198, 200–201, 347, 384

metabolism, 190–192
milk fever, 193–194
relationship to calcium, 192–194
requirements, 23–25, 27, 30–33, 50, 192–193
supplementation, 203–204, 371–372, 384–385, 395–396, 398–399, 403–406
toxicity, 204
vitamin D relationship, 191–192
Phytic acid, 32
Pica, 194–195, 199, 202, 347
Polioencephalomalacia (PEM), 424
Pongolagrass, see Digitaria decumbens
Potassium, 27, 29, 42, 45, 48–50, 53, 77, 88, 93, 105–107, 166–171, 174–176, 180, 214, 228–233, 348–350, 395, 398–399, 403–404, 406
absorption, 229
deficiency, 229–232, 350
forage, 77, 88, 93, 105–107, 166–171, 174, 230, 232
metabolism, 228–229
requirements, 27, 29, 50, 229
supplementation, 233, 395, 398–399, 403–404, 406
toxicity, 233
Poultry waste, 141, 145
Protein, 22–27, 32, 76–77, 80, 84–87, 89–90, 92–93, 96, 99, 101, 114–115, 120–121, 124, 132–146, 150, 153, 361–364
Protein supplements, 139–146, 424
anhydrous ammonia, 142
biuret, 140
blood meal, 145
brewer's grains, 145–146
cassava-leaf meal, 144
cottonseed meal, 143, 158
fish meal, 145, 158
legumes, 144
linseed meal, 144
meat and bone meal, 145
meat meal, 145, 158
nonprotein nitrogen, 27, 139–142
oilseed meal, 142–144
peanut meal, 143
poultry waste, 145
sesame seed meal, 144
soybean meal, 143

sunflower seed meal, 144
urea, 27, 139–140, 151–152, 156
Prothrombin, 422
Pueraria phaseoloides, 97–99
advantages, 99
disadvantages, 99
establishment, 99
growth habit, 99
management and utilization, 99
season of growth, 97–98
soil condition, 98–99

R

Rhodesgrass, see *Chloris gayana*
Rhodopsin, 411
Rice bran, 153
Rice hulls, 153
Rice polishings, 153
Rickets, 194, 420
Ruminant contributions, 12–17
manure, 16–17
meat, 12–14
milk, 12–14
products and services, 16–17
work, 14–16
Ruminant distribution, 3–12
alpaca, 3–4, 9–10
bison, 3–4
buffaloes, 3–6
camel, 3–4, 8–9
camelids, 3–4
cattle, 3–6
goats, 3–4, 7–8
llamas, 3–4, 9–10
sheep, 3–4, 6–7
wild ruminants, 3–4, 10–12
Ruzigrass, see *Brachiaria brizantha*

S

Saccharum officinarum, 89–90, 146–147
advantages, 90
disadvantages, 90
establishment, 89
growth habit, 89
irrigation, 89
management and utilization, 89–90
season of growth, 89
soil condition, 89

Sadabahar, *see Andropogon gayanus*
Salt (NaCl), 29, 42, 45–49, 54, 79,
  166–170, 172, 175, 181, 214–220,
  365–368, 371, 374, 378, 384, 387,
  398–399, 403–406
  absorption, 214–215
  deficiency, 216–218, 365–368, 371, 374, 378
  forage, 166–170, 172, 175, 216–218
  metabolism, 214–215
  requirements, 29, 49–50, 215–216
  supplementation, 218–220, 384, 387, 398–399, 403–406
  toxicity, 220
"Secadera," 372–375, 426
Selenium, 29, 31–32, 49–50, 53, 167–168,
  173–174, 176, 181–182, 184, 260,
  275–286, 334, 348–349, 354–355, 365,
  367–369, 374, 395–396, 401, 403,
  405–406
  absorption, 277
  alkali disease, 276
  blind staggers, 276
  deficiency, 276–281, 354–355, 365, 367, 374
  forage, 167–168, 173–174, 281–282, 355
  metabolism, 276–277
  requirements, 29, 31–32, 50, 277
  stiff lamb disease, 276
  supplementation, 281–282
  toxicity, 29, 283–286, 369, 396, 402
  vitamin E relationship, 277, 280–281, 421–422
  white muscle disease, 276–279, 354
Sesame seed meal, 144
Setaria anceps, 90–91
  advantages, 91
  disadvantages, 91
  establishment, 90
  growth habit, 90
  irrigation, 90
  management and utilization, 90–91
  season of growth, 90
  soil condition, 90
Signalgrass, *see Brachiaria brizantha*
Silage, 148, 154–158
Silicon, 49, 53, 317, 328–330, 334
Siratro, *see Macroptilium atropurpureum*
Sodium, *see* Salt (NaCl)

Soil, as sources of minerals, 176–185
Soil consumption, 176–185
  beneficial effects, 178–182
  detrimental effects, 182–185
  reasons for, 177–178
Solanum malacoxylon, 420
Soybean, perennial, *see Neonotonia wightii*
Soybean meal, 143
Stargrass, *see Cynodon*
Stylosanthes guianensis, 99–101
  advantages, 101
  disadvantages, 101
  establishment, 100
  growth habit, 100
  management and utilization, 100–101
  season of growth, 100
  soil condition, 100
Sugarcane, 144, 146–147, *see also Saccharum officinarum*
Sugarcane bagasse, *see* Bagasse
Sugarcane molasses, *see* Molasses
Sulfur, 27, 29, 32, 49–50, 83, 107, 236,
  251–255, 348–349, 351, 365, 368, 371,
  395, 398–399
  absorption, 251
  deficiency, 252–254, 351, 365, 368, 371
  forage, 107, 253
  metabolism, 251
  relationship to nitrogen, 252
  relationship to copper, 237–238, 254
  requirements, 27, 29, 32, 50, 252
  supplementation, 254, 395, 398–399
  toxicity, 254–255
Sunflower seed meal, 144
Supplemental feeding, animal class, 131–134
  growing males, 131
  mature females, 132
  nonreplacement females, 131
  replacement females, 132

T

Theobromine, 154
Thermal stress, *see* Heat stress
Thiamin, 409–410, 423–426
Thiaminase, 426
Tick clover, *see Desmodium uncinatum*
Tin, 185, 317–318, 328–330, 332–334

Total digestable nutrients (TDN), 23–26, 114–115

U

Uranium, 54
Urea, 27, 139–140, 151–152, 158, 423

V

Vanadium, 317, 328–330, 333–334
  deficiency, 333–334
  toxicity, 333
Veyale, *see Hyparrhenia rufa*
Vitamin A, 23–24, 33, 69–70, 409–419
  β-carotene, 411, 415, 417–418
  deficiency, 411–415
  function, 411–415
  requirements, 23–25, 69–70, 410–411
  sources, 416–417
  storage, 414–416
  supplementation, 418–419
  toxicity, 417
Vitamin B_{12}, 32, 260–262, 410, 423
Vitamin D, 23–25, 32, 191–192, 409–410, 419–421
  deficiency, 419–420
  function, 419
  requirements, 23–25, 32, 419
  supplementation, 420
  toxicity, 420–442
Vitamin E, 32, 409–410
  deficiency, 421–422
  function, 421–422
  requirements, 32, 421
  selenium relationship, 32, 277, 280–281, 421–422
  supplementation, 421–422
  white muscle disease, 276–279, 421–422
Vitamin K, 409–410, 422

W

Water
  effect on feed intake, 40–49, 54, 67
  effect on nutrient digestibility, 44–45
  electrolyte metabolism, 48–49
  hardness, 39–48
  intake, 40–42, 54, 66–67
  requirement, 48–49
  salinity, 39, 45–48
  temperature, 44–45, 55, 67
  turnover, 39–40, 55, 66–67
Water
  as source of essential elements, 39, 49–50
  as source of toxic elements, 50–54
White muscle disease, 276–279, 354, 421–422

Y

Yaragua, *see Hyparrhenia rufa*

Z

Zinc, 28–29, 31–32, 49–51, 91, 166–171, 173–176, 181, 182, 185, 292–293, 304–311, 334, 348–349, 354, 367–368, 373–374, 384, 395–396, 398–399, 401, 404–406
  absorption, 304–305
  deficiency, 304, 307–310, 354, 367–368, 373–374
  forage, 91, 293, 309–310
  metabolism, 304–306
  parakeratosis, 304, 307, 354
  requirements, 28–29, 31–32, 306–307
  supplementation, 310–311, 384, 395–396, 398–399, 401, 404–406
  toxicity, 311
  vitamin A relationship, 304, 411, 414